T0329846

Grid-Integrated and Standalone Photovoltaic Distributed Generation Systems

Grid-Integrated and Standalone Photovoltaic Distributed Generation Systems

Analysis, Design, and Control

Bo Zhao
State Grid Zhejiang Electric Power Research Institute
Hangzhou, China

Caisheng Wang
Electrical and Computer Engineering Department, Wayne State University
Detroit, USA

Xuesong Zhang
State Grid Zhejiang Electric Power Research Institute
Hangzhou, China

Registered Office
John Wiley & Sons, Inc., 111 River Street, Hoboken, NJ 07030, USA
John Wiley & Sons Singapore Pte. Ltd, 1 Fusionopolis Walk, #07-01 Solaris South Tower, Singapore 138628

Editorial Office
1 Fusionopolis Walk, #07-01 Solaris South Tower, Singapore 138628

For details of our global editorial offices, customer services, and more information about Wiley products visit us at www.wiley.com.

Wiley also publishes its books in a variety of electronic formats and by print-on-demand. Some content that appears in standard print versions of this book may not be available in other formats.

Library of Congress Cataloging-in-Publication Data

Names: Zhao, Bo, 1977- author.
Title: Grid-Integrated and standalone photovoltaic distributed generation
    systems : analysis, design and control / Dr. Bo Zhao, State Grid Zhejiang
    Electric Power Research Institute, Hangzhou, China; Dr. Caisheng Wang,
    Electrical and Computer Engineering Department, Wayne State
    University, Detroit, USA; Dr. Xuesong Zhang, State Grid Zhejiang
    Electric Power Research Institute, Hangzhou, China.
Description: Hoboken, NJ, USA : Wiley, 2017. | Includes bibliographical
    references and index. |
Identifiers: LCCN 2017011367 (print) | LCCN 2017026557 (ebook) | ISBN
    9781119187363 (pdf) | ISBN 9781119187356 (epub) | ISBN 9781119187332
    (cloth)
Subjects: LCSH: Photovoltaic power generation. | Interconnected electric
    utility systems. | Distributed generation of electric power.
Classification: LCC TK1087 (ebook) | LCC TK1087 .Z45 2017 (print) | DDC
    621.31/244–dc23
LC record available at https://lccn.loc.gov/2017011367

Cover Design: Wiley
Cover Image: © Petmal/Gettyimages

Set in 10/12pt WarnockPro by SPi Global, Chennai, India

Printed in Singapore by C.O.S. Printers Pte Ltd

10 9 8 7 6 5 4 3 2 1

Contents

Preface

With the progress of technology, human beings have undergone the transition from industrial civilization to ecological civilization, from extensive and inefficient expending to economical and efficient consumption, and from high carbon production to low carbon production. The present fossil-energy-dominated world energy paradigm is gradually changing into a multiple energy structure and will eventually be dominated by nonfossil energy. Against this background, the development of distributed photovoltaic (PV) generation systems, characterized by adaptation to local conditions, clean and efficient, decentralized layout, and local consumption, have experienced phenomenal growth in the past two decades. The world total solar power capacity had reached over 227 GW by 2015 and over 50 GW of the solar capacity were added in that year. Over 28 GW of solar power had been installed in the USA by the end of 2016. The PV capacity in Germany is currently over 40 GW. China is expected to deploy 70 GW PVs by 2017. The International Energy Agency estimates that solar power will become one of the mainstream energy sources by 2050 and contribute about 11% of world electricity generation. The majority of these PV systems have been and will be installed in distribution networks. As a result, the PV penetration level will become unprecedentedly high (e.g., well over 50%) and continue to grow in many distribution networks around the world. The high penetration of PV systems has led to great technical challenges, including voltage problems, harmonics, grid protection, and so on, in the operation and development of modern distribution networks.

In recent years, the authors and their teams have undertaken a series of PV-related projects, such as the implementation of PV systems for Jiaxing PV Science Park, Jianshan New Industrial Zone in Haining, and Hangzhou Bay New Zone PV Science Park. The main characteristic of these projects was to analyze the impact of integration of distributed PVs into the distribution networks at high penetration and develop measures to better accommodate those sources.

This book is the result of over 10 years of research on distributed PVs and integration of PVs in distribution networks and microgrids. It combines the theory, modeling, analysis and control, and the actual implementation of distributed PVs in one place. The book is focused on the operation and control of distribution networks and microgrids with high penetration of distributed PVs, covering the topics of PV hosting capacity analysis, power flow of distribution networks, reactive voltage regulation, short-circuit current calculation, power quality evaluation, methods of integrating distributed PVs into distribution networks at high penetration levels, and actual system implementation experiences. The book is intended to be a resource for all engineers, and for all those interested in designing policies for facilitating renewable energy development. An overview of the chapters covered in the book is now presented.

Chapter 1 introduces the current status and future development trends of PVs around the world. The PV industry development history of different countries, including the USA, Japan, Germany, and China, and the relevant policies, laws, and demonstration projects in these countries are also briefly introduced. Chapter 2 gives a brief coverage of the basic techniques of distributed grid-connected PV systems, with focus on their configurations, components and maximum power point tracking techniques. Chapter 3 presents the load characteristics of a distribution network with and without distributed PVs and provides theoretical foundations for further analysis on PV penetration and power flow in distribution networks.

Chapter 4 covers the concepts of PV power penetration, capacity penetration and energy penetration, analyzes the key challenges in different development stages of distributed grid-connected PV, studies the impact of grid-connected PVs on distribution networks, explores the maximum allowable capacity of grid-connected PVs under the requirement of safe and stable operation of the distribution networks, and presents the methods to increase PV penetration in distribution networks.

Chapter 5 introduces various power flow calculation methods for distribution networks. It then shows the impact of PVs on the power flow in distributed networks, including voltage variation and distribution loss caused by the PVs added. Models of voltage and distribution loss considering PVs are established, which are then used to analyze the impact of different PV capacities and locations.

Chapter 6 addresses one of the important issues caused by high PV penetration: the voltage control of distribution networks with distributed PVs. In this chapter, the impacts on voltage due to distributed PVs are analyzed first and three control strategies (the constant power factor control strategy, variable power factor control strategy, and the reactive voltage control strategy) and relevant modeling methods are then introduced.

Chapter 7 analyzes the short-circuit characteristics of distributed PVs under symmetrical and asymmetrical grid voltage sags, discusses different low voltage ride-through (LVRT) standards for PVs and the LVRT control strategies, and characterizes the fault currents of PVs. An iterative algorithm is given in the chapter for the calculation of fault currents for distribution networks with distributed PVs.

Chapter 8 discusses the impact of grid-connected PVs on power quality. Chinese and international standards on PV integration harmonic requirements are compared and discussed. The differences in power quality terms and the methods for analyzing the impacts due to distributed PVs on distribution network power quality are given.

Chapter 9 discusses various technologies, including energy storage, demand response, and a network partition-based zonal control technology to better accommodate PVs for distribution networks.

Chapter 10 is a good addition to the other chapters in the book by addressing microgrids with PVs. The chapter reviews the configurations of AC, DC, and AC/DC hybrid microgrids, system unit sizing of microgrid components, and the control framework and the implementation of microgrids. The optimal design, planning, and control of microgrids are given in the chapter through a discussion of the implementation examples of two real standalone microgrids. The implementation and operation experiences and lessons learned from the two microgrids are also summarized in the chapter.

The authors are honored to have the privilege to work with their collaborators for this book, which is a result of an exemplar group effort. Mr. Yibin Feng and Dr. Junhui Zhao are the main authors of Chapter 1. Dr. Xuesong Zhang and Dr. Junhui Zhao are the leading authors of Chapter 2, along with Dr. Jinhui Zhou for Chapter 3, Miss Chen Xu and Dr. Jian Chen for Chapter 5, Prof. Li Guo for Chapter 7, Mr. Peng Li and Prof. Zaijun Wu for Chapter 8. Professor Ming Ding and Professor Hongbin Wu of Hefei University of Technology offered a great deal of help to the authors. Professor Saleh A. Al-Jufout from Tafila Technical University kindly helped proofread Chapter 5. The authors also would like to thank team members and students for their help and contributions to this book, including Da Lin, Ke Xu, Xiaohui Ge, Ziling Wang, Ke Wang, Meidong Xue, Xiangjin Wang, Chuanliang Xiao, Zhichen Xu, Haifeng Qiu, Tingli Hu, Chang Fu, Zhongyang Zhao, Michael Fornoff, and Nicholas Lin.

An important feature of this book is that there are case studies accompanied with simulation studies for many chapters. We hope this book and the examples given in the book will be useful to industry professionals, educators, students, and researchers worldwide in developing distributed PVs at high penetration level.

Bo Zhao
State Grid Zhejiang Electric Power
Research Institute, China
Caisheng Wang
Wayne State University, USA
Xuesong Zhang
State Grid Zhejiang Electric Power
Research Institute, China

August, 2017

1

Overview

1.1 Current Status and Future Development Trends of Photovoltaic Generation around the World

With the growing challenges in global resource depletion, global warming, and ecological deterioration, increasing attention has been given to renewable energy generation, especially to photovoltaic (PV) generation. The global market of PVs has experienced a rapid increase since 1998, with a yearly increase of 35% of the installed capacity. The total PV installed capacity was 1200 MW in 2000, and PV installations rose rapidly up to 188 GW in 2014 and is projected to be 490 GW by 2020 [1]. With the rapid development of the PV industry, the market competition is getting increasingly fierce. The investment in the PV market is being boosted in some countries and regions, like the USA, China, Japan, and Europe. By the end of 2014, the global production of PV modules was around 50 GW, in which China increased 27.2% from the previous year to 35 GW, contributing 70% of the global production [2]. The global production of PV modules is expected to reach 85 GW and maintain the momentum of rapid growth [3].

Recently, a number of countries announced their policies and plans to further promote the development of PVs [4, 5]. The US Environmental Protection Agency (EPA) published its Clean Power Plan on June 2, 2014, promising that the usage of renewable energy (including solar energy) will be doubled within 10 years. The US Department of Energy (DOE) will spend $15 million to help families, enterprises and communities develop the solar energy program [6]. The Japanese Government enacted laws, like the *Renewable Energy Special Measure Law* and the *Renewable Portfolio Standard Law*, to identify the development objectives of new energy in Japan and the responsibilities of the participating parties [7]. China has highlighted a few key and crucial demonstration projects of the PV technologies in the *Outline of the National Program for Long- and Medium-Term Scientific and Technological Development (2006–2020)*, the *National 11th Five-Year Scientific and Technological Development Plan* and the *Renewable Energy 12th Five-Year Plan* [4–8].

It is noteworthy that the USA and Japan have both worked through the PVs "Industry Roadmap Through 2030 And Beyond." Japan expects that the future research and development pattern of PVs could be changed from *creating an initial PV market based on the government's guide* to *creating a mature PV market based on cooperation and work sharing among academia, industry, and government*, and targets to have a total PV installation capacity of 100 GW in 2030. The USA anticipates that the development

Grid-Integrated and Standalone Photovoltaic Distributed Generation Systems: Analysis, Design, and Control, First Edition. Bo Zhao, Caisheng Wang and Xuesong Zhang.
© 2018 China Electric Power Press. Published 2018 by John Wiley & Sons Singapore Pte. Ltd.

pattern of the PV industry could be changed from export led to national investment oriented, promoting the industry's significant growth by devoting on the advancement of technologies and market and expansion of the domestic demand. It is projected to install 19 GW of PVs yearly in the USA, with the expectation of a total installed capacity of 200 GW by 2030. By then the cost of the PVs will decline to $0.06/kW, and PVs will make up an important part of the electricity market and become one of the main sources of electricity.

As to the development of the PV industry in China, from the viewpoint of the current status and future trend, the estimated installed capacity was for 300 MW, 1.8 GW, 10 GW, and 100 GW in 2010, 2020, 2030, and 2050 respectively in the *Medium and Long Term Development Plan of Renewable Energies (2007)*, which is apparently lower than actual development and lags behind the trend of the PV industry. Meanwhile, China has not proposed clear goals of the method, direction, and path for developing the critical technologies and devices that has already limited the advancement of the PV industry. In terms of the grid-connected PVs, there is a lack of complete and systematic regulations and policies for operation and management, electricity price, and system maintenance. Therefore, actively promoting the research and practical applications in the Chinese PV industry to follow the main stream of the global PV industry development will be of profound significance in the future.

At present, some developed countries (such as the USA, Germany, Australia, Japan, etc.) are leading the research and development of PV technologies. For example, Australia, represented by Professor Martin A. Green from the University of New South Wales, has made a great contribution to the development of PV cells by leading the research of single crystalline silicon solar cells in the world and proposing the concept of the third-generation PV cells [9]. The USA, the UK, Germany, Spain, Japan, and so on initiated the PV industry and applications early and have experienced rapid development. Although China's PV industry started late, it has experienced exponential growth. Especially after 2004, stimulated by the large demand from the European market, China's PV industry has boomed and saw over 100% yearly growth for five years in a row. In 2007, China became the largest producer of PV cells. China's PV production exceeded 50% of global production in 2010. China has gradually formed an orbicular chain in the PV industry, from silicon material, PV cells, to PV systems and applications [10, 11]. As shown in Table 1.1, China's PV manufacturers now take a dominant role in the world's PV production. Of the world's top 10 PV manufacturers, six are from China and all the top five are from China. Among them, the number one manufacturer Trina Solar produced 3.66 GW in 2014, closely followed by Yingli Green Energy, which yielded 3.36 GW [2].

In 2014, global PV installations increased by 17%, while the total installed capacity reached 47 GW. Figure 1.1 shows the market share of the world's top 10 PV countries in 2014. The top 10 countries were China, Japan, the USA, the UK, Germany, France, South Africa, Australia, India, and Canada with a total installed capacity of 38.3 GW, which accounted for 81.5% of the global increase [12]. As an emerging market, Asia has become the preeminent PV market in the world and took 59% of the global installation in 2014. Although China will maintain its position as the largest PVs market in the world, its development has apparently slowed down recently. Japan has continued its strong

Table 1.1 World's top 10 PV manufacturers in 2014.

Manufacturer	Country	Rank	Production (GW)	(%)
Trina Solar	China	1	3.66	14.6
Yingli Green Energy	China	2	3.36	13.4
Canadian Solar	China	3	3.11	12.4
Jinko Solar	China	4	2.94	11.7
JA Solar	China	5	2	8.0
Sharp	Japan	5	2	8.0
Renesolar	China	7	1.97	7.8
First Solar	USA	8	1.85	7.4
Hanwha SolarOne	South Korea	9	1.45	5.8
Sunpower	USA	10	1.4	5.6

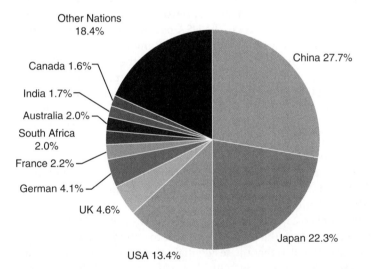

Figure 1.1 Installation percentage of the world's top 10 PV markets in 2014.

growth. The USA surpassed Europe to be the second largest PVs market and took 19.3% of installations in 2014. The European PVs market kept shrinking in 2014 and took only 16.8% of new installations. Spurred by the renewable energy laws, the UK's PVs market flourished in 2014 and exceeded Germany for the first time to be the country with the most new PVs in Europe [2].

1.1.1 USA

Way back to June 26, 1997, President Clinton announced the "Million Solar Roofs Initiative," which planned to install 1 million roof-top solar systems by 2010, including

PV panels and solar thermal collectors. This initiative was driven by the trend of social development and the professionals dedicated to the research and development of PV generation. Two immediate reasons for proposing this initiative were:

- Large greenhouse gas emissions lead to global warming, which requires the reduction of the reliance on conventional energy sources. If the "Million Solar Roofs Initiative" was implemented successfully, the CO_2 emissions would be reduced by more than 3 million tons by the end of 2010.
- In the USA, the technologies of PV panels and solar thermal collectors were mature and implemented in mass production.

At present, the "Million Solar Roofs Initiative" has been carried out in some regions, such as the Civano project in Tucson, AZ. Owing to the huge potential of renewable energy resources in Hawaii, solar power has become the mainstream of the local energy supply and an important part of economic development. In 2001, the California State Government proposed the world-famous "California Solar Initiative Program," planning to install 1 million PV systems in 10 years by investing \$3.2 billion. In September 2004, the US Department of Energy published "Our Solar Power Future: The US Photovoltaics Industry Roadmap Through 2030 And Beyond," revealing an ambitious development plan for the PV industry. In 2006, the USA passed President Bush's Solar Energy Initiative, which increased research funding to \$148 million to strengthen the competiveness of the nation's PV technologies. In April, 2008, the mayor of Philadelphia announced the intention to build the first megawatt-level PV plant in the marine park in Pennsylvania. In May 2008, Duke Energy announced the plan to purchase all of the generation from a 16 MW PV plant in Charlotte, North Carolina. In the middle of June 2008, PEPCO Energy Services signed the contract to build a 2.36 MW PV plant on the roof of the Atlantic City Convention Center in New Jersey. All of the aforementioned projects have been completed.

On March 2015, the US Solar Energy Industries Association (SEIA) released the US Solar Energy Insight Report of 2014 Year in Review. The executive summary of the 2014 report stated that the PV installations in the USA reached 6201 MW_{dc}, up 30% over 2013. The cumulative solar PV installed capacity has reached 18.3 GW_{dc}, and solar accounted for 32% of new generating capacity in the USA, second only to natural gas. It also investigated all kinds of PVs and concluded that the residential solar has soared over the past 3 years, posting annual growth rates over 50% in 2012, 2013, and 2014. In the report of US Solar Industry Year in Review 2009, which was released in April 2010, by SEIA, Lawrence Berkeley National Laboratory, it was estimated that compliance with existing solar and distributed generation carve-outs would require roughly 9000 MW of solar capacity by 2025 [13, 14].

The National Renewable Energy Laboratory (NREL) of the US Department of Energy began its operation in 1977 as the Solar Energy Research Institute and changed its name to NREL in 1991. Nowadays, there are more and more institutions carrying out research on solar energy in the USA. Various incentives have been issued to encourage the development of renewable generation, like feed-in tariffs, investment subsidies, renewable energy certificates, and so on.

1.1.2 Japan

Japan's PV development initially started from a medium- and long-term research and development plan on replacing petroleum by solar-based renewable energy in 1974. The plan, also known as the Sunshine Program, was proposed by the Ministry of Economy, Trade and Industry (METI). After that, the Japanese Government established the New Energy and Industrial Technology Development Organization (NEDO) to take charge of the industrialization of photoelectricity. The industrialization procedure of PV cells was accelerated by large government funds, which dramatically reduced the production cost of PV cells and significantly improved the production technologies. For example, the conversion efficiency of polycrystalline silicon casting substrate increased from 12.7% in 1984 to 15.7% in 1988, while the conversion efficiency of amorphous silicon cells increased from 8.25% in 1985 to 10.1% in 1988.

After the 1992 World Economic and Environmental Conference, more attention was paid to environmental problems, such as climate change due to human activities. Since 1994, METI has implemented preferential policies to subsidy residential PV systems by up to 50% of the total cost, including the cost of inverters, PV panels, grid connection system, and construction. In addition, contractors can also get the same subsidization. The promotion greatly prompts the growth of residential PV installation. Although Japan was falling into the bursting of the bubble economy at that time, the citizens still actively applied for roof-top PV projects. For instance, the total number of applications was 33 000 (121 MW) between 1994 and 1999, while there were 26 000 applications in 2000 (96 MW), which was nearly the summation of the total installation of the previous 5 years.

In 2000, Japan set out the "PVs Industry Roadmap Through 2030 And Beyond," which aimed to transfer the PV research and development from creating an initial PV market guided by government to creating a mature PV market based on the collaboration among academia, industry, and government, and expected the total installed capacity to be 100 GW in 2030. If it can be realized, PV generation can provide approximately 50% of residential electricity consumption, which is 10% of the total electricity demand in Japan. To accelerate the increase of PV generation, on June 28, 2009, NEDO amended the roadmap and predicted the cost of PV generation could decline to $0.14/kW in 2020 and to $0.07/kW in 2030 [15].

As one of the measures taken after the 2011 Fukushima nuclear power plant accident, Japan started to strive to develop PV-based renewable energy generation. On July 1, 2012, the PV Subsidies Act was enacted in Japan. According to this act, Japan's electric companies must purchase the electricity generated by home and industry solar energy, and the feed-in tariff was set as ¥42/kWh (approximately $0.54/kWh) [16]. Although the feed-in tariffs gradually declined in 2013 and 2014, Japan is still the country that has the highest subsidy on PV systems around the world.

1.1.3 Germany

Germany has the largest PV installed capacity in the world so far. Its experience in promoting PV development from the aspects of policymaking, management, and technologies has been the model for several other countries. In 2000, the German Government released the German Renewable Energy Act (EEG-2000) and successfully completed the "100,000 Roofs Programme" in 2003. In 2004, the accumulated PV capacity in Germany surpassed that in Japan and became a leading country in PV generation. The new PV installations in Germany was 1.5 GW in 2008, 3.2 GW in 2009, and peaked at around 7.5 GW between 2010 and 2012. At that time the plan to reach a PVs installed capacity of 10 GW in 2020 had been achieved in advance. However, since 2013, there have been some issues raised related to this renewable energy, such as overload operation and fast growth of electricity price caused by renewable energy subsidies. Therefore, on August 1, 2014, the German Government published the revised German Renewable Energy Sources Act to strictly control the increasing scale of renewable energy generation and reduce funding on newly built projects. Under this new policy, newly installed PVs capacity decreased by 41.3% to around 1.9 GW and kept decreasing for the next 2 years [2]. By the end of 2014, the total German accumulated PV capacity reached 38 GW, and PV generation has turned out to be the largest renewable energy source in Germany: 6.3% of the total energy generated in Germany was from PV generation.

Some effective measures were taken by the German Government to promote PV generation, such as bank loans, feed-in tariffs, and so on. In Germany, if people install PV panels on their own roofs, the generated power can be sold to power grids just like they are running a small power plant, and the country subsidizes up to €0.574/kWh. Currently, the PV industry in Germany has become very active. However, impacted by the widespread European debt crisis and the price decline of PV modules, Germany has begun to repeatedly slash PV subsidies since 2009. The subsidies were cut on July 1 and October 1 of 2010. Compared with 2009, the feed-in tariff of PV installations was decreased by 33–35% in 2011 [17]. According to the latest Renewable Energy Act, the subsidies for renewable energy will be settled through bidding no later than 2017.

While striving to develop its domestic PV industry, Germany also actively expands oversea markets based on its superior technologies. For example, in 2009, Solon SE won the bidding over an 11 MW PV plant project in Spain, which was invested in by Renewable Energies and PVs Spain S.L. (REPS), a majority owned subsidiary of the Norwegian energy company Statkraft AS. In addition, at present, the German company SMA has the largest market share of PV inverters.

1.1.4 China

China has excellent natural resources for exploiting solar energy. Two-thirds of China has considerably good solar irradiance. Provinces like Xizang (Tibet), Qinghai, Xinjiang, Gansu, Ningxia, and Inner Mongolia have the richest solar resources in the country. Eastern, southern and northeastern China are in the second class of the solar irradiance, while the areas of Sichuan Basin and Guizhou Province come last in terms of solar resources.

In the late 1980s, through importing equipment, complete production lines, and manufacturing technologies, the production capability of PV cells reached 4.5 MW in China, and the PV industry was preliminarily set up then. In 2008, the annual production capacity of PV cells reached 2 GW, which took more than 30% of the global market. In 2009, the yearly yield was 7.5 GW, which accounted for 44% of the global production. However, prior to 2008, PV installations in China were very limited, and 90% of PVs production was exported overseas. In 2009, the Chinese Government announced its supporting policies and incentives, such as the Golden Sun Demonstration Project and the Rooftop Project, to promote domestic demand and initiate the massive installation of PV panels in China. The localization in both of the upstream and downstream chains of the PV industry was accelerated, and the production line of PV panels and some key equipment for polycrystalline silicon production was supported by domestic research and development [18]. As Table 1.1 clearly shows, China had six PV manufacturers in the world's top 10 in 2014.

According to the Medium and Long-Term Renewable Energy Development Plan published by the National Development and Reform Commission, the total PV capacity would be 300 MW by the end of 2010, including 150 MW in remote agricultural and pastoral areas, 50 MW on buildings and public facilities, 20 MW large-scale grid-connected PV plants, and 80 MW for the others. The plan also projected to have a total PV capacity of 1.8 GW by the end of 2020, including 300 MW in remote agricultural and pastoral areas, 1 GW on buildings and public facilities, 200 MW large-scale grid-connected PV plants, and 300 MW for the others. This goal was already achieved in 2011, much earlier than the original deadline.

Since several promoting policies and incentives were issued in 2009, like the Solar Rooftop Program, the Golden Sun Demonstration Project, and the Memorandum of Actively Supporting Golden Sun Demonstration Project, new opportunities were opened up for the development of the PV industry [19, 20]. The three programs also showed China's determination on developing the PV industry. In the same year, a number of large-scale PV plants were successively built, such as the Grid-Tied PV Project in Hangzhou Energy and Environment Industrial Park (a total of 20 MW planned and 2 MW completed in the first stage), the 66 MW PV plant in Yulin, Yunan, and China's first 10 MW-scale PV plant in Shizui Mountain. The total PV installed capacity is projected to be 50 GW by the end of 2020, which is equivalent to the capacity of more than 50 large-scale coal-fired power plants.

Figure 1.2 shows Qaidam Basin's million kilowatts solar energy demonstration project, which is the largest grid-connected PV plant in the world. In December 2011, this PV plant's 1.003×10^6 kW generation was successfully interconnected with the Qinghai power grid. It was the first time for the Qinghai grid being penetrated by a million-kilowatt-level of PV power. This project sets five records worldwide: it was the largest group of PV plants, the most concentrated area of PV installations, the largest short-term PV installation in the same region, the largest grid-connected PVs project, and the first time for grid integration of a million-kilowatt-level PV station.

In October 2012, China's State Grid issued the *Suggestions on Supporting Grid-Tied Distributed PV Generation*, in which 15 commitments were made, such as improving the efficiency of the grid-connection service for the distributed PV plant, fee free for some services, and so on. It required all power utilities to fulfill the commitments and to ensure the grid-connection service being carried out in order and smoothly. By the end of January 2013, the utilities had addressed 850 consulting cases in total on grid connection of distributed PV generation, completed 119 grid-connection systems, and installed 338 000 kW of PV panels.

Figure 1.2 Qaidam Basin's million kilowatts solar energy demonstration project.

1.2 Current Research Status of Grid-Connected Photovoltaic Generation

1.2.1 Characteristics of Grid-Connected Photovoltaic Generation

As a distributed generation source, a grid-connected PV system converts the DC output of the PVs to AC power via an inverter; the AC power is then fed into the power grid. Owing to the regional unbalanced, random, and volatile characteristics of the solar resources, the PV generation itself is not dispatchable. Thus, the grid-connection requirements of PV systems vary based on the PV capacity, grid-connection mode, and the target grid. From the perspective of the power grid, on the one hand, as PV generation differs from traditional generation, conventional techniques and calculation methods for grid connection are not suitable for PV systems. On the other hand, at present, the mutual influence between a grid-connected PV system and the grid are still being research, and comprehensive, clear, and operational management standards and technical manuals have not been established yet. It is hard for utilities to do a thorough and credible assessment, covering the aspects of power quality, reliability, stability, security, and standard management. Therefore, connecting PV systems to the power grid can be complex and difficult.

According to the designed capacity, a PV system can be connected to the distribution network at a voltage level of 35 kV, 10(20) kV, or 400 V. The actual operation of a grid-connected PV system has the following characteristics:

1) Currently, a PV grid-connected inverter, normally a voltage source converter (VSC), is primarily controlled by current control to track the voltage of the point of common coupling (PCC) and synchronize with the power grid by controlling the output

current of the VSC. At present, the output PV inverter is pure active power with a unity power factor. However, accompanying the release of relevant standards and regulations, more and more products are equipped with the function of adjusting reactive power output for reactive power support when needed.

2) In order to efficiently explore solar energy, maximum power point tracking (MPPT – see Chapter 2) is normally used for maximizing the output power of the grid-connected PV system.

3) The output of PV generation is heavily dependent on weather. On cloudy days, the PV output can be very volatile.

4) Because of the fast and random fluctuation of the PV output, it is highly possible to have a large peak-to-valley difference of demand from the power grid when the PV penetration level is high. Therefore, it needs the spinning reserve capacity of conventional power units to compensate the power fluctuation, which will increase the overall generation costs.

5) When the output power of a PV inverter is small, the harmonic components become more significant.

6) The correlation between the anti-islanding protection of a grid-connected inverter and the load level is complex. Owing to the low PV/load ratio at present, the protection can identify the islanding by detecting the rapidly declining voltage and frequency when the electric supply from the utility discontinues. However, as the capacity and the number of grid-connected PV generation systems constantly increase, multiple types of inverters with different protection principles will be connected to the power grid and interfere with each other. When the PV generation is close to the load demand level, the anti-islanding detection will take a much longer time or even fails.

7) Overall, the existing PV grid-connection technologies are not grid friendly. With the increase in capacity of grid-connected PV generators the protection technologies of inverters will have a great impact on the safe and stable operation of the power grid and will become a key factor limiting the integration of PV generation. Therefore, it is necessary to study the impact of high-penetration PV generation systems on the power grid.

1.2.2 Impact of High-Penetration Photovoltaic Generations on Distribution Networks

According to the integration methods, grid-connected PV generation can be categorized into centralized and distributed PV systems. A centralized PV plant is usually composed of large-scale PV units and directly feeds the output into a high-voltage grid through inverting. The main problem of centralized PVs is that they need large-scale long-distance transmission to deliver their output power. In addition, since their output is intermittent and random, a large PV plant may bring about severe influences on the frequency and operation of the grid. Distributed PV generation systems are usually connected to low- and mid-voltage distribution networks. A distributed PV generator is close to the load; therefore, it does not require long-distance transmission, and the transmission losses can be significantly reduced. However, the disadvantages of distributed PV generation are low energy density, poor stability, inadequate adjustment, and their output power can be greatly influenced by the surroundings. If distributed PV generation is at a high penetration level, this can induce the following influences [21–23]:

1) ***Bidirectional power flow.*** Different from the one-way power flow in traditional radial distribution networks, distributed PV generation can lead to bidirectional power flow. When the PV output power is higher than the load demand in the distribution network, the extra power is going to be fed back to the transmission grid, which can adversely impact the operation of voltage regulators and the coordination among protection devices.

2) ***Impact on the system voltage.*** The output power of distributed PV generation is heavily influenced by weather and is intermittent. Hence, with high PV penetration, the reverse power flow can increase voltage or even cause overvoltage on some nodes. The rise of voltage is closely related to the position and total PV capacity.

3) ***Impact on the system protection.*** In a distribution network, the short-circuit protection usually takes the approaches of overcurrent protection and fuse protection. As to the distribution network with high PV penetration, the protection may fail to detect the short-circuit faults in the feeder if the PV generation contributes the majority of the short-circuit current. The peak of the short-circuit current is determined by the PV inverter controllers. In addition, when the protection trips to isolate and clear the fault, if the PV generation is still connected to the feeder, an isolated power island powered by the PV generation may be formed, which can lead to the asynchronous reclosing of an automatic reclosing lock and arc restriking at the point of fault.

4) ***Impact on the operation and control of the power grids.*** Because of the uncertainty of PV generation, the accuracy of short-term load forecast is reduced, which makes the planning and operation of traditional generation and the control of exchange power more difficult. Distributed PV generation significantly increases the number of generation in the grid, but these points are dispersed and small, which increases the complexity of the coordination among the power sources and causes the conventional strategies of reactive dispatch and voltage control to be ineffective. In addition, more impacts can also be introduced to peak-shaving, safety reserve, voltage stability, and frequency stability. Therefore, with high PV penetration in the distribution network, the regulation and control of conventional power system is compromised, which brings new challenges for the operation of the power grid.

5) ***Impact on the design, planning and operation of the distribution network.*** With more and more PV generation in the distribution network, the percentage of the power from the central source (i.e., substation) declines, which can result in a significant change in the structure and control patterns of the current grid. The challenges and opportunities due to this call for a fundamental change of the grid in many aspects, such as, but not limited to, design, planning, operation, and control. It is projected that a lot of electricity will be generated in the distribution network; thus, it is markedly important to upgrade and optimize the structure of the distribution network in advance. In addition, when the penetration level of distributed PV generation is high, the customer side can actively participate in energy management and system operation, which will change the traditional operation and sales mode of the distribution network.

1.2.3 The Necessary of Research on Distribution Network with High Photovoltaic Penetration

With the improvement of technologies and the decline of cost, there will be more and more distributed PV generation connected to the distribution network, which brings new issues and. Hence, in order to efficient utilization of solar energy, it is necessary to carry out a comprehensive and deep research on grid-connection technologies of distributed PV generation, as discussed in the following points.

1) *Laboratory development for PV generation–grid interconnection.* In order to study the impacts of distributed PV generation on the safety and power quality of the distribution network, it is necessary to develop research laboratories to test PV systems and investigate different grid connection scenarios. By analyzing the characteristics of PV generation and developing the models of PV control systems, the present power system simulation platform can be updated to have the capability of including the PV generation in the analysis and calculation. Meanwhile, it is also important to establish a database of typical cases of grid-connected PV generation. Therefore, a well-designed and equipped PVs–grid interconnection laboratory is vital for studying the interactions between distributed PV systems and the distribution network.

2) *Research on the mechanism of interaction between PV generation and the grid.* As discussed previously, when the penetration level of PVs is high, it is important to carry out the research on the interaction between PV generation and the grid to discover the mechanisms, which can provide a powerful rationale for updating and improving the technologies for future power systems. Aiming at solving challenges induced by random PV output and power electronic converters, the following mechanisms should be studied by using the simulation platform on the laboratory development for PV grid interconnection. First, the distribution mechanism of power quality distortion. Second, the PVs' response during grid faults and the interaction between the PVs and the grid's control systems. Third, the impacts of PV generation on the voltage, phase, and frequency stability of the grid. The goal of research is to explore the principles of the interaction between the PVs and the grid, develop the relevant theories and methods, and lay out the theoretical framework for the control and stability analysis of the grid with massive PV systems.

3) *Research on collaborative planning for future distribution systems.* When PVs penetration is high, the distribution system will transfer from a system with a single purposed energy-distribution function to one with multiple functions of energy collection, energy transmission, and energy distribution. The integration with PVs and the other distributed generation sources can impact the distribution networks from the aspects of economy, voltage, power flow, short-circuit current, reliability, and power quality. Therefore, the following new challenges are emerging in the planning and design of the power grid.
    - Under high PV penetration, load prediction becomes more difficult. The traditional planning of the generation sources is based on the forecasting of the load distribution; the optimal capacity and location of the sources are then chosen for

balancing the demand and supply. When the PVs are connected, the distribution network becomes active, which makes the traditional source planning methods unsuitable.

- The topology of a traditional distribution network can be radial or networked, which is selected based on the requirements of power supply reliability. A high PV penetration can introduce a significant amount of stochastic power flow in the distribution network. In order to meet this, there needs to be a study of the maximum random power that can be accommodated in the existing distribution network.
- The output of the PV generation changes with the fluctuation of solar irradiance, which causes uncertainty for reactive power optimization of the grid.

4) **Study on the control and operation of power systems with high PV penetration.** On the one hand, when PV penetration is high, stochastic PV generation can lead to uncertain power flows and adversely influence the safe and stable operation of the distribution network. On the other hand, the utilization of PV generation can benefit the distribution network from the aspects of reducing line losses, improving energy structure, increasing energy utilization, and so on. However, how to maximize the benefits by using appropriate economic dispatch is an urgent problem to be solved. Furthermore, extensive research is needed to address the power quality issues, such as the harmonic pollution caused by a large amount of power electronic devices and the grid's three-phase imbalance aggravated by single-phase PVs. Grid disturbances can cause abnormal operations in PV generation, which may in turn intensify the impact of PV generation on the grid. Therefore, research on the operation and control of the power grid with high PV penetration needs to be carried out on the following aspects:

- optimal energy management of the power grid with high PV penetration;
- safety operation, economic dispatch, and optimal control of the grid and the PV generation;
- reactive power dispatch and voltage control strategy;
- load forecasting;
- the impact of PV generation on the safe and stable operation of the distribution network;
- the response of PV generation under grid disturbances.

5) **Research on coordinated protection of distribution networks with PV generation.** Protection is a core function to maintain the safe and stable operation of power systems. When PVs is integrated at a high penetration level, the system faults induced electrical characteristics of the distribution network are altered. Therefore, the traditional fault detection and protection methods need to be updated. To solve the theoretical and technological protection problems faced by a grid with high PV penetration it is necessary to carry out following research:

- To study the characteristics of short-circuit current of grid-connected PV systems and develop simulation models it is required to develop appropriate models of the PVs to accurately characterize external faults.
- It is necessary to develop principles and methods of protection adjustment for different configurations and types of protection devices.

- To ensure the correct operation of protection devices, the research on coordination mechanisms among protection devices plays an important role.

6) ***Research and development of equipment for the monitoring, protection, and control of power systems.*** After analyzing and studying the impacts of PV generation on the distribution network and the mechanism of interaction, it is necessary to develop devices for monitoring, protection, and control of the power system. The main devices need to be developed are:

- *Protection devices considering the integration of PV generation.* New protection methods and devices are required to overcome the PV-induced complex changes of electrical quantities after faults.
- *Islanding detection systems.* For electrical safety and high power quality, there needs to be immediate detection of islanding and taking action on the islanding part and the rest.
- *Real-time monitoring and control systems.* The integration of PV generation leads to more information that needs to be detected and more tasks to be coordinated.
- *New electricity measuring devices.* The PVs may change the power flow to bidirectional in the distribution network. Thus, it is necessary to change the one-way electricity measuring and metering to two-way. Meanwhile, as the PV generation cost is currently higher than the cost of traditional generation, it is also important to study how to reflect this difference in the measurement and metering devices.

7) ***Perfecting the technological standards and regulations for integrating PV generation with power grids.*** At present, the development of distributed generation including PVs is still in the initial stage in China and many other regions. Various technologies are under development, and many standards for integrating PV generation with the grid are immature or even missing. To guide and standardize the integration of renewable distributed generation with the grid and ensure that distributed generation systems and their control devices do not interfere with the grid's safe and stable operation, it is important to improve the technological standards and regulations for integrating PV generation with the power grid, including the standards to identify the technical parameters, control functions, disturbance ride-through capability, grid capacity, voltage level of the PCC, and reactive power and power quality requirement of the PV systems, and the standards to regulate the functions of the distribution network for accommodating distributed generation sources.

1.3 Summary

The current status and future development trends of PV generation around the world have been presented in this chapter. The PV industry development history of the USA, Japan, Germany, and China and the relevant policies, laws, and demonstration projects of these countries have been also introduced.

According to the integration methods, grid-connected PV systems can be categorized into centralized PV plants and distributed PV generation. There are various impacts of high PV penetration on distribution networks, including power flow direction, system voltage, protection, operation, planning, and so on. Therefore, in order to ensure safe and stable operation of the grid, extensive research needs to be carried out from the aspects of the research laboratory development of PV systems, the interaction mechanism between PV generation and the grid, novel planning methods, operation and control, coordinative protection, technological standards, and regulations of distribution networks with high PV penetration.

References

1 Grand View Research. (2012) *Solar PV Market Analysis and Segment Forecasts to 2020*. Grand View Research, Inc.: San Francisco, CA.

2 China Center for Information Industry Development. (April 2015) *White Paper on Photovoltaic Industry Development*. The Ministry of Industry and Information Technology Research Institute CCID: Beijing.

3 http://guangfu.bjx.com.cn/special/?id=818844,2017-04-07.

4 Solangi K., Islam M., Saidur R., *et al.* (2011) A review on global solar energy policy. *Renewable and Sustainable Energy Reviews*, **15**(4), 2149–2163.

5 Timilsina G., Kurdgelashvili L., and Narbel P. (2012) Solar energy: markets, economics and policies. *Renewable and Sustainable Energy Reviews*, **16**, 449–465.

6 Pitt D. and Congreve A. (2016) Collaborative approaches to local climate change and clean energy initiatives in the USA and England. *Local Environment*, 1–18. doi: 10.1080/13549839.2015.1120277.

7 Lau L., Tan K., Lee K., *et al.* (2009) A comparative study on the energy policies in Japan and Malaysia in fulfilling their nations' obligations towards the Kyoto Protocol. *Energy Policy*, **37**, 4771–4778.

8 Liu L., Wang Z., Zhang H., *et al.* (2010) Solar energy development in China – a review. *Renewable and Sustainable Energy Reviews*, **14**(1), 301–311.

9 Khan A., Mondal M., Mukherjee C., *et al.* (2015) A review report on solar cell: past scenario, recent quantum dot solar cell and future trends. *Advances in Optical Science and Engineering*, **166**, 135–140.

10 Sun H., Zhi Q., Wang Y. *et al.* (2014) China's solar photovoltaic industry development: the status quo, problems and approaches. *Applied Energy*, **118**, 221–230.

11 Zhang S., Andrews-Speed P., and Ji M. (2014) The erratic path of the low-carbon transition in China: evolution of solar PV policy. *Energy Policy*, **67**, 903–912.

12 Hanergy. (2015) *The Global New Energy Development Report 2015* (in Chinese). http://www.hanergy.com/upload/contents/2015/04/552e21b0871a3.pdf (accessed February 20, 2017).

13 SEIA. (2014) *U.S. Solar Market Insight Report Overview: 2014 Year in Review. Executive Summary*. http://www.ourenergypolicy.org/wp-content/uploads/2015/03/SEIA-Report.pdf (accessed April 27, 2017).

14 SEIA. (2010) *US Solar Industry Year in Review 2009*. http://www.seia.org/sites/default/files/us-solar-industry-year-in-review-2009-120627093040-phpapp01.pdf (accessed February 20, 2017).

15 Kaizuka I., Ohigashi T., Matsukawa H., and Ikki O. (2010) PV trends in Japan – progress of the PV market by new support measures. In *35th IEEE Photovoltaic Specialists Conference*. IEEE; pp. 136–141.

16 Yamaya H., Corp R., Ohigashi T., *et al.* (2015) PV market in Japan and Impacts of grid constriction. In *2015 42nd IEEE Photovoltaic Specialists Conference (PVSC)*. IEEE; pp. 1–6.

17 Grau T., Huo M., and Neuhoff K. (2012) Survey of photovoltaic industry and policy in Germany and China. *Energy Policy*, **51**, 20–37.

18 Zhao R., Shi G., Chen H., *et al.* (2011) Present status and prospects of photovoltaic market in China. *Energy Policy*, **39**(4), 2204–2207.

19 Zhang S. and He Y. (2013) Analysis on the development and policy of solar PV Power in China. *Renewable and Sustainable Energy Reviews*, **21**, 393–401.

20 Shen J. and Luo C. (2015) Overall review of renewable energy subsidy policies in China – contradictions of intentions and effects. *Renewable and Sustainable Energy Reviews*, **41**, 1478–1488.

21 Chidurala A., Saha T. and Mithulananthan N. (2015) Field investigation of voltage quality issues in distribution network with PV penetration. In *IEEE PES Asia-Pacific Power and Energy Engineering Conference (APPEEC)*. IEEE; pp. 1–5.

22 Nguyen D. and Kleissl J. (2015) Research on impacts of distributed versus centralized solar resource on distribution network using power system simulation and solar nowcasting with sky imager. In *42nd IEEE Photovoltaic Specialists Conference (PVSC)*. IEEE; pp. 1–3.

23 Liu J., Zhang W., Zhou W., and Zhong J. (2012) Impacts of distributed renewable energy generations on smart grid operation and dispatch. In *IEEE Power and Energy Society General Meeting*. IEEE; pp. 1–5.

2

Techniques of Distributed Photovoltaic Generation

2.1 Introduction to Distributed Photovoltaic Generation

2.1.1 Distributed Generation: Definition and Advantages

Distributed generation (DG) refers to relatively small generation systems that are designed, installed, and operated in distribution networks or distributed at the customer side to meet special customers' needs and support the operation of distribution networks based on economic, efficient, convenient, and reliable generation [1–5]. DG capacity is typically between kilowatts and tens of megawatts and is smaller than traditional central power plants. DG is beneficial for improving environment quality, promoting sustainable energy development, and enhancing the competitiveness of green energy sources.

DG can be divided into renewable and nonrenewable types. Electricity is the main output/product of various DG systems, while other forms of output, such as thermal energy, can be obtained by adopting combined heat and power (CHP) generation. CHP generation can make efficient and comprehensive utilization of limited energy resources. DG can operate in two modes: islanded (also called standalone) mode and grid-connected mode. According to the energy conversion technologies used, there are various types of DG systems, including reciprocating generators, Stirling generators, microturbines, gas turbines, fuel cells, PV panels, wind turbines, and microhydro generators. The concept of DG can be extended to distributed energy when distributed energy storage is included.

The IEEE Std 1547.2-2008 provides two classification methods for the DG technologies [6]. According to the types of prime mover, there are rotating types and nonrotating types of DGs. On the other hand, according to the grid interfacing schemes, there can be synchronous, asynchronous, and power electronic-interfaced DG systems.

Compared with the traditional centralized generation, the deployment of DG at mass scale has the following advantages [7–9].

1) *Economic.* As DG systems are normally installed close to load centers, compared with the traditional generation, transmission, and distribution facilities, they can help not only reduce transmission losses, but also greatly reduce the investment required because there is no need of the right of way for transmission lines, which are becoming more and more difficult and expensive to obtain. Compared with the long cycle and large investment of constructing traditional facilities, the benefits of installing DG include the miniaturization and modularization of equipment, short

Grid-Integrated and Standalone Photovoltaic Distributed Generation Systems: Analysis, Design, and Control, First Edition. Bo Zhao, Caisheng Wang and Xuesong Zhang.
© 2018 China Electric Power Press. Published 2018 by John Wiley & Sons Singapore Pte. Ltd.

construction cycle, scalability to closely follow the load increase, economic invest-ment, and low risk.

2) ***Environmentally friendly.*** Renewable (e.g., solar and wind) and clean (e.g., fuel cell) technologies have been widely used in DG systems to reduce the emissions due to electricity/energy development.

3) ***Reliable.*** After DG systems are connected to the grid, the reliability of the distri-bution network can be enhanced by employing appropriate operation schemes. For example, once a large-scale blackout occurs, the DG systems with islanding capability can still support their loads by intentional islanding from the main grid.

4) ***Secure.*** As already mentioned, various generation technologies and energy sources can be used for DG. Therefore, the massive employment of DG systems is an effec-tive approach to power supply diversity, energy safety, and energy crisis mitigation. In addition, the DG dispersibility and miniaturization are helpful in making them resilient in disasters and other emergency situations.

5) ***Flexible.*** DG systems usually utilize small and medium-sized modular devices, which have the merits of fast start-up and shutdown, easy maintenance and man-agement, simple operation and control, and flexible load adjustment. Various DG units in a system can be quite independent of each other, and are capable of pro-viding customizable functions, such as peak shaving, specific supply for remote or important customers, and so on.

2.1.2 Principle and Structure of Distributed Photovoltaic Generation

The development of PV technology has been growing very fast recently. PV genera-tion basically transform solar power into electric power through photo-thermal con-version, photoelectric conversion, or photochemical conversion. At present, PV con-version is the most mature and widely used solar energy application for electricity gen-eration. The PV effect is the creation of voltage and current in the solar cells upon exposure to light, causing excitation of an electron or other charge carrier to a higher energy state after the solar energy is absorbed. PV generation has the advantages of being fuel-less, clean and renewable, free of geographical restrictions, scalable in size, easy to maintain, and safe and reliable in operation, and so on. However, PVs also have the disadvantages of low energy density, high investment, high weather sensitivity, and so on.

Nowadays, PVs are mainly used in the following applications [10, 11]:

1) Solar lights and traffic lights. For example, around 90% of the street lights in the 2008 Beijing Olympic Village were powered by PVs.

2) Standalone PV generation to meet basic household electricity demands in remote areas or hard-to-reach sites.

3) Large-scale PV plants. These are usually built in areas with abundant solar and land resources and connected to high-voltage transmission networks;

4) Distributed, residential small grid-connected PV generation . Installed on the rooftops of residential buildings, these PV generation systems are dispersely connected to the distribution system, and their generated electricity is consumed locally.

5) Power supply for telecommunication cellular base stations.

6) PV-based consumer electronics, such as solar toys, small solar chargers, and so on.

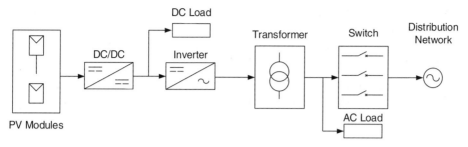

Figure 2.1 A grid-connected PV system.

Grid-connected PV generation can be categorized into centralized PVs and distributed PVs. Distributed PV generation is normally installed at the distribution network level with a voltage level lower than 63 kV and a size normally no greater than 10 MW.

Figure 2.1 shows the generation process diagram for grid-connected distributed PV generation. First of all, the PV modules convert the solar energy into electrical energy. The voltage can be boosted to the required voltage through a DC/DC converter first if needed and then is inverted into AC. In the end, after filtering, the AC power is delivered back into the distribution network or consumed locally. As shown in Figure 2.1, the system mainly consists of PV modules, inverter, controllers and transformer, and maybe batteries and a solar tracking system in some applications [12].

1) **PV module.** PV modules are the core element of any PV generation. The quality and cost of the entire system is mainly determined by the PV modules. A PV module is formed via series and parallel connections of PV cells.

2) **Inverter.** The PV modules produce DC power, which needs to be converted to AC, to meet the AC load needs, via an inverter. In addition to the main function of energy conversion from DC to AC, the inverter usually has other functions, such as MPPT and protection. The MPPT function can be accomplished via the DC/DC converter shown in Figure 2.1.

3) **Charging/discharging controller of battery.** For a PV system with battery storage, the charging/discharging controller of battery manages the battery energy storage system to guarantee proper charging/discharging control of the battery. If the battery is inappropriately managed (charged/discharged), the battery life can be seriously impacted. Thus, to have a well-designed battery charging/discharging controller is essential for the battery energy system.

4) **Transformer.** If the output voltage of a PV system is not high enough, a step-up transformer can be considered to further boost the voltage to meet the load needs and/or deliver the requirements of the grid.

5) **Solar tracking system.** For a specific area, the altitude angle of the sun changes with respect to the time of day and the seasons. A solar tracking system aims to keep the panels always pointing directly at the sun to maximize the electricity generation and improve efficiency.

6) **Battery energy storage.** As the PV output is intermittent during the day and is zero at night, a battery energy storage system may be needed to store the excess electric energy and release the electricity to support the loads when there is not enough

PV generation. In the PV system, the battery energy storage is required to meet the following requirements: low self-discharge rate, strong charging–discharging capability, easy maintenance, long lifespan, low price, and large range of working temperature.

In a grid-connected PV system, the battery energy storage usually is not necessary, and the efficiency of the PV generation mainly depends on the efficiency of the PV cells, inverter, and MPPT control strategy.

2.2 Photovoltaic Cells

A PV cell is usually made from solid semiconductor materials. In general, a PV material should have the following features: appropriate bandgap of semiconductor, high photo-electric conversion efficiency, stable performance, environmentally friendly, and ease of manufacture.

2.2.1 Classification of the Photovoltaic Cells

2.2.1.1 Classification Based on Cell Structure

- *A **homojunction solar cell*** has one or more P–N junctions formed by the same semiconductor material.
- *A **heterojunction solar cell*** is a hetero P–N junction-based solar cell on the interface of two different semiconductor materials with different bandgaps if the two materials have similar structure and are well matched on the interface.
- *A **Schottky junction solar cell*** has a Schottky barrier formed at a metal–semiconductor junction. Its principle is based on the rectifying characteristics of the Schottky effect at the metal–semiconductor junction under certain conditions. This type of PV cell is currently used for metal–oxide–semiconductor PV cells, metal–insulator–semiconductor PV cells, and so on.
- *A **thin-film solar cell*** is a second-generation solar cell that is made by depositing one or more thin layers or thin films of PV material on a substrate. Thin-film technologies largely reduce the amount of semiconductor material in a cell, which is helpful to reduce the cost of solar cells. There are quite a few materials that can be used to make thin-film solar cells, mainly including polycrystalline silicon, noncrystalline silicon, CdTe, copper indium selenide (CIS), and so on, among which the polycrystalline silicon thin film solar cell has the best performance.
- *A **multi-junction solar cell*** is comprised of two semiconductor layers stacked one upon the other, with each layer having different light-wave absorption capabilities. According to the light wave's characteristics, short-wavelength light is usually absorbed by the top wide bandgap material, while the long-wavelength light is absorbed by the bottom narrow bandgap material. In this way, the solar energy can be converted to electricity as much as possible.
- *A **wet PV cell*** is formed by adding electrolyte into conducting glass whose both sides are with light-active semiconductor films. This kind of solar cell not only reduces the use of semiconductor materials, but also provides possibilities to integrate buildings and solar energy applications together.

2.2.1.2 Material-based PV Cell Classification

- *Silicon solar cell.* This group of cells includes monocrystalline silicon cells, poly-crystalline silicon cells, and noncrystalline silicon cells. The monocrystalline silicon cell has a homogeneous crystallinity, has high carrier mobility, small series resistance, high photoelectric conversion efficiency (around 20%), but also high cost. As to the polycrystalline silicon cell, its crystal direction is irregular. The photoelectric conversion efficiency of polycrystalline silicon is lower than that of monocrystalline silicon, but the former has a lower cost. Polycrystalline silicon materials can be divided into multiple types, such as ribbon silicon, foundry silicon, thin-film polycrystalline silicon, and so on. Meanwhile, the solar cells made by the monocrystalline and poly-crystalline silicon materials can also be further divided into two types: the thin-film type and the sheet type. In terms of noncrystalline silicon PV cells, they are made from noncrystalline silicon (a-Si) materials that are short-range ordered but long-range dis-ordered and usually used to make thin-film solar cells. Noncrystalline silicon PV cells have the advantage of low cost, but also with low photoelectric conversion efficiency and stability. At present, they are mainly used in weak light PV cells, like the cells for watches and calculators.
- *Noncrystalline silicon semiconductor solar cell.* These cells can mainly be categorized into CdS solar cells and GaAs solar cells. The CdS cells can be further divided into monocrystalline and polycrystalline and are usually synthesized with other semiconductor materials to make CuS/CdS solar cells, CdTe/CdS solar cells, CIS/CdS solar cells, and so on. As for GaAs, it has appropriate temperature characteristic and high photoelectric conversion efficiency and is suitable for space applications. GaAs can be in either heterojunction form or homogeneous form, in either monocrystalline sheet structure or thin-film structure to make PV cells.
- *Organic solar cell.* These cells are mainly made from organic photoelectric polymer materials [13].

2.2.2 Development History of Solar Cells

The first-generation solar cells were invented by Bell Laboratories in 1954 [14]. Owing to the fossil-fuel energy crisis, renewable energy became increasingly of interest, which promoted the rapid development of solar cells. The first-generation solar cells are made from crystalline silicon materials, including monocrystalline silicon (shown in Figure 2.2) and polycrystalline silicon (shown in Figure 2.3). In 2004, the first-generation solar cells shared 86% of the global market. Although these first-generation solar cells have mature manufacturing technologies, their cost, material usage, and relevant energy consumption are very high. Therefore, in order to utilize solar energy more efficiently and widely, there was a need to reduce the material and manufacturing costs even with the sacrifice of conversion efficiency to develop the second-generation cells based on the thin-film solar cells.

The second-generation solar cells are based on thin-film technology by depositing a very thin layer of only 1 μm PV material on a non-silicon substrate, which significantly reduces the use of semiconductor materials. A thin-film cell has a unit area that is 100 times that of the first-generation solar cells. In addition, it is easy to manufacture, which results in a dramatically reduced cost. There are quite a few materials that can be used to make thin-film solar cells, mainly including a-Si, p-Si, CdTe, and CIS, among

Figure 2.2 Monocrystalline silicon solar cell.

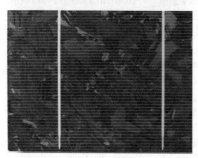

Figure 2.3 Polycrystalline silicon solar cell.

which the polycrystalline silicon solar cell has the best performance. Although the cost of noncrystalline silicon solar cells is low, it was discovered that its material properties can degrade due to the fatigue effect and its photoelectric conversion efficiency can be reduced after exposure to sunlight. Usually, the noncrystalline silicon cell's photoelectric conversion efficiency is under 10%, which is much lower than the 18% of the polycrystalline silicon solar cells. Recent studies show that the photoelectric conversion efficiency of the polycrystalline silicon solar cell has been continuously improved to that of the monocrystalline silicon solar cell and is able to provide stable output.

Currently, the third-generation solar cells, which aim to improve the photoelectric conversion efficiency and reduce the relative production cost, are still at the experimental stage. In order to improve the efficiency, the endeavors are carried out in three directions. First, reducing relevant energy losses such as heat loss; second, improving the effective utilization of the photon; and third, reducing the internal resistance of solar cells. The types of third-generation solar cells are mainly multi-junction cells, multi-energy-band cells, hot PV cells, hot-carrier cells, and quantum dot cells.

Table 2.1 compares the various solar cells in terms of photoelectric conversion efficiency, cost, lifetime, and so on. The table clearly shows that the monocrystalline and polycrystalline silicon solar cells take the biggest shares of the market. Compared with the wafer solar cells, the efficiency and lifetime of the thin-film solar cells needs to be significantly improved.

2.2.3   Model of a Silicon Solar Cell

A solar cell is the basic generation unit in a PV system. Since the output voltage, current, and power of a cell are all very small, multiple cells need to be connected in series and parallel to form a PV module, which has an output voltage up to tens of volts. In addition, the output voltage and power can be further increased by connecting PV modules in series and parallel to form PV arrays.

Table 2.1 Comparison of the solar cells.

Types / Standard environment	Thin film				Wafer	
	a-Si	CdTe	CIS	a-Si/μc-Si	m-Si	p-Si
Cell efficiency (%)	5–7	8–11	7–11	8	16–19	15–16
Module efficiency (%)					13–15	12–14
Module area (m²/kW)	15	11	10	12	7	8
Market share (%)	2.3	4.7	0.5	5.2	42.2	45.2
Lifetime (years)	>10				>20	

m-Si: monocrystalline silicon; μc-Si: microcrystalline silicon; p-Si: polycrystalline silicon.

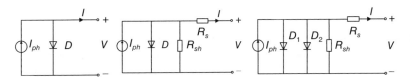

Figure 2.4 The equivalent circuits of solar cells: (a) the solar cell ideal circuit model; (b) single-diode equivalent circuit; (c) double-diode equivalent circuit.

The ideal equivalent circuit of a solar cell, with no internal losses considered, is shown in Figure 2.4a, which is formed by a photogenerated current source in parallel with a diode [15]. It is worth noting that this diode is not an ideal switch that can be switched between the on and off modes and has continuous nonlinearity between its voltage and current. The solar cell's internal losses can be modeled by adding a series resistor R_s and a parallel resistor R_{sh} to the ideal model, as shown in Figure 2.4b. Compared with the single-diode equivalent circuit in Figure 2.4b, one more diode can be added to model the space charge's diffusion effect, shown in Figure 2.4c. The circuit in Figure 2.4c is called a double-diode equivalent circuit, and the circuit can match the output characteristics of polycrystalline silicon solar cells, especially when the solar irradiance is low.

The output volt-ampere characteristic of the double-diode model can be described as

$$I = I_{ph} - I_{s1}[e^{q(V+IR_s)/kT} - 1] - I_{s2}[e^{q(V+IR_s)/AkT} - 1] - \frac{V + IR_s}{R_{sh}} \tag{2.1}$$

For the simplified single-diode model, its output volt-ampere characteristic is

$$I = I_{ph} - I_s[e^{q(V+IR_s)/AkT} - 1] - \frac{V + IR_s}{R_{sh}} \tag{2.2}$$

where V is the output voltage of the solar cell, I is the output current of the solar cell, I_{ph} is the light current, I_{s1} is the saturation current of the diode's diffusion effect, I_{s2} is the saturation current of the diode's compound effect, I_s is the saturation current of the diode, q is the electron charge constant (around 1.602×10^{-19} C), k is Boltzmann's constant (around 1.381×10^{-23} J/K), T is the PV cell's absolute operating temperature, A is the diode fitting coefficient (a variable that needs to be estimated in the single-diode model for different solar cells, while it is a fixed value of 2 in the double-diode model), R_s is the series resistance, and R_{sh} is the shunt resistance.

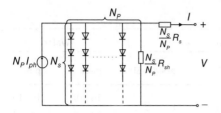

Figure 2.5 Single-diode model-based PV array's equivalent circuit.

Usually, it is assumed that the PV modules that build a PV array have identical characteristics. If the connecting resistances among the PV modules are ignored, the single-diode-model-based PV array's equivalent circuit is as shown in Figure 2.5.

The relationship between the output voltage and current of the PV array in Figure 2.5 can be represented as

$$I = N_\mathrm{P}I_\mathrm{ph} - N_\mathrm{P}I_s\{e^{(q/AkT)[(V/N_S)+(IR_s/N_\mathrm{P})]} - 1\} - \frac{N_\mathrm{P}}{R_\mathrm{sh}}\left(\frac{V}{N_S} + \frac{IR_s}{N_\mathrm{P}}\right) \tag{2.3}$$

where N_S is the number of series cells and N_P is the number of shunt cells.

Furthermore, if the double-diode model-based equivalent circuit is used, shown in Figure 2.6, the relationship between the output voltage and current of a PV array can be written as

$$I = N_\mathrm{P}I_\mathrm{ph} - N_\mathrm{P}I_{s1}\{e^{(q/kT)[(V/N_S)+(IR_s/N_\mathrm{P})]} - 1\} - N_\mathrm{P}I_{s2}\{e^{(q/AkT)[(V/N_S)+(IR_s/N_\mathrm{P})]} - 1\}$$
$$- \frac{N_\mathrm{P}}{R_\mathrm{sh}}\left(\frac{V}{N_S} + \frac{IR_s}{N_\mathrm{P}}\right) \tag{2.4}$$

Typical *I–V* and *P–V* curves of a PV cell/array are shown in Figure 2.7. Three important points shown in this figure are:

1) $(0, I_\mathrm{sc})$: output short-circuit point, where I_sc is the short-circuit current when the PV's output voltage is zero;
2) $(V_\mathrm{oc}, 0)$: output open-circuit point, where V_oc is the open-circuit voltage when the PV's output current is zero.
3) $(V_\mathrm{mp}, I_\mathrm{mp})$: maximum power output point, where $dP/dV = 0$ and the output power $P_\mathrm{mp} = V_\mathrm{mp}I_\mathrm{mp}$ is the maximum power that can be obtained under certain operating conditions. In a practical system, the load should be adjusted to operate the whole system near the maximum power point, which can improve the operation efficiency to the maximum extent.

The output characteristic of PV generation is largely influenced by the solar irradiance and ambient temperature. Figures 2.8 and 2.9 show the impacts of the irradiance and the

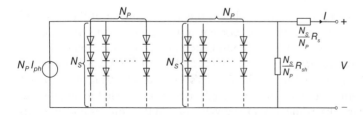

Figure 2.6 Double-diode model-based PV array's equivalent circuit.

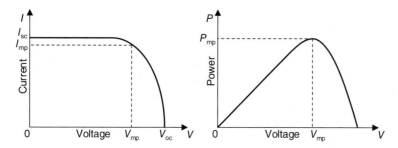

Figure 2.7 Typical *I–V* and *P–V* curves of a PV cell/module.

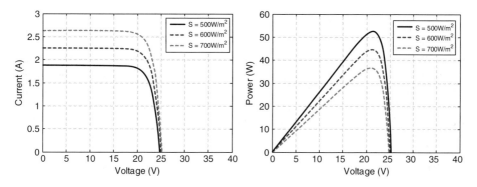

Figure 2.8 The impact of light irradiance on *I–V* curve and *P–V* curve.

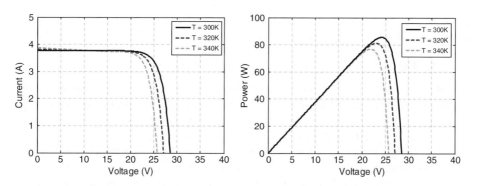

Figure 2.9 The impact of surrounding temperature on *I–V* curve and *P–V* curve.

ambient temperature respectively on the *I–V* and *P–V* curves. Figure 2.9 clearly shows that a rise in temperature results in an increase of the PV cell's short-circuit current and a decrease of the open-circuit voltage. The change in open-circuit voltage with respect to temperature is more significant than that of the short-circuit current. Hence, for a constant light irradiance, the higher the temperature is, the smaller is the maximum power that can be generated and the larger is the voltage change at the maximum power point. On the other hand, an increase in the irradiance can correspondingly boost all of the short-circuit current, open-circuit voltage, and maximum power. In addition, a change in irradiance has little impact on the maximum power point voltage.

It is usual to use Equations 2.1–2.4 to model a PV cell or an array in a distributed generation system simulation study. The $I–V$ curve of a PV array can be obtained by carrying out actual tests on the PV array. Then, based on the measured $I–V$ curve, the parameters of the equations (i.e., the model) can be identified by parameter fitting. In general, the $I–V$ curve provided by manufacturer is obtained under the standard conditions suggested by IEC60904; that is., 1000 W/m² irradiance, 25 °C (298 K) working temperature. In consideration of the actual irradiance and temperature are different from the standard suggestions, the parameters need to be finely tuned. Take Equation 2.2 as an example; the light current I_{ph} and saturation current I_s should be adjusted as

$$I_{ph} = \frac{S}{S_{ref}}[I_{ph,ref} + C_T(T - T_{ref})] \tag{2.5}$$

$$I_s = I_{s,ref}\left(\frac{T}{T_{ref}}\right)^3 e^{(qE_g/Ak)[(1/T_{ref})-(1/T)]} \tag{2.6}$$

where S (W/m²) is the actual irradiance, S_{ref} is the irradiance under the standard condition (i.e., 1000 W/m²), C_T (A/K) is the temperature coefficient, $I_{s,ref}$ (A) is the diode's saturation current under the standard condition, and E_g (eV) is the bandgap width, which is related to the material of the PV cell. T is the PV cell's actual operating temperature (K), and T_{ref} is the cell's operating temperature under standard conditions (298 K).

2.3 Inverter

An inverter is one of the key devices in a PV system, which can track the maximum output power of the PV array with respect to the irradiance level and follow the grid voltage to adjust the amplitude and phase of the output power. The output power and power factor of the inverter can also be adjusted according to dispatch commands.

2.3.1 Topology of Connection between Inverter and Photovoltaic Module

The types of connections between the inverter and PV modules mainly consist of the centralized inverter connection topology (① in Figure 2.10), the series inverter topology (② in Figure 2.10), the module integrated inverter topology (③ in Figure 2.10), and the multiple series inverter topology (④ in Figure 2.10) [16].

PV plants larger than tens of kilowatts usually use topology ①, in which the PV modules are in series and parallel connection to form arrays and connected to a high-voltage DC bus. This topology has low unit price and is suitable for high power applications, but the outputs of the modules in series and parallel can be mutually influenced. In addition, the centralized MPPT strategy used for this topology may largely underperform when some of the modules in the PV array are overshadowed or the photoelectricity conversion efficiency is poor. This can reduce the output of the entire system.

In contrast with the centralized inverter connection topology, every serial module has an individual inverter in the series inverter topology. Therefore, this topology has a higher cost than topology ①, but every serial module can independently track the maximum output power. The typical output power of each serial module is around 2 kW. The topology of a series inverter is shown in Figure 2.11.

In the module integrated inverter topology, the number of cells in each serial module is much less than that in topology ①. Every serial module can also independently

Figure 2.10 Four connection
topologies of the inverter.

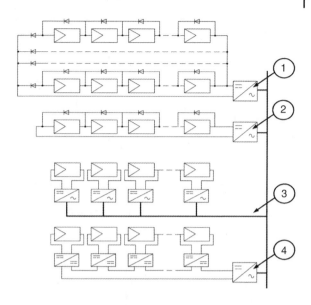

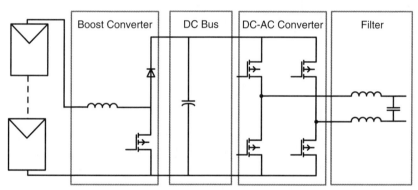

Figure 2.11 The topology of the series inverter.

achieve the MPPT and has the output power varying from tens to hundreds of watts. Therefore, the choice of switches in these inverters is flexible, and even metal–oxide–semiconductor field-effect transistor switching devices can be used. In addition, this topology is small and flexible in integrating with the power grid. However, the overall conversion efficiency of the inverters is lower than that of a large power inverter. The topology is illustrated in Figure 2.12.

By combining the advantages of the centralized inverter connection topology and the module integrated inverter topology, the multiple series inverter topology has individual MPPT for each module and high conversion efficiency of its inverters. The multiple series inverter topology is shown in Figure 2.13.

A PV plant can directly integrate with the power grid or through an isolated transformer. Usually, residential PV generation is directly connected to the grid or through power frequency isolated transformers. Step-up transformers are used when PV generation is connected to the grid at a voltage level of 10 kV or higher.

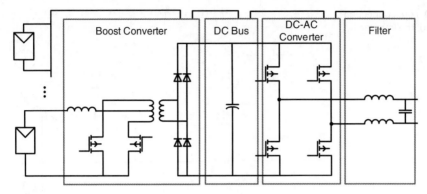

Figure 2.12 The module integrated inverter topology.

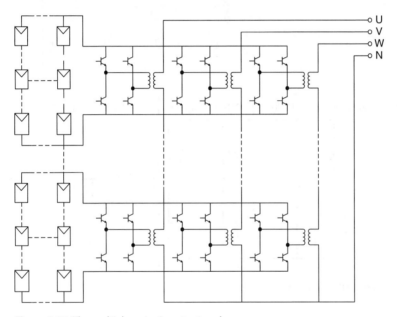

Figure 2.13 The multiple series inverter topology.

2.3.2 The Classification and Characteristics of the Inverter

The inverter roughly makes up 12.5% of the total cost for a typical PV system and is the key power regulator. The classification of inverters is shown in Table 2.2.

Low-power inverters are usually used for single-phase PV generation and can be divided into single-stage inverters and multistage inverters. A single-stage inverter uses one circuit to convert the boosted DC input into AC output. According to the number of switches, the single-stage inverters can be further divided into the topologies of a four-switch bridge and a six-switch bridge. A multistage inverter has a multistage electrical energy conversion circuit, which mainly includes multistage boost, buck, and/or electrical isolation. The multistage inverter can be divided into DC–DC–AC topology, DC–AC–DC–AC topology, and DC–AC–AC topology. The front end of the main circuit converts the DC into AC, while the rear end regulates the DC voltage to

Table 2.2 The classification of PV inverters.

Inverter	
Phase	Single-phase inverter, three-phase inverter, multiphase inverter
Frequency	Power-frequency inverter, intermediate-frequency inverter, high-frequency inverter
Voltage waveform	Square-wave inverter, switch-ladder inverter, sine-wave inverter
Circuit principle	Self-oscillation inverter, pulse-width-modulated inverter, resonance inverter
Circuit topology	Single-ended inverter, half-bridge inverter, full-bridge inverter, push–pull inverter
Power	Low-power inverter, medium-power inverter, high-power inverter
Output power flow	Active inverter, passive inverter

Figure 2.14 A low-power single-phase PV grid-connected inverter with a flyback converter.

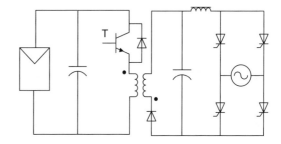

fulfill the MPPT control. The lower power PV inverters usually have a more complex topology.

A single-phase PV inverter usually uses the rear-end converter (boost, buck-boost, or flyback) to realize the MPPT control. Figure 2.14 shows the topology of a low-power single-phase PV inverter which uses a flyback converter (DC–DC–AC topology) to control the output DC voltage of the PV array for generating maximum output power.

High-power conversion is usually carried out through three-phase PV inverters. Series PV modules larger than 10 kW can be connected directly to a 0.4 kV grid. Their output DC voltage is high enough for the MPPT control and a DC–DC circuit is no longer needed on the DC side. For inverters with a DC–DC circuit, the DC voltage generated by the PV array can be adjusted by controlling the duty cycle of the switches in the DC converter. As for the DC–DC-less inverters shown in Figures 2.15 and 2.16, the DC voltage of the PV array must be controlled by regulating the output AC current, which can be achieved by adjusting the modulation index of the sinusoidal pulse width modulation or through other algorithms like hysteresis control, deadbeat control, and so on. The schematic diagram of a three-phase grid-connected system is shown in Figure 2.17.

2.3.3 Requirements of a Grid-Connected Photovoltaic Inverter

The inverter plays a very important role in a grid-connected PV system, converting the DC power generated by the PV cells into AC that can then be injected into the power grid. Compared with inverters that are free from connecting to the grid, the requirements of a grid-connected inverter are stricter, because the grid-connected inverter

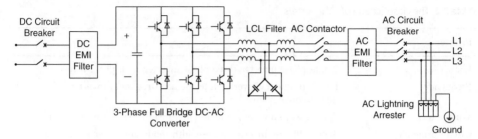

Figure 2.15 A high-power three-phase grid-connected inverter without isolation transformer.

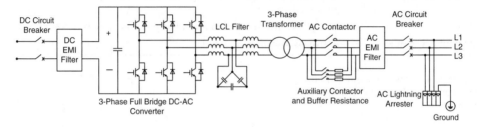

Figure 2.16 A high large-power three-phase grid-connected inverter with a power transformer.

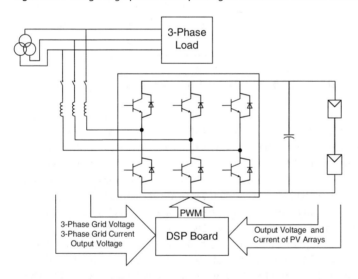

Figure 2.17 The schematic diagram of a three-phase grid-connected system.

needs not only to change the power from DC to AC, but also control the frequency, voltage, current, phase, and power quality of its output to meet the grid connection requirements. Some specific requirements are given as follows:

1) **High efficiency.** As the output voltage and current of the PV cells vary with the irradiance and ambient temperature, the inverter should be able to automatically control the PV cells to generate maximum power output. The inverter is one of the most important components for the efficient operation of a PV system; thus, it is necessary to enhance the efficiency of the inverter.

2) **High reliability.** As PV systems installed in remote areas have high operation and maintenance costs, the inverters in such applications are required to be robust and reliable. In addition, various protection functions are required for the inverters. At present, inverters should have the following protection:
   - protection of overvoltage and undervoltage from the grid;
   - Protection of overfrequency and underfrequency from the grid;
   - short-circuit protection on the AC output side;
   - islanding protection;
   - overheating protection;
   - DC polarity reversed protection;
   - DC overvoltage protection;
   - overload protection;
   - earth-leakage current protection;
   - internal self-checking protection to check lightning protection damage, analog/digital channel damage, insulated gate bipolar transistor damage, and so on.

   In addition, the inverter may have the other protection, like low-voltage ride through (LVRT) protection, protection for overhigh output DC component, protection for overhigh output current harmonics, anti-reverse current protection, and three-phase imbalance protection. It is noteworthy that the LVRT should be well coordinated with the islanding protection.

3) **Accommodation of large fluctuations of input DC voltage.** The terminal voltage of the PV module varies with load, irradiance, and ambient temperature. Even though a battery or capacitor can clamp or mitigate the voltage variation of the PV cells, battery degradation after long-time use or the limitation of capacitor capacity can limit the voltage regulation effect, and the voltage may show fluctuations to some extent. Especially when the battery is aging, its terminal voltage varies significantly; for instance, a 12 V battery can fluctuate between 10 and 16 V. Because of the uncertainties introduced by aging batteries, inverters must be able to adapt to the fluctuations of the input DC voltage and still generate appropriate and stable output AC voltage.

4) **Detection and protection of islanding.** When a power outage occurs, it is hard to detect the outage if the load demand is equal to the PV output, and the PV generation can continuously supply power to the local area. However, this is very dangerous to power system maintenance crews. Therefore, the islanding phenomenon should be avoided on direct grid-tied inverters, and the protection of the inverter must act immediately to separate the inverter from the grid once the outage on grid happens. According to the IEEE Std 929-200 and UL1741 standards, all grid-connected inverters must be designed with islanding protection capability. The maximum islanding detection time defined in the two standards is shown in Table 2.3. The islanding of a PV inverter can be detected by passive or active methods. Passive methods mainly detect the amplitude, frequency, and phase of the voltage and the harmonics in the grid. When power failures happen, certain changes in the amplitude, frequency, and phase of the voltage and in the harmonics in the grid occur that can be used to estimate whether the grid is with power or not. As to active detection methods, they inject a small disturbance signal into the grid at the grid-connected point and then detect the feedback signal to estimate whether a grid outage has happened or not.

Table 2.3 The restriction of the maximum islanding detection time.

State	Grid voltage amplitude after outage	Grid frequency after outage	Allowed maximum detection time
A	$V < 0.5\,V_{nom}$	f_{nom}	12 ms
B	$0.5\,V_{nom} \le V < 0.88\,V_{nom}$	f_{nom}	2 s
C	$0.88\,V_{nom} \le V \le 1.10\,V_{nom}$	f_{nom}	2 s
D	$1.10\,V_{nom} < V < 1.37\,V_{nom}$	f_{nom}	2 s
E	$1.37\,V_{nom} \le V$	f_{nom}	2 s
F	V_{nom}	$f < f_{nom} - 0.7$	12 ms
G	V_{nom}	$f < f_{nom} + 0.5$	12 ms

Note: V_{nom}, the nominal amplitude of the grid voltage; f_{nom}, the nominal frequency of the grid voltage.

5) ***To meet the requirement on power quality.*** The switching harmonics generated by power electronic inverters can "pollute" the grid. In addition, the variation and intermittency of PV generation can cause voltage fluctuations and flickers. Therefore, the capability to restrain power quality issues, like harmonics and flickers, is an important criterion to evaluate the performance of a PV inverter. Taking the harmonic current assessment method in the IEEE 929 standard as an example, the harmonic currents from the second to thirty-third are divided into four classes based on their ratios to the total current, which are 4%, 2%, 1.5%, and 0.6%, and it is restricted that the total harmonic distortion should be less than 5%.

6) ***To meet the requirements on electromagnetic compatibility.*** The influences from the grid on grid-connected PV inverters mainly include voltage rise/drop, frequency drift, three-phase imbalance, and electrical noise. The inverters need to be equipped with functions to endure these interferences. On the other hand, PV inverters also cause disturbances to the grid, consisting of current harmonics, voltage fluctuations, flickers, reactive power consumptions, and so on. The impacts of the inverters on other electrical equipment include conduction interferences, space radiation interferences, and so on. These electromagnetic interferences must satisfy the requirements of standards such as IEEE 1547.

2.4 Maximum Power Point Tracking Control

The output of a PV array is nonlinear and changes with the solar irradiance level, ambient temperature, and load. At a given irradiance and ambient temperature, the PV array can work with different output voltages. However, the PV array only generates the maximum output power at a specific output voltage, which corresponds to the highest point of the PV's *P–V* curve, shown in Figure 2.18. This point is also called the maximum power point (MPP). Hence, in order to obtain efficient PV output an important approach is continuously adjusting the operating point of the PV array to be around the MPP. This process is called MPPT [17–20].

Many MPPT algorithms reach the optimal point by regulating the DC output voltage of the PV array. For the two-stage inverter (DC–DC–AC), the DC voltage can be

Figure 2.18 Output power characteristic curve of a PV array.

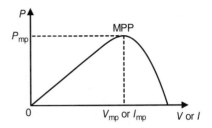

controlled by adjusting the duty cycle of the DC–DC converter (boost, buck-boost, Ćuk or flyback). On the other hand, for a single-stage three-phase grid-connected PV inverter without the DC–DC stage, the control of the DC voltage can directly affect the MPPT efficiency and the dynamic performance of the inverter.

Several common MPPT algorithms are introduced in the following subsections.

2.4.1 Hill Climbing/Perturb and Observe

The hill climbing/perturb and observe (P&O) method [21] determines the MPP by observing the change in direction of the PV output power after periodically increasing or decreasing the output voltage. If the output power increases, the control system further changes the operating point in that direction; otherwise, the operating point will be changed in the opposite direction.

Figure 2.18 clearly shows that, if the operating point is on the left side of the MPP, the output power changes in the same direction as the voltage; otherwise, the output power changes in the opposite direction. Hence, if the power variation is positive, the voltage change will be kept the same as the perturbation direction to reach to the MPP; and if the power variation is negative, the voltage will be changed to the opposite direction until the power no longer increases.

The P&O method is easy to implement, and the accuracy of the sensors does not need to be very high. The disadvantage is that it may result in oscillations of power output near the MPP. The oscillation amplitude is determined by the perturbation step. The larger the perturbation step is, the faster the tracking is, but the more severe the fluctuation. Therefore, the design needs to consider the compromise between the tracking accuracy and the speed of response. According to the literature, this contradiction can be eased by adopting a variable-step algorithm that is to determine the length of the next step by using some adaptive method, such as a fuzzy control strategy [22].

In addition, if the solar irradiance changes dramatically, the P&O method may become invalid. As shown in Figure 2.19, given the PV array operates at point A on power curve I, when the P&O method works normally, the operating point moves to A′ after a perturbation ΔV. However, if the solar intensity drops dramatically at that moment, the power curve I shifts to power curve II, and the operating point moves to point B. Owing to $\Delta V > 0$ and $\Delta P < 0$, the P&O method estimates the PV array works on the right side of the MPP, and then it determines the next perturbation step ΔV should be negative and the operating point of the PV array will improperly move to the MPP's left side. If the solar irradiance continuously drops, the algorithm keeps $\Delta V < 0$ and the operating point will repeatedly move toward the left, which takes the operating point increasingly far away from the MPP. Finally, it loses the ability to track the MPP. This problem can

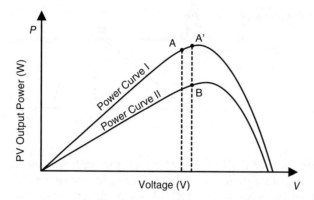

Figure 2.19 The P&O method becomes invalid when the solar irradiance changes dramatically.

be solved by increasing the perturbation frequency as well as decreasing the perturbation step.

2.4.2 Incremental Conductance

Incremental conductance (IncCond) [23] is an upgrade to the P&O. According to the PV array's $P–V$ characteristic:

$$\begin{cases} \dfrac{\mathrm{d}P}{\mathrm{d}V} = 0 & \text{MPP} \\[2mm] \dfrac{\mathrm{d}P}{\mathrm{d}V} > 0 & \text{on left of MPP} \\[2mm] \dfrac{\mathrm{d}P}{\mathrm{d}V} < 0 & \text{on right of MPP} \end{cases} \tag{2.7}$$

Hence

$$\frac{\mathrm{d}P}{\mathrm{d}V} = \frac{\mathrm{d}(IV)}{\mathrm{d}V} = I + V\frac{\mathrm{d}I}{\mathrm{d}V} \approx I + V\frac{\Delta I}{\Delta V} \tag{2.8}$$

So

$$\begin{cases} \dfrac{\mathrm{d}I}{\mathrm{d}V} = -\dfrac{I}{V} & \text{MPP} \\[2mm] \dfrac{\mathrm{d}I}{\mathrm{d}V} > -\dfrac{I}{V} & \text{on left of MPP} \\[2mm] \dfrac{\mathrm{d}I}{\mathrm{d}V} < -\dfrac{I}{V} & \text{on right of MPP} \end{cases} \tag{2.9}$$

According to this equation, the MPPT can be achieved by comparing the incremental conductance ($\Delta I/\Delta V$) and instantaneous conductance (I/V). The corresponding algorithm is shown in Figure 2.20.

In Figure 2.20, V_{ref} is the output reference voltage of the PV. The output voltage of the PV's rear-end converter is controlled by a closed-loop controller to track V_{ref}. When the PV system is not operating at the MPP, V_{ref} is increased or decreased by the IncCond algorithm to drag the system to the MPP. When the system operates at the MPP (i.e.,

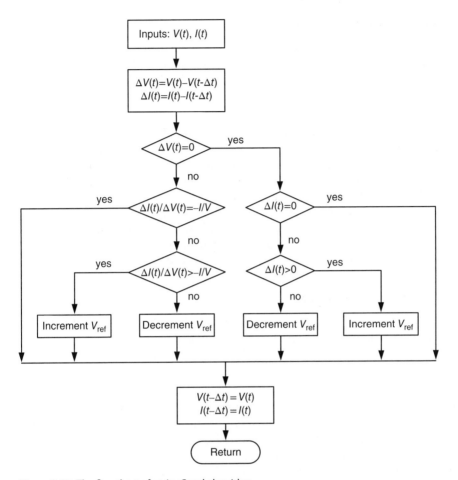

Figure 2.20 The flowchart of an IncCond algorithm.

$V_{ref} = V_{mp}$), the inverter can maintain the PV array working at this state. The response speed of the IncCond is determined by the incremental step, which should be carefully tuned considering the compromise between steady-state accuracy and dynamic speed.

Compared with the P&O method, IncCond keeps the system at the MPP, which nearly eliminates the power oscillation. However, IncCond requires accurate sensors and fast response speed, which results in a high cost of hardware implementation.

In addition, there is another IncCond-based control algorithm that takes the summation of incremental conductance and instantaneous conductance as an error signal; that is:

$$e = \frac{I}{V} + \frac{dI}{dV} \tag{2.10}$$

where e is the error signal, I/V is the instantaneous conductance, and dI/dV is the incremental conductance.

Based on the aforementioned analysis, when the error signal is zero, the system reaches the MPP. Usually, a proportional–integral control is adopted to regulate the error to be zero.

2.4.3 Open-Circuit Voltage Method

Although the PV output curve changes with respect to the solar irradiance and ambient temperature, numerical analysis and experiments show that the PV array's MPP voltage is linearly proportional to its open-circuit voltage V_{oc}; that is:

$$V_{mp} \approx k_1 V_{oc} \tag{2.11}$$

where k_1 is a constant and known as the voltage factor. The value of k_1 depends on the PV array characteristics and is usually within 0.71–0.78. The V_{oc} can be measured through periodically shutting down the back-end converter to open circuit the PV array. The value of V_{mp} can then be calculated based on Equation 2.11. When the PV output voltage is close to the V_{mp}, the MPPT is achieved.

However, the open-circuit voltage method [24] needs the PV array to periodically open circuit and generate zero output power, which will result in extra power losses. The open-circuit voltage method is simple and easy to implement, but it has low control accuracy and may not work if a part of the PV array is shaded.

2.4.4 Short-Circuit Current Method

Similar to the principle of the open-circuit voltage method, the MPP current of the PV array is also proportional to the short circuit current; that is:

$$I_{mp} \approx k_2 I_{sc} \tag{2.12}$$

where k_2 is a constant whose value is within 0.78–0.92 depending upon the PV array's characteristics. Compared with the open-circuit voltage method, the short-circuit current method [25, 26] needs to measure the short-circuit current of the PV array. In order to repeatedly short circuit the PV array, an extra switch is needed in the converter, which greatly increases the control complexity and cost. However, if a boost circuit is used in the DC–DC stage, the PV array can be shorted without adding an additional switch.

Similar to the open-circuit voltage method, the low control accuracy and periodic short circuit operations in the short-circuit current method can result in large power losses. The solution for improving the control accuracy and compensating the power losses is to update k_2 by using complex algorithms.

2.4.5 Ripple Correlation Control

Ripple correlation control (RCC) [27, 28] achieves the MPPT by analyzing the voltage, current, and power ripples generated by converters. The switching operations of power electronic converters can induce ripples in the PV output voltage and current, which correspondingly result in ripples in the output power. The RCC method determines the duty cycle of the power converter through studying the derivatives of power, voltage, and current over time. The duty cycle is controlled to keep the gradient of the PV output power zero to maximize the output power.

As shown in Figure 2.18, if the output voltage V or current I increases ($dV/dt > 0$ or $dI/dt > 0$) and the output power P also increases ($dP/dt > 0$), the operating point can be estimated on the left of the MPP; otherwise, if the output voltage V or current I increases ($dV/dt > 0$ or $dI/dt > 0$) and output power P decreases ($dP/dt < 0$), the operating point can be estimated on the right of the MPP; that is:

$$
\begin{cases}
\dfrac{dP}{dt}\dfrac{dV}{dt} > 0 \quad \text{or} \quad \dfrac{dP}{dt}\dfrac{dI}{dt} > 0 & \text{left of MPP} \\[2ex]
\dfrac{dP}{dt}\dfrac{dV}{dt} < 0 \quad \text{or} \quad \dfrac{dP}{dt}\dfrac{dI}{dt} < 0 & \text{right of MPP} \\[2ex]
\dfrac{dP}{dt}\dfrac{dV}{dt} = \dfrac{dP}{dt}\dfrac{dI}{dt} = 0 & \text{MPP}
\end{cases}
\tag{2.13}
$$

Given the back-end stage of the PV generation is a boost converter, the input inductance current of the boost converter is equal to the PV output current. It is known that if the duty cycle of the boost converter increases, the inductance current also increases, but the output voltage decreases.

Therefore, the RCC strategy of the boost converter is

$$
d(t) = -k_3 \int \left(\frac{dP}{dt}\frac{dV}{dt} \right) dt
\tag{2.14}
$$

or

$$
d(t) = k_3 \int \left(\frac{dP}{dt}\frac{dI}{dt} \right) dt
\tag{2.15}
$$

where k_3 is a positive constant. The MPPT can be achieved in real time by continuously adjusting the duty cycle $d(t)$. Take current control, for example; if the $dP/dt \cdot (dI/dt)$ is positive, the operating point is on the left side of the MPP, and the integration enlarges the duty ratio $d(t)$, increases the inductance current, and finally moves the operating point toward the MPP.

The advantages of the RCC are the perturbation can be obtained without adding extra circuit or control and dI/dt can be easily obtained by measuring inductance voltage.

However, small errors can be introduced to the RCC method because of the phase angle among the current, voltage, and power ripples induced by inductive/capacitive components, such as the input inductance, the parallel parasitic capacitance, and so on.

2.4.6 Load Current or Load Voltage Maximization Method

According to the load characteristics, the loads can be mainly divided into four types, i.e. voltage source loads, impedance loads, current source loads and mixed impedance and voltage source loads, which are shown in Figure 2.21. From the figure, for the voltage source load, the MPPT can be achieved by maximizing the current; for the current source load, the maximum power can be obtained by maximizing the voltage; for the other loads, only if the load does not show any negative impedance characteristics, either maximizing the load current or voltage can extract the most PVs. This is termed the load current or load voltage maximization method.

The load current maximization is widely applied for battery loads, in which a positive feedback is used to maximize the output current to achieve the MPPT. It is noteworthy that the Load Current/Load Voltage Maximization method is not a real MPPT method because it is infeasible to implement lossless converters.

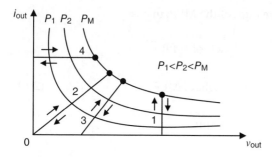

Figure 2.21 The load type: (1) voltage source load; (2) impendence load; (3) impedance and voltage mixed load; (4) current source load.

2.4.7 dP/dV or dP/dI Close-Loop Control

With the advancement of digital control technology and the increase in the calculation frequency of digital signal processors or microcontrollers, massive calculations can be completed in real time. In MPPT, a straightforward method is to control the slope of the power curve, dP/dV or dP/dI, to be zero by implementing feedback control [22].

2.4.8 Maximum Power Point Tracking Efficiency

In normal conditions the MPPT algorithm test is quite time consuming and hard to calibrate. The best way to test MPPT performance is to use PV array simulators. The requirements of the PV array simulator are that a PV array with a given number of serial and parallel PV modules can be easily set up; that the V–I curves of monocrystalline silicon cells, polycrystalline silicon cells, and noncrystalline silicon cells can be accurately simulated; that the short-circuit characteristics can be simulated; and that the power capacity can reach 100 kW. At present, the standard on evaluating the performance of MPPT is EN 50530-2010: Overall Efficiency of Grid Connected PV Inverters. This standard defines three MPPT efficiency values: MPPT efficiency η_{MPPT}, static MPPT efficiency η_{MPPTstat}, and dynamic MPPT efficiency η_{MPPTdyn}.

1) **MPPT efficiency.** This refers to the ratio of DC energy obtained from the PV modules to the PV modules' theoretical MPP output in the testing time T_{M}. The efficiency is calculated as

$$\eta_{\mathrm{MPPT}} = \frac{\displaystyle\int_0^{T_{\mathrm{M}}} p_{\mathrm{DC}}(t)\,\mathrm{d}t}{\displaystyle\int_0^{T_{\mathrm{M}}} p_{\mathrm{MPP}}(t)\,\mathrm{d}t} \tag{2.16}$$

where $p_{\mathrm{DC}}(t)$ is the instantaneous power obtained from the PV arrays and $p_{\mathrm{MPP}}(t)$ is the PV array's instantaneous maximum output power in theory.

2) **Static MPPT efficiency.** This refers to the accuracy of tracking the MPP on the given static characteristic curve. The η_{MPPTstat} is calculated as:

$$\eta_{\mathrm{MPPTstat}} = \frac{1}{P_{\mathrm{MPP,PVS}} T_{\mathrm{M}}} \sum_i U_{\mathrm{DC},i} I_{\mathrm{DC},i} \Delta T \tag{2.17}$$

where $P_{\mathrm{MPP,PVS}}$ is the maximum output power of the PV array simulator at the MPP, $U_{\mathrm{DC},i}$ is the sampling input voltage, $I_{\mathrm{DC},i}$ is the sampling input current, T_{M} is the total testing time, and ΔT is the sampling time step. In addition, there are the other two static MPPT efficiency definitions: Europe static MPPT efficiency $\eta_{\text{MPPstat,EUR}}$ and California Energy Commission static MPPT efficiency $\eta_{\text{MPPstat,CEC}}$. They are calculated as follows:

$$\eta_{\text{MPPTstat,EUR}} = 0.03\eta_{\text{MPP5\%}} + 0.06\eta_{\text{MPP10\%}} + 0.13\eta_{\text{MPP20\%}} + 0.1\eta_{\text{MPP30\%}}$$
$$+ 0.48\eta_{\text{MPP50\%}} + 0.2\eta_{\text{MPP100\%}} \tag{2.18}$$

$$\eta_{\text{MPPTstat,CEC}} = 0.04\eta_{\text{MPP10\%}} + 0.05\eta_{\text{MPP20\%}} + 0.12\eta_{\text{MPP30\%}} + 0.21\eta_{\text{MPP50\%}}$$
$$+ 0.53\eta_{\text{MPP75\%}} + 0.05\eta_{\text{MPP100\%}} \tag{2.19}$$

where $\eta_{\text{MPP5\%}}$, $\eta_{\text{MPP10\%}}$, $\eta_{\text{MPP20\%}}$, $\eta_{\text{MPP30\%}}$, $\eta_{\text{MPP50\%}}$, $\eta_{\text{MPP75\%}}$, and $\eta_{\text{MPP100\%}}$ are the efficiencies of the PV array when the actual MPP power is respectively 5%, 10%, 20%, 30%, 50%, 75%, and 100% of its rated power $P_{\text{DC,r}}$.

3) **Dynamic MPPT efficiency.** The variation of solar irradiance results in the change of the output curves. The efficiency in dynamically tracking the MPP on varying output curves is called the dynamic MPPT efficiency η_{MPPTdyn}. The calculation is

$$\eta_{\text{MPPTdyn}} = \frac{1}{\sum_j P_{\text{MPP,PVS},j}\Delta T_j} \sum_i U_{\text{DC},i}I_{\text{DC},i}\Delta T_i \tag{2.20}$$

where ΔT_j is the duration length of $P_{\text{MPP,PVS},j}$ and ΔT_i is sampling step of $U_{\text{DC},i}$ and $I_{\text{DC},i}$.

In addition, an overall dynamic MPPT efficiency $\eta_{\text{MPPTdyn,t}}$ is used to get the average dynamic MPPT efficiency during the testing sequence, and is calculated as

$$\eta_{\text{MPPTdyn,t}} = \frac{1}{N}\sum_{i=1}^{N} a_i\eta_{\text{MPPTdyn},i} \tag{2.21}$$

where $\eta_{\text{MPPTdyn},i}$ is the dynamic MPPT efficiency at testing point i, N is the number of the testing points in the sequence, and a_i is the weight factor for testing point i and $a_i = 1$ for $i = 1,\ldots,N$.

2.5 Summary

This chapter has introduced the basic techniques of distributed grid-connected PV generation, with focus on the configurations and components. The PV array and the inverter are the core components in any PV system.

The PV cells are classified based on their structures, materials used, and their development history. The main characteristics and applications of different PV cells have been briefly introduced and compared. The single-diode model and double-diode model to simulate PV cells have been given. The impacts of solar irradiance and ambient temperature on the PV output characteristics have been presented.

The common topologies of grid-connected PV inverters have been compared; and the inverters are categorized into different groups. The basic requirements of a PV inverter have been discussed to achieve high efficiency, high reliability, accommodation of large fluctuations of input DC voltage, function of islanding detection, high power quality, and electromagnetic compatibility.

Finally, several MPPT technologies are discussed, including P&O, IncCond, open-circuit voltage method, and short-circuit current method. The assessment index and testing method for evaluating MPPT efficiencies are also introduced.

References

1 Guan F., Zhao D., Zhang X., *et al.* (2009) Research on distributed generation technologies and its impacts on power system. In *International Conference on Sustainable Power Generation and Supply*. IEEE; pp. 1–6.

2 Chiradeja P. (2005) Benefit of distributed generation: a line loss reduction analysis. In *IEEE/PES Transmission & Distribution Conference & Exhibition: Asia and Pacific*. IEEE; pp. 1–5.

3 Ault G., McDonald J., Burt G. (2003) Strategic analysis framework for evaluating distributed generation and utility strategies. *IEE Proceedings – Generation, Transmission and Distribution*, **150**(4), 475–481.

4 Kirubakaran A., Jain S., and Nema R. (2011) DSP-controlled power electronic interface for fuel-cell-based distributed generation. *IEEE Transactions on Power Electronics*, **26**(12), 3953–3864.

5 Nissen M. (2009) High performance development as distributed generation. *IEEE Potentials*, **28**(6), 25–31.

6 1547.2-2008. (2009) *IEEE standard for interconnecting distributed resources with electric power systems*. IEEE: Piscataway, NJ.

7 Gu Y., Xiang X., Li W., and He X. (2014) Mode-adaptive decentralized control for renewable DC microgrid with enhanced reliability and flexibility. *IEEE Transactions on Power Electronics*, **29**(9), 5072–5080.

8 Xu X. and Zha X. (2007) Overview of the researches on distributed generation and microgrid. In *2007 International Power Engineering Conference (IPEC 2007)*. IEEE; pp. 966–971.

9 Kundu D. (2008) An overview of the distributed generation (DG) connected to the grid. In *2008 Joint International Conference on Power System Technology and IEEE Power India Conference*. IEEE; 1–8.

10 Mekhilef S., Saidur R., and Safari A. (2011) A Review on solar energy use in industries. *Renewable and Sustainable Energy Reviews*, **15**, 1777–1790.

11 Pratap N., Sharma S., Singh P., and Rani R. (2014) Solar energy: trends and enabling technologies. *International Journal of Education and Science Research Review*, **1**(3), 21–30.

12 Gawad H. and Sood V. (2014) Overview of connection topologies for grid-connected PV systems. In *IEEE 27th Canadian Conference on Electrical and Computer Engineering (CCECE)*. IEEE; pp. 1–8.

13 Forrest S. (2005) The limits to organic photovoltaic cell efficiency. *MRS Bulletin*, **30**, 28–32.

14 Green M. (1982) *Solar Cells: Operating Principles, Technology, and System Applications*. Prentice-Hall, Englewood Cliffs, NJ.

15 Das N., Wongsodihardjo H., and Islam S. (2013) Photovoltaic cell modeling for maximum power point tracking using MATLAB/Simulink to improve the conversion efficiency. In *2013 IEEE Power & Energy Society General Meeting*. IEEE; pp. 1-5.

16 Kouro S., Leon J., Vinnikow D., and Franquelo L.G. (2015) Grid-connected photovoltaic systems: an overview of recent research and emerging PV converter technology. *IEEE Industrial Electronics Magazine*, **9**(1), 47–61.

17 Faranda R., Leva S., and Maugeri V. (2008) MPPT techniques for PV systems: energetic and cost comparison. In *2008 IEEE Power and Energy Society General Meeting – Conversion and Delivery of Electrical Energy in the 21st Century*. IEEE; pp. 1–6.

18 Subudhi B. and Pradhan. (2012) A comparative study on maximum power point tracking techniques for photovoltaic power systems. *IEEE Transactions on Sustainable Energy*, **4**(1), 89–98.

19 Esram T. and Chapman P. (2007) Comparison of photovoltaic array maximum power point tracking techniques. *IEEE Transactions on Energy Conversion*, **22**(2), 439–449.

20 De Brito M.A.G., Sampaio L.P., Junior L.G., and Canesin C.A. (2011) Evaluation of MPPT techniques for photovoltaic applications. In *2011 IEEE International Symposium on Industrial Electronics*. IEEE; pp. 1039–1044.

21 De Brito M.A.G., Galotto L., Sampaio L.P., *et al.* (2013) Evaluation of the main MPPT techniques for photovoltaic applications. *IEEE Transactions on Industrial Electronics*, **60**(3), 1156–1167.

22 Jiang Y., Qahouq J. and Haskew T. (2013) Adaptive step size with adaptive-perturbation-frequency digital MPPT controller for a single-sensor photovoltaic solar system. *IEEE Transactions on Power Electronics*, **28**(7), 3195–3205.

23 Hsieh G., Chen H.L, Chen Y., *et al.* (2008) Variable frequency controlled incremental conductance derived MPPT photovoltaic stand-along DC bus system. In *2008 Twenty-Third Annual IEEE Applied Power Electronics Conference and Exposition*. IEEE; pp. 1849–1854.

24 Ahmad J. (2010) A fractional open circuit voltage based maximum power point tracker for photovoltaic arrays. In *2010 2nd International Conference on Software Technology and Engineering (ICSTE 2010)*. IEEE; pp. 247–250.

25 Sher H.A., Murtaza A.F., Noman A., *et al.* (2015) A new sensorless hybrid MPPT algorithm based on fractional short-circuit current measurement and P&O MPPT. *IEEE Transactions on Sustainable Energy*, **6**(4), 1426–1434.

26 Sandali A., Oukhoya T., and Cheriti A. (2014) Modeling and design of PV grid connected system using a modified fractional short-circuit current MPPT. In *2014 International Renewable and Sustainable Energy Conference (IRSEC)*. IEEE; pp. 224–229.

27 Barth C. and Pilawa-Podgurski R. (2015) Dithering digital ripple correlation control for photovoltaic maximum power point tracking. *IEEE Transactions on Power Electronics*, **30**(8), 4548–4559.

28 Bazzi A. and Krein P. (2014) Ripple correlation control: an extremum seeking control perspective for real-time optimization. *IEEE Transactions on Power Electronics*, **29**(2), 988–995.

3

Load Characteristics in Distribution Networks with Distributed Photovoltaic Generation

3.1 Introduction

In recent years, as one of the main renewable energy sources, PV generation has been growing very rapidly. Large-scale grid-connected PV generation has substantially changed the power flow in distribution networks and changed the networks from passive to active. Meanwhile, the ever-increasing PV capacity also influences the load level because of the coincident nature of the peak PV output relative to the diverse nature of the peak load. Grid-connected PV generation can further cause severe voltage issues, impacting the stability and safety of the distribution network.

This chapter aims to study the load characteristics of a distribution network with and without distributed PV generation and provide theoretical foundations for further analysis on PV penetration and power flow in the distribution network. In order to correctly describe load variations in time and accurately estimate the load variations, this chapter first introduces some common load characteristic indices and analyzes the time-sequence characteristics of typical loads and PV generation. The impacts on the system load levels and load fluctuations from PV generation are explored. Finally, the effective output power and equivalent electricity of the PV generation in a distribution network are derived.

3.2 Load Characteristics of a Distribution Network

3.2.1 Load Types and Indices

The study on load characteristics is important for connecting large-scale distributed PV generation to a distribution network. Meanwhile, the classification of loads is the cornerstone of analyzing the load characteristics. Currently, in distribution network planning and load data statistics, loads can be mainly divided into four types: industrial loads, agricultural loads, commercial loads, and residential loads [1, 2]. Among them, the industrial loads can be further divided into light-industrial loads and heavy-industrial loads, while the commercial loads include business load and finance service load, and so on. These four load types have their own variations over time.

In order to describe the load characteristics in a scientific and normative way and to predict the load variations, a number of parameters and curves are required to build

Grid-Integrated and Standalone Photovoltaic Distributed Generation Systems: Analysis, Design, and Control,
First Edition. Bo Zhao, Caisheng Wang and Xuesong Zhang.
© 2018 China Electric Power Press. Published 2018 by John Wiley & Sons Singapore Pte. Ltd.

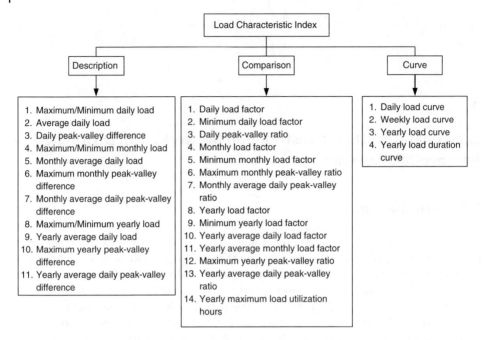

Figure 3.1 The classification of load characteristic indices.

indices for understanding the load characteristics. As shown in Figure 3.1, the load characteristic indices can be divided into description indices, comparison indices, and curve indices [3]. The first load index type reflects the overall load situation in specific areas; the second index type outlines the cross-region comparison; and all kinds of load curves to intuitively reflect the load variation are included in the third index type.

Some of the key load characteristic indices are as follows.

1) ***Load factor.*** This is used to represent load imbalance, including average and minimum load factors, which can be further subdivided with respect to different time scales, such as daily, monthly, or yearly load factors [4]. For example, the daily load factor and the minimum daily load factors can be calculated as

$$\text{Daily load factor } (\%) = \frac{\text{Average daily load}}{\text{Maximum daily load}} \times 100 \tag{3.1}$$

$$\text{Minimum daily load factor } (\%) = \frac{\text{Minimum daily load}}{\text{Maximum daily load}} \times 100 \tag{3.2}$$

The monthly and yearly load factors can also be obtained by using similar equations. In addition, the monthly and yearly average daily load factors are defined as the average of the daily load factor in that month or year.

2) ***Peak–valley difference.*** This is defined as the difference between the maximum and the minimum loads, and the peak–valley difference factor is the ratio of the peak–valley difference to the maximum load [5]. In the daily active power curve, the maximum load is termed the peak, while the minimum consumption is called the valley. The peak–valley difference reflects the capability of the grid to regulate the peak demand, which greatly depends on the load constitution and seasonal consumption.

3) ***Yearly maximum load utilization hour.*** This is used to represent the efficiency of time utilization of the load, which is defined as [6]

$$\text{Yearly maximum load utilization hours} = \frac{\text{Yearly electricity consumption}}{\text{Yearly maximum load}}$$
$$= 8760 \times \text{Yearly load factor} \qquad (3.3)$$

According to this definition, the yearly maximum load utilization hour is a comprehensive index and is usually determined by the major load type in the distribution network. Generally speaking, for a region with a large portion of heavy-industrial loads, its yearly maximum load utilization hour is usually more than 6000 h. On the other hand, the yearly maximum load utilization hour is relatively low for the regions mainly with commercial and residential loads.

Similarly, the monthly maximum load utilization hour can be calculated as the ratio of the monthly electricity consumption to the monthly maximum load.

4) ***Monthly load duration curve.*** The monthly load duration curve [7] shows load values versus their corresponding duration hours in a month. Different from a conventional load curve, the monthly load duration curve cannot illustrate the load variation in a time sequence, but it depicts the duration of different load levels. In fact, this index represents the relationship between the load magnitudes and their durations, which can contribute to scheduling generation sources and evaluating system reliability.

3.2.2 Time-Sequence Characteristics of Typical Loads

In analysis of load characteristics, to gain a better understanding of the intrinsic procedure of load evolvement, the analysis should include not only the total load demand, but also the level and change rules for different types of load in the distribution network. In this section, the daily time-sequence characteristics of four load types are analyzed [8, 9].

The time-sequence profiles of the four different loads in four seasons in a distribution network are shown in Figure 3.2. The typical daily variation profile of the industrial load is shown in Figure 3.2a. The figure shows that the variations almost follow the same pattern in four seasons. Especially in spring and autumn, the curves almost overlap each other. The industrial load is seen to be highest in summer and lowest in winter. The industrial electricity demand mainly lasts from 09:00 to 21:00, among which the demand peaks at 12:00 and bottoms between 00:00 and 07:00 in a day.

Figure 3.2b clearly shows that the agricultural consumption has significant seasonal variations, with the order from the most to least being winter, autumn, summer, and spring. Two intervals of agricultural electricity demand peak are between 08:00 and 11:00 and between 15:00 and 22:00. The global peak load occurs at around 10:00, while the load level is relatively low between 00:00 and 07:00.

The 24-h variations of commercial load in the four seasons are nearly the same, as shown in Figure 3.2c. The duration of business load in summer is a little longer than in the other three seasons. The highest demand happens at 11:00 and is around zero between 00:00 and 08:00.

Figure 3.2d shows a daily residential load profile. As shown in the figure, the highest demand happens in summer and the lowest is in spring. The peak residential load occurs between 17:00 and 22:00.

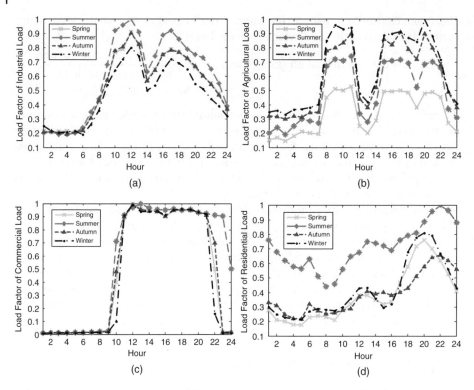

Figure 3.2 Four typical daily load profiles in four seasons: (a) industrial load; (b) agricultural load; (c) commercial load; (d) residential load.

3.2.3 Case Study

This case study analyzes the load demand of a 110 kV SA substation in 2014 in City J in China's east coastal province. In this area, the load consists of a large portion of light-industrial loads and a small portion of commercial loads and residential loads. According to the measured data in 2014, the maximum load in this area was 56.2 MW, the maximum daily load factor was 75.8%, the minimum load factor was 55.1%, the average daily load factor in that year was 78.1%, the maximum peak–valley difference of the whole year was 48.9 MW, and the average monthly maximum load utilization hour was 483.5 h. The monthly maximum load and maximum load utilization hours are listed in Table 3.1. The table shows that, in this region, the load level is relatively high and the load changes gently.

In consideration of the weather conditions in this region, the loads of four typical months, April, July, October, and January of 2014, were selected to represent the load features of spring, summer, autumn, and winter respectively. The daily load profiles in different seasons are then obtained from typical days. Figure 3.3 shows that the load in this area varied significantly over the seasons; that is, the load peaks in the summer, bottoms in the winter and lies in the middle in the spring and autumn. The daily load profiles almost have the same pattern in the four seasons. The electricity consumption mainly lasts from 09:00 to 21:00, and the peak load appears approximately between 11:00

Table 3.1 The maximum load and the maximum load utilization hour in a 110 kV SA substation of City J in 2014.

Month	Maximum load (MW)	Maximum load utilization hours (h)	Month	Maximum load (MW)	Maximum load utilization hours (h)
January	47.8	361.68	July	53.5	531.56
February	54.1	434.09	August	54.3	502.05
March	56.2	509.86	September	48.3	470.04
April	51.7	500.87	October	44.9	485.77
May	54.8	510.85	November	46.7	503.40
June	55.2	479.25	December	50.1	512.73

Figure 3.3 Time-sequence load curves in the SA substation area.

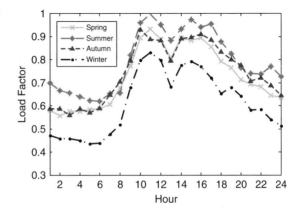

and 15:00. The comparison of the load variation between Figures 3.2a and 3.3 shows that the light-industrial load dominates in this area.

Take July, which has the highest load, as an example, its monthly load duration curve is shown in Figure 3.4. If the peak load duration is defined as the duration when the load is over 85% of the maximum load, the peak load duration in July shown in Figure 3.4 is 154 h, which takes up 21% of the month. If distributed PV generation is connected to the distribution network in this area, the peak load duration can obviously be reduced, significantly relieving the stress on the power supply.

The electricity demand in the grid is influenced by multiple factors and shows certain random variations that can be described by using statistical methods. The probability density function (PDF) is a function to describe the relative likelihood for a random variable to take on a given value. The probability of the random variable falling within a particular range of values is given by the integral of this variable's density over that range; that is, it is given by the area under the density function but above the horizontal axis and between the lower and the upper bounds of the range. The PDF is nonnegative everywhere, and its integral over the entire space is equal to one.

The PDF of the 5 min load fluctuation in July is shown in Figure 3.5, where the horizontal and vertical axes represent the percentage of 5 min power variations and the probability density of the load variations respectively. Figure 3.5 clearly shows that 99.6%

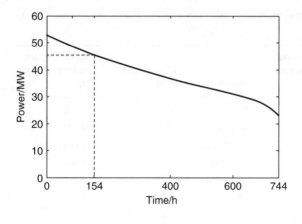

Figure 3.4 Monthly load duration curve in July.

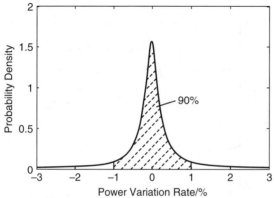

Figure 3.5 PDF of 5 min load fluctuation in July.

of the 5 min load variations are no larger than 2% of the peak load, and 90.0% of the variations are within 1% of the peak load. Therefore, from the perspective of load fluctuations, the distribution network in this area can, to a certain amount, adopt highly random and fluctuant distributed generation sources, such as PV generation.

3.3 The Output Characteristics of Photovoltaic Generation

3.3.1 Regulations on Grid-Connected Photovoltaic Generation

At present, the intermittency and volatility of PV output are the key hurdles for developing distributed PV generation. The impacts from the volatile PV output on the distribution network are outlined in the following [10–12].

- *The impact on distribution network planning.* The intermittent and random PV output makes it hard to meet the strict requirements on grid stability, continuity, and reliability, which requires large spinning reserves to compensate for the fluctuations. In addition, the fluctuations can result in serious waste of system resources and bring new challenges on distribution network planning.
- *The impact on the operation and dispatch of the distribution network.* In order to guarantee the stable operation of the grid, the fluctuations of PV output should be

limited by using well-designed operation and control of the distribution network. The fluctuations also introduce difficulties in load prediction and power dispatch.

- *The impact on power quality.* The fluctuations of PV output could be high enough to cause significant fluctuations of grid voltage. When the PV output declines, the interface inverters are lightly loaded, leading to increased current harmonics.

In order to mitigate the negative impacts of distributed PV generation on the distribution network, a number of standards have been established to regulate the capacity and induced voltage fluctuation range of grid-connected distributed generation. For example, in IEEE Std 1547 [13] and Germany's *mid-* and *low-voltage grid-connected guides* [14, 15], the capacity of a distributed generation source is restrained by limiting the maximum voltage change rate at the point of common coupling (PCC). Moreover, IEEE Std 1547 requires the voltage fluctuation at the PCC be less than ±5% of the rated voltage, while the voltage change should be within 2% of the nominal voltage when no distributed generation is connected as required in Germany's *mid-voltage grid-connected guide* [14], and the requirement is slightly relaxed to 3% in the *low-voltage grid-connected guide* [15].

China's GB/T 29319-2012 standard [16] requires that the voltage deviation at the PCC needs to meet the GB/T 12345; that is, the three-phase voltage deviation should be less than ±7% of the rated voltage that is less than 20 kV, and the single-phase voltage deviation should be within [−10%, +7%] of the 220 V rated voltage. This standard further specifies that the PV generation must operate appropriately with 90–110% of the rated voltage at the PCC and should be able to follow reactive dispatch orders to regulate the grid voltage by adjusting its reactive power output. China's NB/T 32015-2013 [17] requires that the operation voltage needs to be kept within 85–110% of the rated voltage when the distributed generation connects to a 380 V grid. In addition, in China State Grid's standards, the power fluctuation ranges of grid-connected PV generation are regulated based on the guidance shown in Table 3.2.

Compared with China, the standards in some other countries may put tighter restrictions on grid-connected distributed generation systems, such as on frequency deviation and phase deviation. Some of the regulation requirements are listed in Table 3.3.

3.3.2 Time-Sequence Characteristics of Photovoltaic Generation

As introduced in Chapter 2, PV output power is affected by solar irradiance and ambient temperature and has obvious seasonal time-sequence characteristics. To study the PV output characteristics in an area, one should collect the meteorological data from this area to obtain the corresponding solar irradiance and temperature curves and derive the time-sequence curves of the PV generation.

Table 3.2 China State Grid's technical standards on power fluctuation of grid-connected PV generation [18].

PV interconnection voltage level (kV)	Maximum active power change in 10 min (MW)	Maximum active power change in 1 min (MW)
0.38	Installed capacity	0.2
10–35	Installed capacity	Installed capacity/5

Source: Wu 2014 [3.18]. Reproduced from Modern Electric Power.

Table 3.3 Standards for connecting distributed generation to distribution network [19].

Power of the distributed generation (MW)	Frequency deviation (Hz)	Phase deviation (deg)	Standard of Nation
0–0.5	0.3	20	
0.5–1.5	0.2	15	IEEE 1547
1.5–10	0.1	10	
Small scale	0.1	10	France
Small scale	0.5	(5.6%)	Italy
0–1	0.1	10	Spain
>1	0.2	20	Spain
Mid-voltage grid-connected	0.5	10	Germany

Source: Yang 2012 [19]. Reproduced from Tsinghua Tongfang Knowledge Network Technology Co., Ltd.

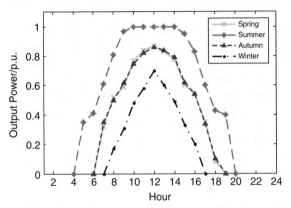

Figure 3.6 Time-sequence characteristic curves of PV generation in four seasons.

Typical time-sequence characteristic curves of the PV generation in four seasons are presented in Figure 3.6 [20]. The figure shows that the PV generation has the largest output power in summer due to the long duration of sunshine and strong irradiance, while it has the smallest output in winter and yields moderately in spring and autumn. The daily output power peaks around noon, and remains zero at night.

The seasonal time-sequence characteristic curves of the PV generation are presented in Figure 3.7 [20]. The figure shows that the weather conditions have great impacts on the PV output. On sunny days, the fluctuations of PV output are relatively small. The PV system starts generating electricity at dawn, and the output power increases as the solar irradiance gets stronger and reaches a peak around noon. In the afternoon, the PV output decreases with the decline in irradiance and drops to zero when the sun sets. The time-sequence characteristic of PV generation in overcast conditions or on rainy days is similar to that on a sunny day but with larger fluctuations. The PV output power at the same time, in the order from most to least, is sunny days, cloudy days, overcast days, and rainy days. On cloudy days, as the sunshine may be blocked by clouds from time to time, the fluctuation of the output power can be quite significant, which may exceed 50% of the installed capacity occasionally. Therefore, cloudy weather may induce a very volatile voltage profile of the distribution network with distributed PV generation.

Figure 3.7 Time-sequence characteristic curves of PV generation under different weather conditions.

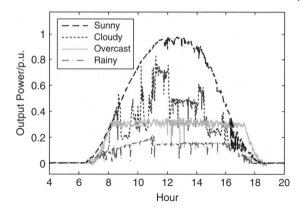

3.3.3 Case Study

In this section, the output characteristics of a distributed PV generation will be analyzed based on the data from the SA substation area. The distributed PV generation, with 24 arrays and a total of 480 modules, was installed on the rooftop of the buildings in a plant. The first-stage 120 kW PV generation came into operation on February 28, 2013.

A field photograph of the system is shown in Figure 3.8. The generation of this PV system is mainly consumed by the customers in the buildings, and the rest of the PV power is fed back to the local distribution network through a step-up transformer.

Similarly, the PV outputs in April, July, October, and January of 2014 were selected to represent the generation in spring, summer, autumn, and winter respectively. The maximum daily output curves for these four months are shown in Figure 3.9. The figure shows that the maximum daily output of the PV generation fluctuates significantly. The average maximum daily output power in the four seasons was 69.8 kW, 73.4 kW, 68.5 kW, and 52.0 kW in spring, summer, autumn, and winter respectively. The whole year's average maximum daily output power is 65.9 kW, about 54.9% of the rated capacity. Therefore, it could be concluded that the irradiance condition in this area is not very high.

The daily operation hours and daily effective operation hours of the distributed PV generation in four seasons are shown in Table 3.4, where the former parameter refers to the total time in 1 day that there is PV generation has output power and the latter is defined as the total daily PV-generated energy divided by its installed capacity. According to the table, both the daily operation hours and the effective operation hours are

Figure 3.8 A field photograph of the distributed PV generation in the SA substation area.

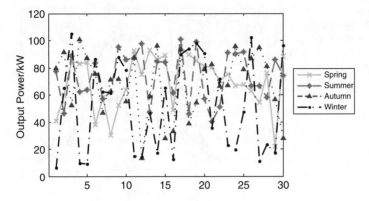

Figure 3.9 The maximum daily output power curves of the distributed PV generation in four seasons.

Table 3.4 The daily operation hours and the daily effective operation hours of PV generation.

Season	Daily operation hours/h			Daily effective operation hours/h		
	Maximum	Minimum	Average	Maximum	Minimum	Average
Spring	12.5	0	7.2	6.7	0	3.3
Summer	13.8	0	8.1	7.4	0	4.0
Autumn	11.7	0	5.8	5.6	0	2.5
Winter	9.6	0	4.2	4.5	0	1.6
Yearly	13.8	0	6.3	7.4	0	2.9

relatively longer in spring and summer and shorter in winter and autumn. In general, the operation hours and the effective hours are short and the annual average generation capacity is not high. The generation capability in winter is particularly limited and it may not be suitable for PV generation to serve as the main power supply in the season.

Figure 3.10 shows the PV output power duration curves in the four seasons. The figure clearly shows that the distributed PV generation has the longest time to generate electricity in summer and the shortest in winter. The generation time lies in the middle in spring and autumn. The maximum power utilization hours of this distributed PV generation in spring, summer, autumn, and winter are 376.1 h, 427.0 h, 307.7 h, and 245.6 h respectively. Therefore, the maximum power utilization hours is relatively low as the PVs can only operate in the daytime and its output is low at dawn and dusk.

The PDF of the 5 min output variations of the distributed PV generation for the four seasons in 2014 is shown in Figure 3.11, where the horizontal axis represents the percentage and the vertical axis denotes the probability density. The figure shows that the output fluctuates similarly in the four seasons. Owing to the weak irradiance in winter, the fluctuation of PV output is relatively small. The 5 min output power variations are mainly within ±20% of the rated power, and in rare cases some can reach 50% of the rated power. Therefore, the fluctuation of the PV output sometimes affects the peak load regulation of the power system.

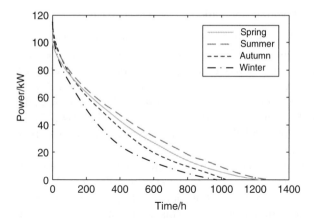

Figure 3.10 The duration output power characteristic curves of the distributed PV generation in the four seasons.

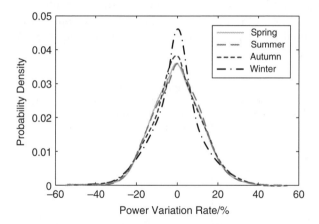

Figure 3.11 The PDF of 5 min output power variation of the PV generation for the four seasons in 2014.

3.4 Characteristics of the Net Load in a Distribution Network with Distributed Photovoltaic Generation

Based on the aforementioned analysis, the maximum load of a distribution network does not appear in the same time frame as the peak hour of output of distributed PV generation. When the distribution network is connected with the PVs, the peak value of net load can still be reduced, where the net load is defined as the net value of the load demand subtracted by the PV system's output power. However, if the capacity of the PV generation is large, the randomness and uncertainties of the PV output can induce issues of overvoltage and reverse power flow in the distribution network and even impact the operation and dispatch of in the system. In this section the net load characteristics in a distribution network with distributed PV generation will be discussed.

3.4.1 Influence of Distributed Photovoltaic Generation on System Load Level

Similarly, the load and distributed PV generation in the SA substation area in April, July, October, and January of 2014 are again selected as the example for discussion. Given that the output power of distributed PV generation is directly proportional to its capacity and assuming that the capacity of the PV system accounts for 15% of the maximum load in that season, the system net load can be obtained by superposing the PV output with the system load. Figure 3.12 shows the monthly load duration curves of the original load and the corresponding net load in July, and Tables 3.5 and 3.6 show the parameters of the load characteristics in those four months.

According to Figure 3.12 and Table 3.5, comparing with the original load, the maximum monthly net load decreases slightly after the PV generation is connected. As the maximum load utilization hours is determined by the load consumption and the maximum load demand but the PV integration decreases the maximum load and the power consumption simultaneously, the variations of the maximum net load utilization hours depend on actual conditions. As to the maximum load utilization hours, compared with the original data, it declines in January and July and increases in April and October. The net load duration hours are significantly reduced after connecting the distributed PV generation. For example, the duration of the heaviest load in July drops from 154 to 126 h, which means the distributed PV generation has the obvious function of peak shaving for the distribution network.

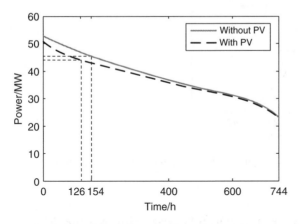

Figure 3.12 The monthly load duration curves of the SA substation area with/without the PV generation in July 2014.

Table 3.5 The maximum load, the maximum load utilization hours, and the peak load duration hours of the system with/without the distributed PV generation.

Month	Original load			Net load		
	Maximum load (MW)	Maximum load utilization hours (h)	Peak load duration hours (h)	Maximum load (MW)	Maximum load utilization hours (h)	Peak load duration hours (h)
January	47.8	361.7	44	47.2	355.0	32
April	51.7	500.9	133	48.7	514.4	117
July	53.5	531.6	154	51.7	530.1	126
October	44.9	485.8	100	43.5	486.7	82

Table 3.6 Load factor and peak–valley difference of the system with/without the distributed PV generation.

Month	Original load			Net load		
	Minimum load factor (%)	Average load factor (%)	Maximum peak–valley difference (MW)	Minimum load factor (%)	Average load factor (%)	Maximum peak–valley difference (MW)
January	15.1	48.6	30.4	15.5	47.7	28.6
April	38.0	69.6	35.6	40.3	71.4	34.2
July	43.2	71.4	40.6	44.7	71.3	40.2
October	20.7	65.3	32.1	21.4	65.4	29.0

According to Table 3.6, the minimum load factor increases while the maximum peak–valley difference declines after the distributed PV generation is connected. The reasons for this are as follows. The load in the selected SA substation area consists mainly of the light-industrial load, whose time-sequence characteristic curve is the same as that shown in Figure 3.3. The load usually bottoms at night, and the PV output at night is zero. If the capacity of the distributed PVs is not large, the minimum load of the system may not be influenced. In addition, based on the previous analysis, the distributed PV generation can shave the peak demand. Therefore, all of these afore-mentioned factors analyzed together determine the variation trend of the minimum load factor and maximum peak–valley difference in this system. As for the average load factor, since it is related to the average and maximum load of the distribution network and both of these two figures decline after the PV generation is connected while the variation of the average load factor should be analyzed on a case-by-case basis. From Table 3.6, the average load factor of the system net load declines in January and July and increases in April and October.

The links between the capacity of the distributed PV generation and the system's net load duration curves are studied further. Taking the heaviest loaded July as an example, the monthly duration curves under different PV capacities are shown in Figure 3.13, where the PV capacity is presented as the percentage over July's maximum load. The figure shows that the distributed PV generation can reduce both the peak load and the

Figure 3.13 The monthly load duration curves of the net load in July in the SA substation area.

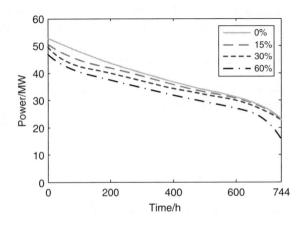

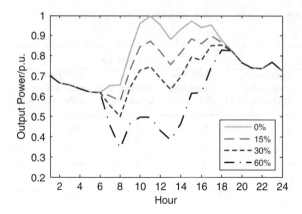

Figure 3.14 The daily time-sequence characteristic curves of the system net load under different sizes PVs.

valley load of the system. When the PV capacity reaches a certain level, the peak load reduces slowly, but the valley load drops faster, which is because the peak load in the distribution network appears at a different time from the PV peak hour. In addition, with the increase of the PV capacity, the time for the peak net load shifts. When the PV output reaches a peak at noon, the system net load may, on the contrary, be at the bottom in the daily load curve, which can be observed from Figure 3.14.

The peak load duration hours of the system net load in January, April, July, and October of 2014 under different PV capacities are listed in Table 3.7. In the table, the peak load duration hours can be effectively reduced by connecting the PV generation; and the larger the PV capacity, the lower the peak load duration hours will be.

Based on the analysis here, distributed PV generation has the benefit of shaving the peak demand, increasing the power supply capacity, and reducing the reserve generation in the power system.

3.4.2 Influence of Distributed Photovoltaic Generation on Load Fluctuation

This section takes July as an example to study the influences of the distributed PV generation on load fluctuation. The maximum variation and the PDF of 5 min system net load under different PV penetration levels are shown in Table 3.8 and Figure 3.15 respectively.

Table 3.8 and Figure 3.15 show that, with the increasing penetration of distributed PV generation, the fluctuation range of the system net load becomes larger, which increases the uncertainty in load forecasting and the demand for more reserve capacity.

Table 3.7 The peak load duration of the net load in January, April, July, and October of 2014 in the SA substation area.

PV capacity/maximum load (%)	Peak load duration (h)			
	January	April	July	October
0	44	133	154	100
15	32	117	126	82
30	25	104	78	60
60	24	70	44	53

Table 3.8 The maximum fluctuation of system net load under different PV capacities.

PV capacity (%)	0	15	30	40	50	60
Maximum fluctuation $(\Delta P\%)_{max}$ (%)	5.1	8.1	16.4	21.9	24.6	29.4

Figure 3.15 The PDF of the variations of system net load under different PV penetration levels.

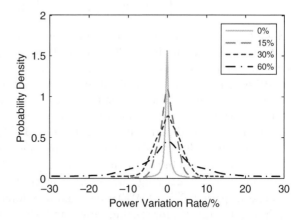

In addition, in order to meet the requirements on peak load regulation, the fluctuation between two consecutive 5 min loads should be less than 25% of the maximum load. In other words, the maximum load variation rate should be less than 25% [21]. As shown in Table 3.8, when the PV penetration level is no greater than 50%, the load variation rate is much less than 25% in most cases and the influence on the peak load regulation is very limited. However, if the distributed PV capacity increases to 60% of the maximum load, a few times the power variation rate of 5 min net load can be around 30%. This means that overly high PV penetration can cause adverse impacts on the peak load regulation and automatic generation control, though the possibility is low.

3.5 Power and Energy Analysis of Distributed Photovoltaic Generation

3.5.1 Effective Power and Equivalent Energy of Distributed Photovoltaic Generation

Based on the previous analysis, distributed PV generation can cause difficulties in the forecasting of the maximum load and further influence peak load regulation. For a distributed PV generation, its capability of shaving the peak load is related to its average output power and the coincidence between its generation and the load.

For a certain period, the effective output power of a PV system, which is defined as the equivalent capability for shaving the peak load in the chapter, can be calculated by Equations 3.4–3.7:

$$P_{g\,max} = \sum_{i=1}^{N_g}(\eta_i P_{gN,i}) \tag{3.4}$$

$$\Delta P_{cut\text{-MLDday}} = k_1 k_2 P_{g\,max} \tag{3.5}$$

$$\Delta P_{\text{Ldmax}}^{\text{d-n}} = P_{\text{Ldmax}}^{\text{day}} - P_{\text{Ldmax}}^{\text{night}} \tag{3.6}$$

$$\Delta P_{\text{cut}} = \begin{cases} 0, & \text{if} \quad P_{\text{Ldmax}}^{\text{day}} \leq P_{\text{Ldmax}}^{\text{night}} \\ \min\{\Delta P_{\text{cut-MLDday}}, \Delta P_{\text{Ldmax}}^{\text{d-n}}\}, & \text{if} \quad P_{\text{Ldmax}}^{\text{day}} > P_{\text{Ldmax}}^{\text{night}} \end{cases} \tag{3.7}$$

where $P_{\text{g max}}$ (MW) is the total of the maximum output power of all the distributed PV systems, N_{g} is the number of distributed PV systems, η_i is the maximum generation efficiency of the ith PV system ($i = 1, 2, 3, \ldots, N_{\text{g}}$), $P_{\text{gN, }i}$ (MW) is the rated capacity of the ith PV system, $\Delta P_{\text{cut-MLDday}}$ (MW) is the effective output power of all the distributed PV systems during the peak load period in the daytime on the day with the maximum load, k_1 is the correction coefficient of the effective output power of distributed PV systems during the peak load period in the daytime on the day with the maximum load, k_2 is the correction coefficient of the average output power of distributed PV systems during the peak load period in the daytime on the day with the maximum load, $P_{\text{Ldmax}}^{\text{day}}$ (MW) is the maximum forecasting value of the conventional load in the daytime during the forecasting period, $P_{\text{Ldmax}}^{\text{night}}$ (MW) is the maximum forecasting value of conventional load in the night during the forecasting period, and ΔP_{cut} (MW) is the effective output power of all the PV systems.

The impact of distributed PV generation on power forecasting is determined by the maximum power utilization hours and the maximum generation efficiency of the PV system. The equivalent electricity of the distributed PV generation within a certain period of time can be calculated as

$$W_{\text{g},i} = k_3 \eta_i P_{\text{gN},i} T \tag{3.8}$$

$$W_{\text{g}} = \sum_{i=1}^{N_{\text{g}}} W_{\text{g},i} \tag{3.9}$$

where $W_{\text{g},i}$ (MW h) is the equivalent electricity of the ith PV generation, k_3 is the correction coefficient of the equivalent electricity, T (h) is the total hours, and W_{g} (MW h) is the equivalent electricity of all the distributed PV systems.

3.5.2 Calculation Methods of the Correction Coefficients

The correction coefficients k_1, k_2, and k_3 in Equations 3.5 and 3.8 can be obtained based on historical data of the load and the PVs for a given area.

First, the steps to calculate k_1 and k_2 are as follows:

1) ***Total output power of all in-service distributed PV systems.*** Given that the distributed PV systems of the same type in the same area share similar output curves, a typical PV output profile can be generated by statistical calculation based on the PV recorded historical time-sequence power data in the area. The typical PV output profile reflects the local PV output characteristics effectively. So with the typical output profile, the output power of all same type in-service PV systems in the area is obtained as follows:

$$P_{\text{g},d,t}^{\text{on}} = \frac{P_{\text{g},d,t}^0}{P_{\text{g max}}^0} \sum_{i=1}^{N_{\text{g}}^{\text{on}}} (\eta_i P_{\text{gN},i}^{\text{on}}) \tag{3.10}$$

where $t = 1,2,\ldots, T_d$, $d = 1,2,\ldots,D$, T_d is the number of records of the daily PV output power (since the PV output power is recorded every 5 min, the number of the records in 1 day is $12 \times 24 = 288$), D is the number of days in the calculation period, $P^{on}_{g,d,t}$ (MW) is the total active power of all the distributed PV systems in the tth record on the dth day, $P^0_{g,d,t}$ (MW) is the active power of a typical distributed PV system in the tth record on the dth day, $P^0_{g\,max}$ (MW) is the maximum active power of a typical distributed PV system, N^{on}_g is the number of all the in-service distributed PV systems, and $P^{on}_{gN,i}$ is the rated capacity of the ith in-service distributed PV system.

2) **Correction of the original load power.** Given the measured load in the specific area includes the total output power of in-service distributed PV systems, the original load demand can be obtained by superposing the measured load with the PV output power; that is:

$$P_{Ld,d,t} = P_{Ldg,d,t} + P^{on}_{g,d,t} \tag{3.11}$$

where $t = 1,2,\ldots, T_d$, $d = 1,2,\ldots,D$, $P_{Ld,d,t}$ is the conventional load demand in the tth time slot on the dth day, and $P_{Ldg,d,t}$ is the measured net load demand in the tth time slot on the dth day.

3) **Selection of the day with the maximum load and the period in the daytime with the peak load.** Theoretically, the PV maximum effective output power may be equal to its maximum generation power. However, as the PV output power curve is not always coincident with the load curve, the actual maximum effective output power is less than the maximum output power. In order to obtain suitable effective output power, the day with the maximum load is selected based on the ratio of the maximum effective PV output power to the maximum original load.

$$P^{on}_{g\,max} = \sum_{i=1}^{N^{on}_g} P^{on}_{g\,max,i} \tag{3.12}$$

$$P_{Ld\,max,d} = \max\{P_{Ld,d,t}|t = 1,2,\ldots,T_d\}; \quad d = 1,2,\ldots,D \tag{3.13}$$

$$P_{Ld\,max} = \max\{P_{Ld,d,t}|t = 1,2,\ldots,T_d; \quad d = 1,2,\ldots,D\} \tag{3.14}$$

$$\alpha = 1 - \frac{P^{on}_{g\,max}}{P_{Ld\,max}} \tag{3.15}$$

$$\varphi_{MLD} = \{d|P_{Ld\,max,d} \geq \alpha P_{Ld\,max}; \quad d = 1,2,\ldots,D\} \tag{3.16}$$

$$\psi_{MLDday} = \{t|P_{Ld,d,t} \geq \alpha P_{Ld\,max,d}; \quad t = T^S_d, T^S_d + 1, T^S_d + 2, \ldots, T^E_d; \quad d \in \varphi_{MLD}\} \tag{3.17}$$

where $P^{on}_{g\,max}$ (MW) is the maximum output power of all in-service distributed PV systems, $P^{on}_{g\,max,i}$ (MW) is the maximum output power of the ith in-service distributed PV system, $P_{Ld\,max,d}$ (MW) is the maximum power of the original load on the dth day, $P_{Ld\,max}$ (MW) is the maximum power of original load in the calculating period, α is the threshold ratio to select the maximum load day, φ_{MLD} is the set of the maximum load days, ψ_{MLDday} is the set of the periods in the daytime with peak load in the maximum load day, T^S_d is the start time of the distributed PV systems in the daytime, and T^E_d is the stop time of the distributed PV systems in the daytime.

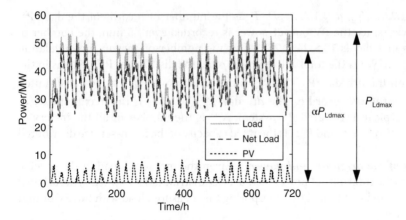

Figure 3.16 The selection of the maximum load day.

The procedure to select the maximum load day is shown in Figure 3.16, where $\alpha P_{\text{Ld max}}$ is the minimum threshold to determine the maximum load days during the calculation period; that is, the days with their maximum loads higher than the threshold are identified as the maximum load days. The steps are as follows:

i) The original load power without distributed PV systems can be obtained based on the given historical data of PV output power and the system's net load.

ii) The ratio of the maximum PV output power to the maximum original load is used to find the minimum threshold and further identify the maximum load days.

4) *Average output power of the in-service distributed PV systems during the periods in the daytime with peak load on the maximum load days.* That is:

$$\overline{P}^{\text{on}}_{\text{g-MLDday}} = \frac{1}{N_{\text{MLDday}}} \sum_{\substack{t \in \psi_{\text{MLDday}} \\ d \in \varphi_{\text{MLD}}}} P^{\text{on}}_{g,d,t} \tag{3.18}$$

where $\overline{P}^{\text{on}}_{\text{g-MLDday}}$ is the average output power of the in-service distributed PV systems during the periods in the daytime with peak load on the maximum load days and N_{MLDday} is the number of peak load periods in the daytime on the maximum load days.

5) *Average effective output power of the in-service distributed PV systems during the period with peak load in the daytime on the maximum load days.*

$$P^{\text{MLD}}_{\text{Ld max},d} = \max\{P_{\text{Ld},d,t} | t \in \psi_{\text{MLDday}}\}, \quad d \in \varphi_{\text{MLD}} \tag{3.19}$$

$$P^{\text{MLD}}_{\text{Ldg max},d} = \max\{P_{\text{Ldg},d,t} | t \in \psi_{\text{MLDday}}\}, \quad d \in \varphi_{\text{MLD}} \tag{3.20}$$

$$\Delta\overline{P}^{\text{on}}_{\text{cut-MLDday}} = \frac{1}{N_{\text{MLD}}} \sum_{d \in \varphi_{\text{MLD}}} \left(P^{\text{MLD}}_{\text{Ld max},d} - P^{\text{MLD}}_{\text{Ldg max},d}\right) \tag{3.21}$$

where $P^{\text{MLD}}_{\text{Ld max},d}$ (MW) is the maximum original load on the maximum load days, $P^{\text{MLD}}_{\text{Ldg max},d}$ (MW) is the maximum system net load on the maximum load days, $\Delta\overline{P}^{\text{on}}_{\text{cut-MLDday}}$ is the average effective output power of the in-service distributed PV systems during the period with peak load in the daytime on the maximum load days, and N_{MLD} is the number of maximum load days.

6) **Correction coefficients k_1 and k_2 of the effective PV output power.**

$$k_1 = \frac{\overline{\Delta P}^{on}_{cut-MLDday}}{\overline{P}^{on}_{g-MLDday}} \qquad (3.22)$$

$$k_2 = \frac{\overline{P}^{on}_{g-MLDday}}{P^{on}_{g\,max}} \qquad (3.23)$$

The steps to calculate the correction coefficient k_3 of the PV equivalent electricity generation are as follows. First, the total electricity generation of a typical in-service distributed PV system in the calculating period is calculated as

$$W^0_g = \sum_{d=1}^{D} \sum_{t=1}^{T_d} \left(P^0_{g,d,t} \Delta \tau_{d,t} \right) \qquad (3.24)$$

where $\Delta \tau_{d,t}$ (h) is the time slot and W^0_g (MW h) is the total electricity generation of the typical in-service distributed PV system in the calculation period.

The maximum power utilization hours of the typical in-service distributed PV system is obtained as

$$T^0_{g\,max} = \frac{W^0_g}{P^0_{g\,max}} \qquad (3.25)$$

where $T^0_{g\,max}$ (h) is the maximum power utilization hours of the typical in-service distributed PV systems.

Finally, the correction coefficient of the distributed PV equivalent electricity generation is

$$k_3 = \frac{T^0_{g\,max}}{T} \qquad (3.26)$$

3.6 Summary

The time-sequence characteristic curves of four typical types of loads, namely industrial load, agricultural load, commercial load, and residential load, in a distribution network have been introduced in this chapter. The case study on a 110 kV substation area (called the SA substation) in City J in the east of China has been carried out based upon the following load characteristic indices: load factor, peak–valley difference, the maximum load utilization hours, and load duration curves.

According to the standards for grid-connected distributed PV generation, the time-sequence characteristics of typical distributed PV output power have been analyzed in four seasons and under different weather conditions. The SA substation area has been taken as an example to analyze and compare the characteristics of the system net load, including the system load level and system load fluctuations, before and after the distributed PV generation is penetrated. The results show that the distributed PV generation has obvious benefits of shaving the peak load of the distribution network. However, with increasing PV capacity, fluctuation of the net load can be aggravated.

At the end of this chapter the equations to calculate the output power and energy of distributed PV generation are introduced, including how to get the effective output power and equivalent electricity generation. The calculations of the correction coefficients of the effective PV output power are deduced. These calculations provide theoretical foundations for the analysis in the following chapters.

References

1 Singh D., Misra R. K., and Singh D. (2007) Effect of load models in distributed generation planning. *IEEE Transactions on Power System*, **22**(4), 2202–2212.

2 Liu J. (2015) Distribution network planning including distributed generation. Master dissertation, Southwest Jiaotong University.

3 Chen W., Yue L., Cui K., and Li J. (2011) Analysis on power load and consumption characteristics of representative consumers. *Energy Technology and Economics*, **23**(9), 44–49.

4 Chang R.F. and Lu C.N. (2002) Feeder reconfiguration for load factor improvement. In *IEEE Power Engineering Society Winter Meeting*. IEEE; pp. 980–984.

5 Tang Y.D., Song H.K., Hu F.N., and Zou Y. (2005) Investigation on TOU pricing principles. In *IEEE/PES Transmission & Distribution Conference & Exposition: Asia and Pacific*. IEEE; pp. 1–9.

6 Gomez J.C. and Morcos M.M. (2003) Impact of EV battery chargers on the power quality of distribution systems. *IEEE Transactions on Power Delivery*, **18**(3), 975–981.

7 Poulin A., Dostie M., Fournier M., and Sansregret S. (2008) Load duration curve: a tool for technico-economic analysis of energy solutions. *Energy and Buildings*, **40**(1), 29–35.

8 Li L., Tang W., Bai M.K., *et al.* (2013) Multi-objective locating and sizing of distributed generators based on time sequence characteristics. *Automation of Electric Power System*, **37**(3), 58–63.

9 Xu X. (2013) Distribution network planning with micro-grids. Master dissertation, Shanghai Jiaotong University.

10 Chidurala A., Saha T., and Mithulananthan N. (2015) Field investigation of voltage quality issues in distribution network with PV penetration. In *IEEE PES Asia-Pacific Power and Energy Engineering Conference (APPEEC)*. IEEE; pp. 1–5.

11 Nguyen D. and Kleissl J. (2015) Research on impacts of distributed versus centralized solar resource on distribution network using power system simulation and solar nowcasting with sky imager. In *42nd IEEE Photovoltaic Specialists Conference (PVSC)*. IEEE; pp. 1–3.

12 Liu J., Zhang W., Zhou W., and Zhong J. (2012) Impacts of distributed renewable energy generations on smart grid operation and dispatch. In *2012 IEEE Power and Energy Society General Meeting*. IEEE; pp. 1–5.

13 1547-2003. (2003) *IEEE standard for interconnecting distributed resources with electric power systems*. IEEE: Piscataway, NJ.

14 BDEW. (2008) *Technical Guideline: Generating Plants Connected to the Medium-Voltage Network*. Bundesverband der Energie- und Wasserwirtschaft e.V.: Berlin.

15 VDE-AR-N 4105: 2011-08. (2011) *Power generation systems connected to the low voltage distribution network*. VDE: Frankfurt.

16 GB/T 29319-2012. (2012) *Technical requirements for connecting photovoltaic power system to distribution network*. Standards Press of China.

17 NB/T 32015-2013. (2013) *Technical rule for distributed resources connected to distribution network*. China Electric Power Press.

18 Wu Z.W., Jiang X.P., Ma H.M., and Ma S.L. (2014) Study on fluctuations characteristics of photovoltaic power output in different time scales. *Modern Electric Power*, **31**(1), 58–61.

19 Yang D.W., Huang X.Q., Yang J.H., *et al.* (2012) Introduction to and analysis on IEEE 1574 standard series for microgrids and distributed resources. *Southern Power System Technology*, **6**(5), 7–12.

20 Huang F.S.. (2015) Research on optimal allocation of distributed generator based on time-sequence characteristic. Master dissertation, Southwest Jiaotong University.

21 Wang J.X., Wang X.F., Zhang X., and Hu Z. (2003) Bidding model of reserve capacity in electricity market. *Automation of Electric Power System*, **27**(2), 7–11.

4

Penetration Analysis of Large-Scale Distributed Grid-Connected Photovoltaics

4.1 Introduction

The past 10 years have witnessed the rapid development of distributed PVs all over the world as well as significant PV cost reduction. In some areas, the increasing PVs has already affected the local utilities' price policy and revenue. With continuous technical progress and cost reduction, it is foreseen that PVs will become one of the major energy sources in the future. When the PV cost is lower than the retail electricity price, electricity end users might prefer to use more PV energy instead of grid power, which means a significant load decrease and a big challenge for utilities [1, 2].

On the other hand, owing to the intermittent characteristic of PV output, the capacity of PVs that could be accommodated by a grid is limited. If not dealt with properly, excessive PVs may lead to severe voltage problems and other issues. Therefore, it is important to investigate the PV penetration of a given distribution network and develop advanced control and operation techniques to maximize PV power utilization while keeping system stability and reliability. This chapter focuses focus on PV penetration evaluation and the relevant techniques in the development of the distribution networks with high PV penetration.

Generally, a distribution network experiences three stages in the growth of PV penetration [3]:

- **Stage 1.** Low- or middle-level PV penetration. In this stage, the PV output can be totally absorbed by the local load; no reverse power flows occur, with only minor impact on the distribution network.
- **Stage 2.** High-level PV penetration. As PVs increase, the local load demand may be less than the PV output. The surplus PV power will be delivered back to the grid and affect the operation, protection, and power quality of the distribution network. However, at this stage, utilities still play the role of major power providers for local consumers.
- **Stage 3.** While the PV capacity keeps increasing, the majority of power for local load will be from PVs and utilities then act as a supplemental power provider. At this stage, reverse power flow may occur frequently and make a great impact on the existing distribution network.

In the three different stages, the distribution network will face different economic and technical problems. It needs to coordinate renewable energy sources and traditional energy sources, keep in accordance with load demand management, and overcome the

Grid-Integrated and Standalone Photovoltaic Distributed Generation Systems: Analysis, Design, and Control, First Edition. Bo Zhao, Caisheng Wang and Xuesong Zhang.
© 2018 China Electric Power Press. Published 2018 by John Wiley & Sons Singapore Pte. Ltd.

technical limitations of grid operation when large amounts of intermittent solar power are injected [3, 4]. In this chapter, several concepts of PV penetration are introduced. By analyzing the key challenges in different PV development stages, methods to increase the PV hosting capacity of distribution networks are explored.

4.2 Economic Analysis of Distributed Photovoltaic Systems

At low or middle level of penetration (stage 1), PVs have a minor impact on the distribution networks. The main consideration in this stage is the benefit from the PV investment. If the benefit is too low to cover costs, customers will directly purchase power from the utilities. To protect and support the PV industry, governments in different countries have enacted a series of subsidy policies for PV development. However, those subsidies will be ended eventually. Therefore, the most urgent problem to be solved in stage 1 is to reduce the investment and operating costs of distributed PVs and to establish a self-sustainable business and industry.

4.2.1 Cost/Benefit Analysis of Distributed Grid-Connected Photovoltaic Systems

In the whole PV life cycle, cost efficiency is determined by the total expenditure and the total income. Expenditure includes the initial investment in the initial stage of the project, as well as the operation and maintenance (O&M) cost and the replacement cost in the project cycle, and so on, while the income is mainly derived from the saved electricity and electricity sales from PVs. In most countries, the government also offers different tax subsidies for PVs. The final profit of the project is obtained by deducting tax and all the costs from the total income [5].

4.2.1.1 Cost Composition

The investment costs of PVs can be divided into five parts: capital cost, O&M cost, replacement cost, remaining cost, and other cost [5]. Capital cost refers to the cost needed at the initial stage of project construction, which mainly includes the system components cost and installation cost; O&M cost refers to the expenditure needed in the whole life cycle; replacement cost refers to the cost of replacing system components in the whole life cycle; and the remaining cost is the salvage value of each component, which means the remaining value of components at the end of the project.

1) **Capital cost:**

$$C_{cap} = W_{pv}c_{pv} \tag{4.1}$$

where C_{cap} is the initial investment cost of PV systems, W_{pv} is the capacity, and c_{pv} is the unit PV installed capacity cost.

2) **O&M cost:**

$$C_{op} = W_{pv}c_{op} \tag{4.2}$$

where C_{op} is the total annual O&M cost and c_{op} is the annual O&M cost per unit. Under normal circumstances, the annual maintenance only takes several hours to clean the dust and dirt from PV modules and check the operating performance of the inverter. As such, the O&M cost is mainly the local labor fee.

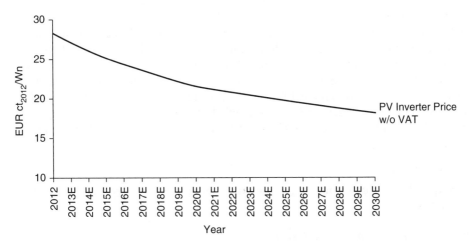

Figure 4.1 PV inverter price 2013–2030. *Source*: CREARA Analysis.

3) **Replacement cost:** denoted C_{re} in this chapter. This refers to the cost of replacing components after a certain number of years. Generally, PV modules have a long life (over 25 years), less than 20% degrading within 25 years. Inverters are the major component in PV installations that need to be replaced. According to research by CREARA [6], the price trend of inverters is shown in Figure 4.1, where the price unit is euro cents per watt. So the total annual cost C_t of a distributed PV project in its operating period can be expressed as follows:

$$C_t = C_{op} + C_{re} \tag{4.3}$$

4) **Remaining cost:** this refers to the salvage value of each component. It is calculated using

$$S = C_{or} \frac{R_{rem}}{R_{comp}} \tag{4.4}$$

where S is the salvage value, R_{rem} is the remaining life of the component, R_{comp} is the total life of the component, and C_{or} represents the original value of the component.

5) **Other cost:** denoted C_{el}, this refers to the cost reserved for emergencies and is mainly used to deal with uncertain situations such as tax rate fluctuations, fluctuations of labor costs, and possible distribution network equipment replacement and line reconstruction in the construction process, and so on. In addition, the initial investment may not be finished in one time step, as bank loans may be used. In this situation, the interest rate of loans should be included in the investment cost.

4.2.1.2 Income Composition

The income from a PV investment consists of two parts: policy subsidy for PVs and electricity consumption saving. Electricity consumption saving includes the saved cost of electricity purchased from the grid and the revenue of PV electricity sold to the grid.

4.2.1.2.1 Subsidy

In China, the government subsidy has two forms. One is the feed-in tariff that is issued by the National Development and Reform Commission; the other is the one-time construction investment subsidy, which generally accounts for a certain proportion of the initial construction investment (e.g., 50% of the construction investment cost of the Golden Sun demonstration project is from the national subsidy). Here only the feed-in tariff is discussed. At present in China, the full electricity subsidy policy is adopted in terms of the distributed PVs. Namely, all distributed PV systems could apply for the subsidy:

$$I_b = c_b E_{pv} \tag{4.5}$$

where I_b is the subsidy income, E_{pv} is the total generated PV energy, and c_b is the unit electricity subsidy.

4.2.1.2.2 Electricity Consumption Saving

According to the existing policy, the PV electricity delivered back to the power grid is priced using the power price of a benchmark desulfurized coal-fired generator:

$$I_{gr} = c_{gr} E_{gr} \tag{4.6}$$

where I_{gr} is the electricity consumption saving, E_{gr} is the sent-to-grid PV energy, and c_{gr} is the unit price of desulfurized benchmark coal-fired power.

In most cases, PVs have the priority to generate power to supply local load and send the surplus power back to the grid. Therefore, the PV power consumed by users can be considered as income since it reduces the purchase of electricity from the grid. In other words, the money saved that was used to purchase electricity should be included in the PV income:

$$I_{sa} = c_{sa} E_{sa} \tag{4.7}$$

where I_{sa} is the income from the saved electricity, E_{sa} is the saved electricity, and c_{sa} is the unit price of electricity purchased from the grid.

Finally, the present value of the investment cost and the present value of the income can be expressed as

$$C_p = C_{cap} + C_{el} + \sum_{t=1}^{N} (C_{op} + C_{re})(1 + i_c)^{-t} \tag{4.8}$$

$$I_p = \sum_{t=1}^{N} (I_b + I_{gr} + I_{sa})(1 + i_c)^{-t} + S(1 + i_c)^{-N} \tag{4.9}$$

where t means the tth year in the PV life cycle. When $t = 0$, it is the construction year of the PV project, and the income starts when $t = 1$. N is the total life years of the project. And i_c is the discount rate, which is used to calculate the present value of assets.

4.2.2 Grid Parity

Levelized cost of energy (LCOE), defined by the National Renewable Energy Laboratory of America, or levelized cost of electricity as defined by Fraunhofer-ISE, is an economic assessment used to analyze various issues concerning the cost of power generation technology. LCOE offers investors a clear view of the unit electricity cost. LCOE is used to evaluate the benefit and cost of PVs in this chapter. Only when the LCOE of PVs falls to or below the retail electricity price from the grid is PVs competitive enough and becomes attractive.

Conceptually, LCOE refers to the ratio of the total cost to build and operate a PV system in its life cycle over the total PV energy throughput [7–13]. In practice, LCOE has two forms: a real LCOE and a normal LCOE. As a base price, the real LCOE is fixed at a certain time point; however, the normal LCOE does consider the time value of money which is often caused by currency inflation and other factors. In this chapter, the normal LCOE is used.

It is assumed that a residential consumer is planning to install a PV system; the normal LCOE can then be obtained through the following equations:

$$\sum_{t=1}^{N} \left[\frac{\text{LCOE}}{(1+i_c)^t} E_{\text{pvt}} \right] = C_{\text{cap}} + C_{\text{el}} + \sum_{t=1}^{N} \frac{C_t}{(1+i_c)^t} \tag{4.10}$$

where E_{pvt} is the total amount of PV electricity generated in the tth year, C_t is the total expenditure in the tth year, and N is the total number of years. Assuming that the price is constant during the whole year, it is derived that

$$\text{LCOE} = \frac{C_{\text{cap}} + C_{\text{el}} + \sum_{t=1}^{N} \frac{C_t}{(1+i_c)^t}}{\sum_{t=1}^{N} \frac{E_{\text{pvt}}}{(1+i_c)^t}} \tag{4.11}$$

It can be seen that in the calculation of the LCOE the government tax incentives and subsidies are not included. As a result, the LCOE of solar PV can only indicate the comparison of the per kilowatt-hour cost of PV electricity with the retail electricity price without considering any external incentives.

In the past 10 years, owing to the high cost of PVs, the price of PV electricity was much higher than the retail electricity price and so made PVs insufficiently attractive to customers. Therefore, many governments have used subsidies to stimulate the PV market demand. With technical progress, the cost of PVs dropped rapidly and in some places it has been close to and even lower than the electricity price from the grid. In this case, PVs have a strong economic advantage to attract investors even without government subsidies. In this situation, PV grid parity is happening, as shown in Figure 4.2. It can

Figure 4.2 Simplified diagram of grid parity of PV electricity.

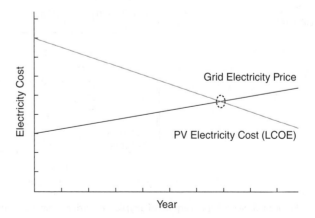

be seen from the diagram that once the PV grid parity is achieved, more electricity end consumers are willing to install distributed PV systems.

4.3 Large-Scale Photovoltaic Penetration Analysis

4.3.1 Further Explanation of Some Concepts

As mentioned previously, high PV penetration happens in a distribution network at stage 2 or stage 3. If the PV capacity is not limited, the net load may drop to below zero, which will lead to reverse power flow. In this chapter firstly the conceptual analysis will be performed where the distribution network is treated as a load aggregation point including local power sources, network, and loads. It interconnects with other distribution networks and power supplies, forming the total power system. The schematic diagram of the power system is shown in Figure 4.3.

Traditional electrical units/power sources include hydropower units, nuclear power units, and fossil-fuel-based (coal or natural gas fired) thermal power units. Among them, thermal power units dominate the electricity generation in China and worldwide. According to the operation and economic efficiency requirement, a thermal power unit must run above its minimum power output level, which is also called the minimum technical output or the minimum stable output. When the load is lower than the minimum technical output, the thermal power unit will be forced to shut down. Large-capacity generator outage often brings serious economic losses (such as load shedding) and stability issues (such as system vibration) to power systems. Owing to

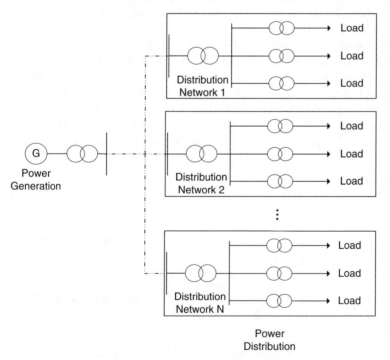

Figure 4.3 Schematic diagram of a typical electrical power distribution system.

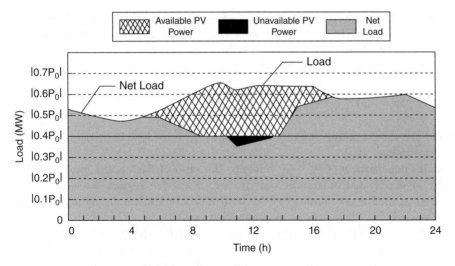

Figure 4.4 Schematic diagram of unavailable PV output.

this and the relatively low price, nuclear and thermal units are normally dispatched to cover the base load to satisfy the minimum technical outputs of nuclear and thermal units.

When the PV penetration level in a power system is high, its net load may fall below the minimum technical output required by the traditional generation units. In this situation, those units will be forced to be taken down from the system and result in substantial economic losses and security issues. To avoid this, it is assumed that the net load is not allowed to be lower than the minimum technical output. As shown in Figure 4.4, the minimum technical output of the system is assumed to be $0.4P_0$, the net load curve is lower than the minimum technical output at 12 h (i.e., the dark section) due to the high PV output. This means that the system cannot afford to have such large PV output and the PV power must be curtailed.

Furthermore, for the distribution network, when PV output exceeds local load demand, excess PV power would flow back to the higher level grid and lead to a lot of problems, such as protection discoordination and failures, overvoltage issues, challenges in system power dispatch, and so on. In order to get high investment cost efficiency and avoid other accompanying issues, high PV penetration with an advanced control approach while keeping the original system structure and configuration unchanged will be analyzed. It should be noted that the hosting capacity discussed here refers to the hosting capacity of the distribution network. For the distribution network, the excess PV power that is not able to be accommodated locally either flows back to the higher level power grid or curtailed, which hence all regarded as unavailable PV power.

4.3.2 Concepts and Assumptions

4.3.2.1 Basic Concepts

First of all, the related basic concepts are defined as follows [14]:

- **Power penetration (PP).** PP refers to the maximum ratio of the PV output to the regional load in the same time point of 1 year in a given region. As given in

Equation 4.12, PP reflects the maximum load support ability by PV output in 1 year. At other moments of the year, the ratio of PV output to load is always lower than PP. PP may be higher than 100%, which means the PV can exceed the load at the moment and reverse power flow on a distribution system upstream of PVs will occur:

$$PP\ (\%) = \max\left\{\frac{P_{pv}}{P_{load}}\right\} \times 100 \tag{4.12}$$

where P_{pv} is the PV output at a certain time and P_{load} is the load in the same region at the same time.

- **Capacity penetration (CP).** CP refers to the ratio of the annual maximum PV output to the annual maximum system load in a given region. As is shown in Equation 4.13, CP is very similar to PP as both of them are the ratios of the PV output to the system load. However, the difference between CP and PP lies in that PP is the ratio of the PV output to load at the same time while CP is the ratio of the PV output to load at different time intervals. CP has been widely used by power companies to assess the capacity of the distributed PV output against the regional load:

$$CP\ (\%) = \frac{P_{pv\ max}}{P_{load\ max}} \times 100 \tag{4.13}$$

where $P_{pv\ max}$ is the maximum PV output power in a year and $P_{load\ max}$ is the maximum load in a year.

- **Energy penetration (EP).** EP refers to the ratio of the annual available PV electricity to the annual load power consumption as shown in Equation 4.14. Even if the PV output far exceeds the load demand at some time points, there will be only a small impact on EP because EP focuses on the long-term energy usage:

$$EP\ (\%) = \frac{E_{pv}}{E_{load}} \times 100 \tag{4.14}$$

where E_{pv} is the annual available PV electricity at and E_{load} is the annual power consumption.

- **PV utility ratio (PUR).** PUR refers to the ratio of the actual annual available PV electricity to the annual generated PV electricity:

$$PUR\ (\%) = \frac{E_{pv}}{E_{pv\text{-}ge}} \times 100 \tag{4.15}$$

where $E_{pv\text{-}ge}$ is the annual generated PV electricity.

- **PV cost (PC).** According to the analysis of the PV economic benefits in Section 4.2, the cost of PV is affected by many factors. However, for investment fixed PVs, a higher PUR indicates a higher recovery efficiency and a lower cost of PV electricity. Therefore, the reciprocal of the PUR is used to represent the PV cost:

$$PC = \frac{1}{PUR} \tag{4.16}$$

4.3.2.2 Basic Assumptions

In order to further discuss the relationship between PP, PUR, PC, and the maximum PV capacity allowed in distribution network, the following assumptions are made in this chapter [15]:

1) The power loss is not considered. In other words, it is assumed that the PV output is directly connected to the load side, and that the PV output can directly meet the load demand.
2) Unified efficiency curves are used in the PV system (including the PV inverter). The difference in the manufacturer and environmental factors is ignored.
3) The scope is confined within the distribution network while its capability of accommodating PVs is studied. The generation dispatch capability of the upper bulk power system is not considered for accommodating PVs in the distribution network.
4) PV systems operate at unify power factor, taking no consideration of reactive power.
5) The distributed deployment of PV systems can reduce the impact caused by cloud dynamics. Hence, the transient performance of PVs is ignored in this chapter and the minimum time interval of analysis is set to be 1 hour.

4.3.3 Power Penetration Analysis

Once connected to the distribution network, the PV output is dispatched to first satisfy the nearby load to reduce the power supply from the grid. The PV output is greatly affected by solar irradiance, which might be inconsistent with the load profile, as shown in Figure 4.5. The data were measured at the SA substation in J City in 2014 in China.

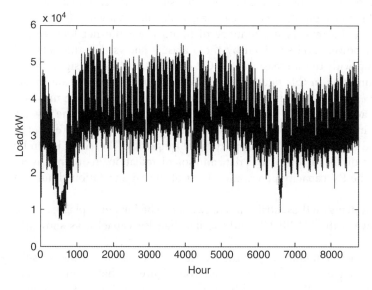

Figure 4.5 Annual load curve of the SA substation.

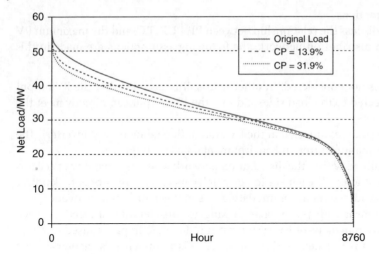

Figure 4.6 Net load duration curve.

The annual maximum load in the region of the SA substation is 56.2 MW. As the 2014 Spring Festival fell in February, the load in February dropped significantly. Spring Festival is the most important festival in China, normally coming with a long holiday break. In general, the annual load profile of the SA substation shows no prominent seasonal characteristics and the annual load is quite stable.

According to the definition, PP and CP are closely related to the installed capacity of grid-connected PV systems. The impact of the PV power on the net load grows as the installed capacity increases. In the SA region, by increasing the PV installed capacity gradually the annual net load duration curve can be calculated, as shown in Figure 4.6.

The x-axis in Figure 4.6 reflects the number of hours when the net load equals or exceeds a certain power level specified by the y-axis. It shows that the net load demand is decreasing with the increase of PV CP. The net load drops significantly in the heavy load condition in this figure. When the PV capacity penetration in the system reaches 31.9%, the net load at one certain time becomes zero, which means if more PV capacity is added, the PV PP would exceed 100% and accordingly the reverse power flow appears in the distribution network. When CP continues to increase, the negative net load and reverse power flow appear more frequently. Therefore, it can be concluded that, based on the historical load data and solar irradiation, in order to avoid reverse power flow, the PV capacity penetration limit of the SA substation should be less than 31.9%.

The impact of PV systems on the distribution network can be further explained using the relationship between the PV PP, CP, and the installed PV capacity, as shown in Figure 4.7.

The actual power generation efficiency of the distributed PV systems at the SA substation region is in the range 78–83%. It can be seen from Figure 4.7 that when a 23 MW PV system is installed, the PV PP will exceed 100%, which means during some moments in a year the PV output power exceeds the load demand at that time. Beyond this point, the PV generated power will exceed the load of the distribution network during some periods of time and reverse power flow would occur, resulting in a possible undesirable

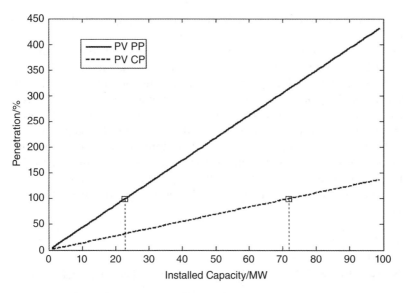

Figure 4.7 Relationship between PV PP, CP, and grid-connected PV capacity.

impact on the power system dispatch and operation. Serious PV power reverses will appear when the installed PV capacity is about 72 MW and the PV PP is higher than 300% at 100% of CP. The distribution of the ratio of the annual PV output to the load with CP reaching 100% is shown in Figure 4.8.

In Figure 4.8, the horizontal axis represents the PV PP while the vertical axis represents the frequency of ratio in a year (8760 h). It can be seen that the PV PP ratio mainly falls between 0 and 100% while the PV PP could be higher than 300%. Although this high ratio seldom appears, special attention should be paid to it, and the possible

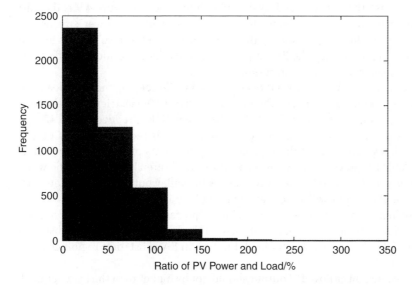

Figure 4.8 Frequency distribution of PV output to load when CP = 100%.

impact of the high reverse power on the distribution network equipment also needs to be investigated.

It should be noted that although the PV PP and PV CP are similar to each other, as both of them describe the mutual impacts of the PV power generation on the distribution network during a certain period of time, they have different focuses. The PV PP reflects the relative matching degree between the PV output and load demand of certain moments in 1 year, while CP reflects the limit of possible impact of the PV power generation on the load of the distribution network with such an installed capacity.

The PP and CP indices evaluate the impact of the PV output throughout the year, but they are not a good measure to assess the cumulative impact. For this purpose, the EP index is used to analyze the cumulative effect of PV output during a given period; that is, 1 year. The available PV power, considering the PV power regulation capability, can be expressed as

$$P_{avai} = \begin{cases} P_{out} & P_{out} \leq P_{load} \\ P_{load} & P_{out} > P_{load} \end{cases} \tag{4.17}$$

where P_{out} is the PV output and P_{load} is the load. The annual available PV energy can be expressed as

$$E_{avai} = \int_0^t P_{avai} \, dt \tag{4.18}$$

Accordingly, the PV energy penetration is defined as the ratio of available PV energy to the annual load consumption in the system. As an example, the cumulative effect of solar energy is analyzed using the real data of 1 day at the SA substation. As shown in Figure 4.9, the ratio of the installed PV capacity versus the annual maximum load is 43%; however, the PV electricity only contributes 27.6% of the total load in that day; the PV output could be completely accommodated and the PV power utilization rate on that day was 100%. As in Figure 4.9, the original load is supported by two sources: the grid and the PV systems. The part covered by dotted oblique lines in Figure 4.9 is the part of the load supported by the PV systems, the so-called "available PV power"; the part in grey is the part of the load supported by the power grid, which is also the net load. It can be seen from Figure 4.9 that the PV output during the midday period can effectively reduce the daily peak load of the distribution network.

It is now assumed that PV capacity has been doubled while keeping the same load and solar radiation conditions unchanged. The ratio between the PV electricity and the total load on the same day is increased to 55.2%, but the actual PUR is reduced to 76.47%, as shown in Figure 4.10. By comparing Figures 4.9 and 4.10 it is seen that the PV output under the zero lines fails to meet the regulation requirements of the system, which is called "unavailable PV power." The "unavailable PV power" often happens at noon when the PV output is very high and the net load reaches the valley bottom. It is obvious that more unavailable PV output will appear after more PV capacity is installed.

In the case of high PV penetration, the available PV power and installed capacities will no longer be a linear positive correlation. Based on the load data of the SA substation in 2014, Figure 4.11 shows that the net load duration in SA substation region changes along with the EP.

To prevent reverse power flow, PV output should not be higher than the load demand. Figure 4.11 shows that, along with higher PV EP, the time that the system net load

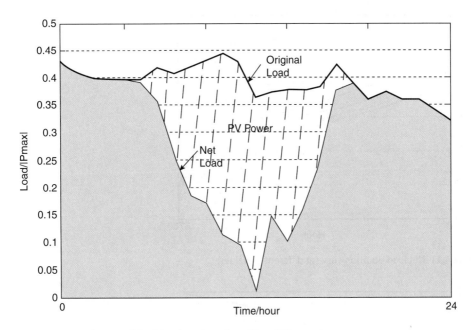

Figure 4.9 Load curve of the SA substation when CP = 43%.

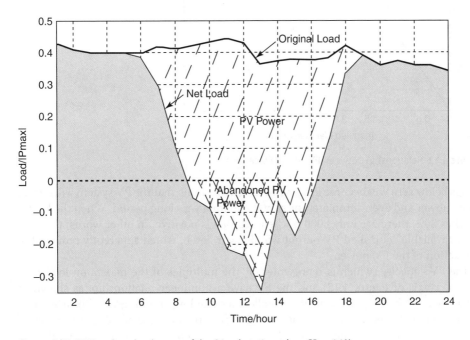

Figure 4.10 PUR and net load curve of the SA substation when CP = 86%.

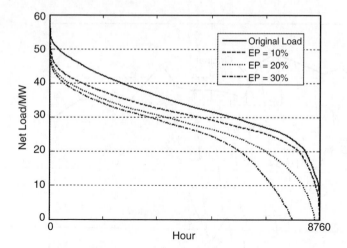

Figure 4.11 Net load duration curve at different EP values.

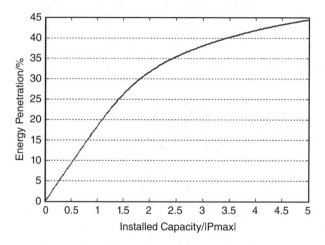

Figure 4.12 Relationship between PV EP and PV capacity.

drops below zero will become longer and longer, meaning that the PV system will curtail more power and even completely be shut down for a very long period. When the PV EP reaches 30%, there are only 8000 h when net load is positive. In other words, the time of "unavailable"/curtailed PV output will be near 760 h, which apparently reduces the utilization of the PV power.

The PV capacity (which is represented by the multiples of the maximum load P_{max} on the x-axis of Figure 4.12) and the EP have a nonlinear relationship, as shown in Figure 4.12. When the PV capacity installed is quite low, the EP grows linearly. After the PV energy penetration arrives at a certain value, the increase of PV energy penetration slows down. The upper limit of PV energy penetration depends on when the PV output is sufficient to support load demand at any time.

Figure 4.13 shows the relationship between the PV PP, CP, and EP. It is concluded that at the beginning of PV development, the PV capacity penetration and PV penetration

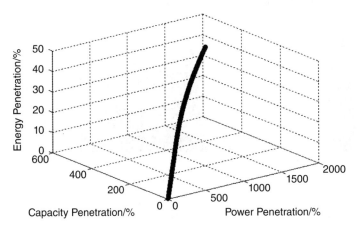

Figure 4.13 Diagram of relationship between PP, CP, and EP.

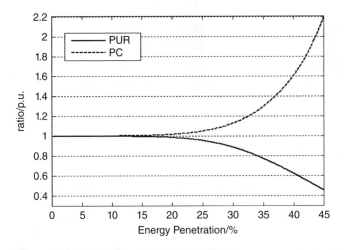

Figure 4.14 Relationship of PUR, PC, and PV EP.

increase and the PV energy penetration increases fast as well. However, when PV energy capacity approaches its upper limit, the curtailed PV power will increase rapidly.

Figure 4.14 shows the curves of PUR and PC versus the change of EP. It can be seen that the PV available hours decrease with the PUR. When the EP is low, the PUR is high and most of the PV output can be utilized. When the PV EP is high, the PUR drops rapidly, resulting in a big loss of solar energy. When the PUR drops, the unit cost of PV electricity will increase. With the increase of EP, the PV electricity cost will constantly increase. It is also known that, for a distribution network with high PV penetration, a higher PV EP indicates a less effective utilization of PV capacity and higher cost.

4.3.4 Photovoltaic Penetration with Different Types of Load

The correlation coefficient between the PV PP and the PV CP is affected by the load distribution. In Chapter 3, loads are classified into four categories: industrial load,

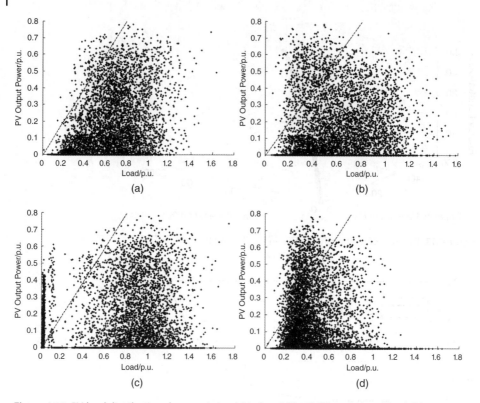

Figure 4.15 PV-load distribution characteristics: (a) industrial load; (b) agricultural load; (c) commercial load; (d) residential load.

agricultural load, commercial load, and residential load. In this chapter, the different kinds of loads, the annual irradiance curve at the SA substation and the typical load daily curve in Chapter 3 (see Figure 3.2) are still adopted to discuss the PV penetration for different types of loads, the fitting curve of which is obtained by using HOMER software (http://www.homerenergy.com/). The PV capacity installed is set to be 100% of the average load. The distributions of the ratio of the annual PV output over load demand are shown in Figure 4.15, which are divided by different typical loads.

In Figure 4.15, the diagrams show the distribution of the ratio of the PV output to load at each time point in 1 year. It can be seen from Chapter 3 that the characteristic of the four types of load are quite different from each other. The dashed lines starting from (0, 0) in Figure 4.15 represent the 100% ratio of PV output over the load. The dots above the 100% line mean the PV output has exceeded the load, which will lead to the reversing of power in the system; the dots below the 100% lines mean the PV output is smaller than load demand and can be completely absorbed.

If the same PV capacity is kept in Figure 4.15, the PV PPs of all types of load will exceed 100%. In other words, all the four cases experience reverse power; this happened more seriously in the commercial load. The PV CPs of all types of load also vary. Details are given in Table 4.1.

It can be seen from Table 4.1 that the commercial load has the largest fluctuation and the PV PP reaches 3907.11%. The PV PP reaches its maximum in the morning when the load demand is very low even though the PV output is not very high. However, the

Table 4.1 Typical daily PV penetration of different types of load.

Load	PV PP (%)	PP CP (%)	Periods power reversed	Correlation
Industrial	216.75	47.80	126	0.48
Agricultural	598.47	49.41	560	0.21
Commercial	3907.11	44.79	1145	0.38
Residential	392.30	47.46	939	−0.10

installed PV capacity is not very high compared with the commercial load as its PV CP is only 44.79%, which is the lowest number among the four types of load.

The last column in Table 4.1 is the correlation between the PV output and load. The big number indicates that the PV output curve and the load curve have consistent change and the PV output can match load well. The calculation formula of the correlation coefficient is

$$
r = \frac{\displaystyle\sum_{i=1}^{N}(x_i - \bar{x})(y_i - \bar{y})}{\sqrt{\displaystyle\sum_{i=1}^{N}(x_i - \bar{x})^2 \sum_{i=1}^{N}(y_i - \bar{y})^2}}
\tag{4.19}
$$

where x represents the PV output, y represents the load, and N is 8760 h (the number of hours in 1 year). The range of r is between −1 and +1. When $r = -1$ the two variables are negatively linearly related, and when $r = 1$ they are positively linearly related. It is noted that the variables in Equation 4.19 have been normalized to fully reflect the correlation of the two variables and provide a better view of the matching degree between the distributed PV output and the load.

As shown in Table 4.1, the industrial load can well match the PV generation, but the residential load only has a weak correlation with the PV output. In addition, the correlation coefficient between the PV output and the residential load is negative, which means the two are negatively correlated. It can be seen from Figure 4.15d that when the load is heavy the PV output is low, and when the load is light the PV output is high, meaning that the reverse of power can be quite serious. Generally speaking, it is encouraged to install more distributed PVs to a distribution network primarily with a major industrial load; and for a distribution network with major residential load, a small PV capacity could be effectively accommodated.

Figure 4.16 shows the PV EP versus installed PV capacity curves for different types of load. It can be seen from the figure that the all the curves have a pretty steep slope at the beginning. This means that the PV power can be well absorbed when the installed PV capacity is small for all four types of load, particularly for the residential type of load. However, with the increase of PV capacity the growth of the PV EP of the residential load rapidly slows down; the PV EPs of the commercial load and industrial load still maintain a good trend, and the EP upper limits of these two types of load are also much higher than the EP limits of the other two types of load. Generally, it is more suitable to integrate low PV capacity into the distribution network primary with residential load while the distribution network with industrial load and commercial load is more suitable for high

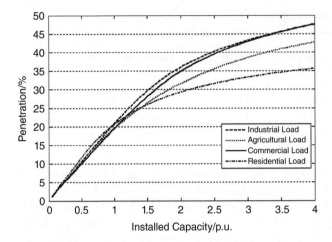

Figure 4.16 Installed PV capacity–EP curves of different types of load.

PV penetration. However, advanced control measures should be taken to prevent the negative impact if overvoltage occurs due to high reverse power.

4.4 Maximum Allowable Capacity of Distributed Photovoltaics in Distribution Network

The PV capacity and penetration in a distribution network from the perspective of power balance and energy balance has been discussed above. According to the previous analysis, the cost of PV electricity will increase and PUR will decrease in distribution networks with PV penetration higher than certain extent. Hence, the attempt to install more PV capacity in the distribution network is not always a good solution. PV electricity utilization is affected by many factors, including the actual distribution network transmission capacity, protection settings, and grid configuration. Especially for high PV penetration cases, the system topology and operation are the key factors limiting the further expansion of grid-connected PVs.

Power quality and voltage stability are other safety issues in high PV penetration networks. Especially when a traditional distribution network is evolving from original radial feeders to a networked system, there will be many other issues in voltage regulation, protection setting, power quality, and other technical problems. Building a new distribution network or upgrading is of much expense. So the capacity of the existing distribution network should be fully utilized. Quantitative analysis is important in order to safely integrate more power into the distribution network [16]. In this chapter, three methods of calculating the maximum allowable capacity will be investigated: the static characteristic constraint method, the constraint optimization algorithm, and the digital simulation method.

4.4.1 Static Characteristic Constraint Method

The static characteristic constraint method aims at PV capacity constraints in consideration of a single characteristic, such as voltage characteristic, harmonic characteristic, and protection characteristic, for stable and safe operation of the distribution network.

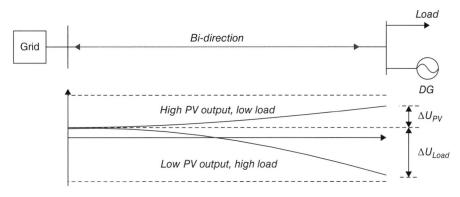

Figure 4.17 Impact of distributed PV system connection on local voltage.

4.4.1.1 Voltage Constraint

The traditional radial feeder only has one power source; the voltage magnitude gradually decreases from the source to the load when there is no distributed generation. After connection of the PV system, reverse power flow and overvoltage may appear when the PV output is high. These phenomena may frequently happen due to the intermittent characteristic of PV output. Figure 4.17 describes the impact of the PV output on voltage. Because overvoltage may damage power system equipment, it is an important factor limiting the allowable PV capacity.

Intermittent PV output also brings voltage fluctuation and flicker issues to the power grid. Once coupled with the load change, serious voltage fluctuations and flickers may be seen, which is also worth paying attention to in the design of PV systems (http://www.wusuobuneng.com/archives/7437).

4.4.1.2 Protection

Typical protection equipment refers to circuit breakers, reclosers, fuses, and control systems. Nondirectional three-phase overcurrent protection is the main protection and backup protection equipped for feeders. The recloser is also equipped for important feeders, which is used to remove temporary faults and restore power. Faults can also be removed and isolated using fuses. The fuses will be blown off when the line current exceeds the unallowable limit for a certain time. Circuit breakers, reclosers, and fuses work together by properly setting the circuit breaker action current values and action delays, recloser operation procedures, and selecting proper melting currents of fuses [17].

However, when a large amount of distributed PVs are connected to a distribution network, the original distribution network changes from a passive, unidirectional, radial network to an active, interconnected network. The fault current value and direction can be changed, which makes the problem more complicated. Coordinating different protection devices is very difficult and sometime impossible. Traditional protection devices are barely able to meet the requirements of sensitivity, selectivity, rapidity, and reliability at the same time. The following is an example to show the maximum allowable capacity of the distributed PVs without the need for changing previous current protection settings.

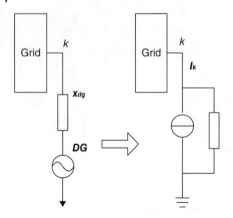

Figure 4.18 Equivalent circuit of distribution network with distributed PV system.

When a fault occurs, the PV system will provide an additional short-circuit current to the fault point that may worsen the fault consequence. As shown in Figure 4.18, the distributed PV system is simplified as a current source plus a reactance.

This example focuses on the short-circuit current analysis flowing through the line where the protection device is located. First of all, the following assumptions are made: (1) ignore the impact of the nonstandard transformation ratio of the transformer; (2) ignore the impact of load; (3) the potential of the current source and the subtransient potential are both 1.0 p.u., and the phase angle is zero.

If the node impedance matrix of the distribution network without PVs is Z, the system voltage satisfies

$$Z\dot{I} = \dot{V} \tag{4.20}$$

where $\dot{I}$ is the injection current of the system and $\dot{V}$ is the bus voltage.

It is assumed that the distributed PV system is connected to the grid at node k. The injected current is calculated as

$$I_k = \frac{1}{jx_{dg}}\frac{S_{kdg}}{S_B} - V_k\frac{1}{jx_{dg}}\frac{S_{kdg}}{S_B} \tag{4.21}$$

where S_{kdg} is the installed PV capacity, x_{dg} is the subtransient impedance (in p.u.) of the PVs, S_B is the system power base, V_k is the voltage (in p.u.) of the node k after the fault occurs, and I_k is the injected current (in p.u.).

By solving Equation 4.20, node voltages can be expressed by a function of S_{kdg}, and according to Equation 4.21), the voltage of node k is

$$V_k = f_k(S_{kdg}) \tag{4.22}$$

The short-circuit current of any branch $m–n$ can be calculated as

$$I_{mn} = \frac{U_m - U_N}{Z_{mn}} = \frac{f_m(S_{kdg}) - f_n(S_{kdg})}{Z_{mn}} \tag{4.23}$$

where I_{mn} is the short-circuit current of the branch $m–n$ and Z_{mn} is the positive sequence impedance of the branch $m–n$.

Equation 4.23 gives the relationship between the capacity of the distributed PV system and the short-circuit current. Further establish the mathematical model to maximize allowable PV capacity with no need to change the current protection settings, which is

$$\begin{cases} \max\ f(x) \\ \text{s.t.}\ \ g_i(x) < b_i;\ \ i = 1,2,\dots,m \end{cases} \tag{4.24}$$

where $g_i(x)$ is the short-circuit current through node i, $f(x)$ is the total PV capacity, b_i is the current protection setting value, and x is the column vector representing the capacities of all the PV systems.

Traditionally, synchronous generators provide short-circuit current and the fault line can be isolated by the line protection. However, owing to the low energy density of PV generation and the limited overcurrent capacity of the PV system, the PV system may not be able to provide enough short-circuit current. One example can be found in Barker and de Mello [18], where distributed generators can only provide small fault current. Experimental and dynamic simulation studies show that the short-circuit current in the PV inverter is allowed to be 25% higher than the rating value at most. As a result, it may be hard to detect a fault in the distribution network with high PV penetration because the PV ssystem can only provide and allow a small short-circuit current, which leads to the protection device refusing to act. Therefore, the impact of PVs on protection should be investigated in distribution system planning and design when the PV system is connected.

4.4.1.3 Harmonic Limit

Harmonics mainly refers to current harmonics caused by power electronic components, such as PV inverters. Flicker refers to the fast fluctuation of voltage perceived by electricity end users. PV-related flicker is due to the high-speed switch in inverters. Harmonics and flicker can influence voltage quality, which is dependent on the system short-circuit capacity at the PV connection bus and the total installed PV capacity under the same step-up transformer.

For instance, consider one PV system to be connected to a 10 kV node in a distribution network. According to China's GB/T 14549-1993 standard, the PV harmonic current component (root-mean-square value) must be no greater than the specified allowable value [19]. Taking the fifth harmonic as an example, it is assumed that the minimum short-circuit capacity of this node is 100 MWA and the allowed fifth harmonic current is 20 A. Meanwhile, it is assumed that the fifth harmonic distortion rate of the PV power is 4% [20]. In this case, the capacity of the single inverter at the PCC, S_{pv}, should meet the following requirement:

$$\frac{S_{pv}}{\sqrt{3} \times 10\,\text{kV}} \times 4\% \leq 20\,\text{A} \tag{4.25}$$

The maximum PV installation capacity is then 8.66 MW with respect to the fifth harmonic current constraint. In the same way, the PV capacity limit of other orders of harmonic is calculated [21].

When multiple inverters are connected to a node in parallel, harmonic amplitude and phase dispersion need to be taken into account. Crucq and Robert [22] give a harmonic current summation rule based on statistical data; the resultant current of the hth harmonic is

$$I_h = k\left(\sum_i I_{hi}^a\right)^{1/a} \tag{4.26}$$

where I_h is the resultant value of the hth harmonic current of all harmonic sources and I_{hi} is the value of the hth harmonic current of the ith single harmonic source. The values of coefficient k and coefficient a are determined by two factors: the chosen probability that the actual value exceeds the calculated value, and the random variation degree of both the current amplitude and the phase of all the harmonics.

4.4.2 Constrained Optimization Method

A variety of optimization methods (such as the deterministic optimization method, the probabilistic optimization method, and the intelligence algorithm) are used to decide the maximum capacity of PV systems under certain constraints. The maximum allowable capacity of the distributed PV systems should satisfy the power balance and voltage constraints and meet the requirements for the safe and stable operation of the distribution network. The mathematical expression is as follows:

$$\begin{cases} \max f = (c^T y) \\ g(u, x, y) = 0 \\ h(u, x, y) \le 0 \end{cases} \tag{4.27}$$

where u and x are the control and state variables of each component in the system. y is the column vector composed of the PV active power (which is proportional to the installed PV capacity) and reactive power, which is separately listed as it is not a control variable or a state variable, and c is the coefficient vector representing the working status of the PV system; $c = 1$ when the PV active power is not zero, and is zero in other cases.

The equality constraints to satisfy power flow balance are given by the power flow equation [23]:

$$\begin{cases} P_i^{sp} - U_i \sum_{j\in i} U_j(G_{ij}\cos\theta_{ij} + B_{ij}\sin\theta_{ij}) = 0 \\ Q_i^{sp} - U_i \sum_{j\in i} U_j(G_{ij}\cos\theta_{ij} - B_{ij}\sin\theta_{ij}) = 0 \\ \qquad\qquad i = 1, 2, \ldots, N - 1 \end{cases} \tag{4.28}$$

where N is the number of nodes in the distribution network, P^{sp} and Q^{sp} are the column vectors of the active power and reactive power injected into the grid at all nodes. P_i^{sp} and Q_i^{sp} are the ith elements of P^{sp} and Q^{sp} respectively, which are composed by

$$\begin{cases} P^{sp} = P^G + P^D + P^S \\ Q^{sp} = Q^G + Q^D + Q^S \end{cases} \tag{4.29}$$

where P^G, P^D, and P^S are the column vectors of the power generation, load, and PV active power respectively at the node (taking the power injection as positive direction,

P^D is usually negative); Q^G, Q^D, and Q^S are the column vectors of the power generation, load, and PV reactive power respectively at the node (excluding the balance node N).

The inequality constraint $h(u, x, y) \leq 0$ includes

1) line constraints (such as current or transmission line power constraints);
2) node voltage constraint;
3) harmonic current constraint.

4.4.3 Digital Simulation Method

The digital simulation method analyzes the impact of a series of grid-connected PVs with given capacities on the power quality, safety, and stability in several typical operation modes through steady-state and dynamic simulations, in order to maximize the allowable PV capacity in the system [23]. Commonly used operation modes include maximum load operation mode and minimum load operation mode. Both of them must satisfy the static and dynamic safety and stability constraints of the distribution network. In other words, with the maximum allowable PV capacity installed, the PVs should meet the following conditions:

1) The voltage of each node and the power flow in the system does not exceed the predefined limit in the steady-state operation.
2) The system can maintain safe and stable operation when facing PV output disturbance because of solar radiance fluctuations.
3) When a grounding fault happens, the system can remain stable.
4) Other specific requirements.

4.4.3.1 Maximum Allowable Photovoltaic Capacity in Static Simulation

The $P–V$ curve analysis method is used to study the PV static capability in system steady-state operation in which the PV output constantly increases until the system loses voltage stability. It is worth mentioning that the maximum PV capacity is not only decided by the "critical point" on the $P–V$ curve, but also by the voltage boundaries allowed in the current distribution network.

4.4.3.2 Maximum Allowable Photovoltaic Capacity in Dynamic Simulations

In dynamic simulations, the PV output is constantly increased and system stability will be observed under the predefined disturbance. The maximum acceptable PV capacity is obtained when the system loses dynamic stability under disturbance.

The main forms of disturbance mainly include:

- *System fault disturbance.* A grounding fault is assumed to happen on the connection line between the PVs and the distribution network. A transient stable condition needs to be satisfied in the fault cleaning process by protection devices and the following oscillation period. Line faults can be divided into two types of PV system: (1) a small PV system with only one line connecting the PV system to the distribution network; the instantaneous three-phase short circuit and the fault cleaning protection should be considered under this circumstance; (2) a large PV system, in which at least two lines are used to integrate the PV system to the distribution network; faults on each one of the lines also need to be investigated in this case.

- **PV power disturbance.** That is, the impact of the PV output fluctuations on the system stability due to the dramatic changes of the solar radiance. The PV output could reduce from its rated value output to zero in a short period of time and then increase from zero to the rated output in another short period of time, and so on.

4.5 Maximum Allowable Capacity of Distributed Photovoltaics Based on Random Scenario Method

Till now, three methods for the calculation of maximum PV capacity have been introduced. The constrained optimization method has high flexibility, low computational cost, and high accuracy. It can be easily used in consideration of a variety of constraint conditions and objectives. Meanwhile, a bunch of optimization algorithms can be used to solve different optimal models. Therefore, the constrained optimization method is a good candidate to decide the maximum allowable PV capacity. In this section the idea of a "random scenario" is introduced. Also, the data of SA substation is employed to discuss the method.

4.5.1 Algorithm Introduction

A distribution network could have many nodes, loads, and various PV connection plans, which all make it difficult to decide the maximum PV allowable capacity. For a certain distribution network, the load distribution and the network structure are determined, but the PV integration plans vary. If only a few interconnection plans are taken into account, the evaluation result may not be very accurate. On the other hand, if a large number of plans are considered, it may be time consuming and cost great efforts with fewer effects and results. The idea of "random scenario" [24] is introduced to solve this

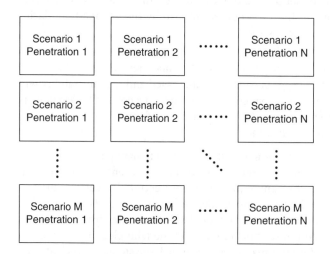

Figure 4.19 Schematic diagram of the "random scenario."

issue. As shown in Figure 4.19, to evaluate the maximum PV allowable capacity, the grid integration strategies of distributed PV systems and the load characteristics of the connection point [25] are key to determining the maximum PV allowable capacity.

In this method, various representative plans should be established first; then, based on the plans, the maximum PV allowable capacity is evaluated. It is assumed that, with a certain load distribution, the quantity of integrated PV systems, PV location, and PV installed capacity are the key variables. In Figure 4.19, M random scenarios determined by the quantity and integration location of the PV systems are listed. The installed PV capacity in each scenario is then randomly and constantly changed so that various kinds of possible situations are simulated to obtain the maximum allowable capacity of the PV systems. Assuming that N PV integration plans can be applied for each scenario, there are thus a total of $M \times N$ integration plans. The N installed capacities in each scenario could be derived from the initial allowable PV capacity with certain rules.

The details of the algorithm are given in Figure 4.20, and the steps are given as follows:

Step 1. k nodes are selected as the candidates to place PV systems; namely, $B_{avai} = \{b_1, b_2, \ldots, b_k\}$. The choice is made based on the load characteristics and planning requirements.

Step 2. Use an heuristic method to determine the appropriate number of random scenarios M (in this chapter the value of M is finally set to be 300).

Step 3. The initial allowable capacity of each location in B_{avai} is calculated in this step, which is also recorded in the set $L_{ref} = \{l_1, l_2, \ldots, l_k\}$.

Step 4. M integers a_i ($0 < a_i < k, 1 \leq i \leq M$) are generated in this step representing the number of connecting PV systems in M scenarios. These integers are recorded as the set $A = \{a_1, a_2, \ldots, a_M\}$.

Step 5. Randomly select a_i PV connection locations in each scenario from the set of B_{avai} as the PV integration locations in scenario i.

Step 6. Constantly increase the PV output power to determine the maximum PV capacity for scenario i. This is realized by scaling the initially given PV capacity; that is, by multiplying the PV initial power by βN times (usually, β is set to be between 0.1 and 0.3 and N is set between 10 and 30).

Step 7. Repeat step 5 and step 6 to obtain M groups of PV interconnection plans.

Step 8. Choose the maximum allowable capacity of the PV systems from the waiting list of PV integration plans.

4.5.2 Case Study

Now a 10 kV feeder at the SA substation in J City is taken as an example to perform the PV installation study. The topology of the feeder is shown in Figure 4.21. For this feeder, the most important factor affecting the integration of distributed PVs is the overvoltage limit. To obtain the maximum allowable PV capacity of this feeder, iterative simulations, analysis, and revisions are done to find the largest PV capacity that will not lead to the violation of the voltage constraint for safe and stable operation.

The radial feeder is composed of 12 types of conductors, including DL-150, JKLY-185, and DL-70, and 36 load nodes. The total load power in 2014 was around 3 MW and the 1-year load profile is shown in Figure 4.22.

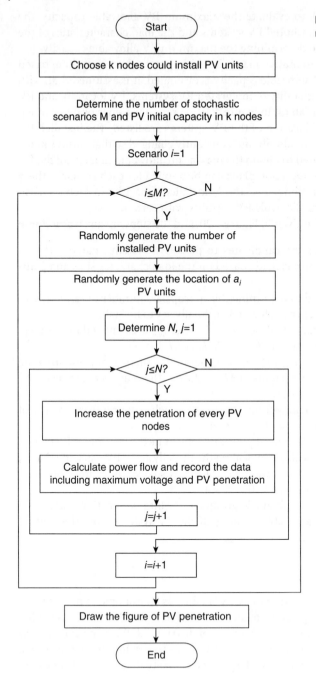

Figure 4.20 Evaluation process of maximum allowable PV capacity.

Figure 4.21 Topology of the 10 kV feeder of the SA substation.

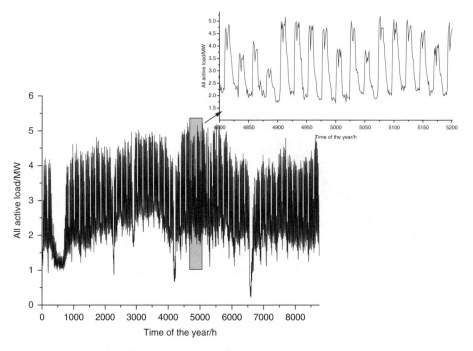

Figure 4.22 Load profile of the 10 kV feeder in the SA substation in 2014.

In Figure 4.22, the maximum load is 5.239 MW and the minimum load is 0.237 MW; the annual average value of the active load is 2.719 MW; the peak load appears in June and July, while the load valley appears in January and February. In addition, it can be seen from the zoom-in window that the load in this region is mainly industrial load as the daily peak appears in two periods between 09:00 and 11:00 and between 13:00 and 17:00.

According to the planning requirements, the PVs integrated into the distribution network at a single node should be higher than 0.25 MW. In the initial step, a total of 10 nodes are selected as the candidate locations (where the loads all exceed 0.25 MW) for PVs installation (labeled as node 1 to node 10), as shown in Figure 4.23.

The voltage at the step-down transformer in the substation is set to be 1.04 p.u. The solar radiance and ambient temperature reference are the actually measured data. The

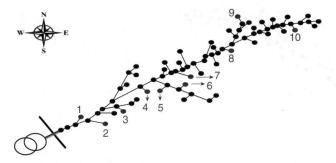

Figure 4.23 Possible PV connection locations.

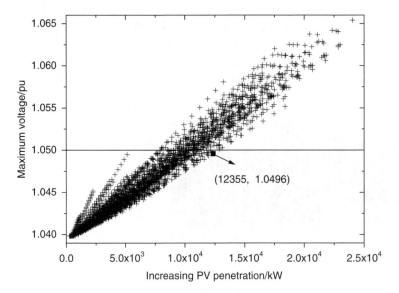

Figure 4.24 Distribution of PV capacity of distribution network.

PV inverter operates at unity power factor, which means the PVs only generate active power; the PV capacities obtained based on random scenarios are given in Figure 4.24.

A total of 3200 PV integration plans were analyzed, as shown in Figure 4.24. All the plans in the figure can be divided into groups by the 1.05 p.u. line. The data points below the line represent the feasible plans that can be accepted, while those plans above the 1.05 p.u. line might lead to overvoltage. The first number in the parentheses is the maximum PV capacity of the distribution network, which is 12.355 MW; at this point, the voltage reaches 1.0496 p.u. when the PVs are installed at nodes 1, 2, 3, 5, 7, and 8.

For the integration plan of the maximum 12.355 MW PVs, the PV penetration indices introduced in Section 4.3 are calculated. The PV PP reaches 1564%, which means the PV output is much higher than the load demand at some time and the power will be reversed back to the 10 kV system. The PV EP reaches 51%, which means the PV can support 51% of the total load consumption in 1 year. The PV CP is 227%, which shows that the feeder is able to accommodate the PV capacity more than twice the maximum load without causing overvoltage problems. It is noted that only the voltage constraint has been considered in this example. When more constraints are considered, the maximum PV capacity will decrease accordingly, and so do the CP, PP, and EP indices.

4.6 Photovoltaic Penetration Improvement

In this subsection, various methods in power generation, power distribution, and final power consumption at end users will be briefly discussed in order to improve PV penetration. A more detailed analysis will be made in Chapter 9.

4.6.1 Full Utilization of the Reactive Power Regulation Capability of a Distributed Photovoltaic System

Based on advanced control and communication techniques, the performance of PVs could be significantly improved and highly effective PV system could be realized. One of those techniques is PV reactive power regulation. When a distribution network is integrated with high PV capacity (i.e., real power), the grid voltage is affected by the intermittent PV power outputs. If an overvoltage issue happens, the system safety and security will be compromised. Therefore, the PV output should be limited and controlled within a certain range. It is well known that the PV inverter can have a certain reactive power regulation ability so that it can play an active role in voltage regulation. This is an important method to keep the system safe and stable, and to avoid frequent reactive transmission and exchange whilst increasing the PV penetration.

4.6.2 Distribution Network Upgrade

A system upgrade is another efficient way to improve the capability of handling a high PV penetration situation. Bidirectional power flow cased by PV systems will change the voltage distribution in the distribution network. In severe cases, PV power rationing or curtailment has to be taken in order to ensure the power system safety and stability. A large amount of PV curtailment will lead to severer economic loss and energy waste. If this occurs frequently, it is strongly suggested to upgrade the current distribution network. Specific reconstruction involves upgrading primary equipment, such as the distribution lines and protection system, and enhancing the grid operation control system, including the enhancement of the PV generation forecasting technique and the economic dispatching function in the energy management system.

4.6.3 Demand Response (DR)

The demand response, which can be used to change electricity users' behavior, is a great alternative to improve the distributed PV penetration. Partially using a price lever and other incentives, load may transfer from low PV output periods to high PV output periods to achieve the best match between the load characteristic and the PV output curve. In this way, the allowable PV penetration can be effectively increased.

4.6.4 Energy Storage Technologies

Energy storage can be used widely for peak load shifting, load characteristic modification, and power quality improvement. It can also play an important role in improving the PV hosting capability of the distribution network. Energy storage devices are often installed accompanying distributed PVs. When the PV output exceeds the load demand, the energy storage system can store the extra PV energy. The stored electricity can be discharged back to the system during an energy shortage period. Hence, energy storage can effectively alleviate the system safety pressure and improve PV penetration. In addition, owing to the intermittence and fluctuation of the PV power, not only the energy storage device can be used as a buffer between the PV systems and the distribution network, but also can effectively solve the electrical power quality problems, such as voltage fluctuations and harmonics.

4.7 Summary

The process of integration of large-scale distributed PVs can be divided into three stages. At first, when PV penetration is low, the cost effectiveness of PV projects is the key factor affecting PV development. In this chapter, several useful indices have been introduced to evaluate the benefits of PV systems. In the future, when grid parity comes true, more electricity customers may transfer to using the clean solar power.

PV power contribution to a system is not always linearly correlated with the PV installed capacity. Through the definitions and the analysis of PV PP, CP, and EP, the impact of distributed PVs on the power flow and energy management of the distribution network has been described.

When the PV penetration reaches a high level, the distributed PVs will lead to great changes in the operation of the distribution network. The existing distribution network might not have the capacity to accommodate such large-scale PVs. Based on the constraints on the safety and stability of the distribution network, the methods for calculating the maximum PV allowable capacity were summarized. In addition, based on the real data, a method for calculating the maximum PV integration capacity based on the random scenario method has been introduced.

Finally, other methods that can be used to improve PV penetration have also been discussed. Those methods are involved in all aspects of power generation, power distribution, and power consumption and are suggested to be considered in PV integration planning.

References

1 Bronski P., Creyts J., Lilienthal P, *et al.* (2014) *The Economics of Grid Defection.* Rocky Mountain Institute: Boulder, CO.

2 Bronski P., Creyts J., Crowdis M., *et al.* (2015) *The Economics of Load Defection.* Rocky Mountain Institute: Boulder, CO.

3 Stetz T., Rekinger M., and Theologitis I. (2014) *Transition from Uni-Directional to Bi-Directional Distribution Grid: Management Summary of IEA Task 14 Subtask*

2 – Recommendations Based on Global Experience. IEA PVPS T14-3:2014. International Energy Agency: Paris.

4 IEA. (2014) *High Penetration of PV in Local Distribution Grids: Subtask 2: Case-Study Collection.* IEA PVPS T14-2:2014. International Energy Agency: Paris.

5 Su J., Zhou L., and Li R. (2013) Cost–benefit Analysis of distributed grid-connected photovoltaic power generation. *Proceedings of the Chinese Society of Electrical Engineering*, **33**(34), 50–56.

6 Briano J.I., Báez M.J., and Moya Morales R. (2015) *PV Grid Parity Monitor: Residential Sector – 3rd Issue.* Creara Energy Experts. http://www.creara.es/actualidad/pv-grid-parity-monitor-residential-sector-3rd-issue (accessed February 25, 2017).

7 REN21. (2008) *Renewables 2007 Global Status Report.* REN21 Secretariat/Worldwatch Institute: Paris/Washington.

8 REN21. (2010) *Renewables 2010 Global Status Report.* REN21 Secretariat: Paris.

9 Black & Veatch Corporation. (2010) *Renewable Energy Transmission Initiative Phase 2B: Draft Report.* RETI Stakeholder Steering Committee: Sacramento, CA.

10 Grana P. (2010) Demystifying LCOE. http://www.renewableenergyworld.com/ugc/articles/2010/08/demystifying-lcoe.html (accessed February 25, 2017).

11 Darling S.B., You F., Veselka T., and Velosa A. (2011) Assumptions and the levelized cost of energy for photovoltaics. *Energy & Environmental Science*, **4**, 3133–3139.

12 Short W., Packey D.J., and Holt T. (1995) *A Manual for the Economic Evaluation of Energy Efficiency and Renewable Energy Technologies.* National Renewable Energy Laboratory (NREL): Golden, CO.

13 Chen R., Sun Y., Chen S., and Shen H. (2015) The LCOE analysis of PV grid-connecting projects. *Renewable Energy Resources*, **33**(5), 731–735.

14 Anderson K., Coddington M., Burman K., *et al.* (2009) *Interconnecting PV on New York City's Secondary Network Distribution System.* National Renewable Energy Laboratory: Golden, CO.

15 Zhao B., Hong B., and Ge X. (2010) Study on energy permeability after many distributed photovoltaic power supplies connected to grid. *East China Electrical Power*, **38**(9), 1388–1392.

16 Cheng S. (2014) *Research on Penetration Ratio and Coordinated Optimization Algorithm of Distribution Power System with Distributed Generations.* China Electric Power Research Institute.

17 Gong Y. (2012) *A Study on Protect Effect and Admittance Capacity of Distributed Generation Connected to Power Grid.* Yanshan University.

18 Barker P.P. and de Mello R.W. (2000) Determining the impact of distributed generation on power systems, part 1: radial distribution systems. In *2000 IEEE Power Engineering Society Summer Meeting (Cat. No.00CH37134).* IEEE. doi: 10.1109/PESS.2000.868775.

19 GB/T 14549-1993. (1993) *Quality of electric energy supply – harmonics in public supply network.* China Zhijian Publishing House.

20 GB/T 19939-2005. (2005) *Technical requirements for grid connection of PV system.* Standards Press of China.

21 Zhong Q., Gao X., Yu N., *et al.* (2014) The adopt capacity and connecting modes for distributed PV system considering the harmonic constraints. *Automation of Electric Power System*, **38**(24), 1–7.

22 Crucq J.M. and Robert A. (1989) Statistical approach for harmonics measurements and calculations. In *10th International Conference on Electricity Distribution, 1989. CIRED 1989.* IET.

23 Tang B. (2012) *Modeling and Research on the Maximum Installed Capacity of Grid-Connected PV Station.* Xinjiang University of China.

24 EPRI. (2012) *Stochastic Analysis to Determine Feeder Hosting Capacity for Distributed Solar PV.* Electric Power Research Institute: Palo Alto, CA.

25 Zhao B., Wei L., Xu Z., *et al.* (2015) Stochastic analysis to determine the allowable distributed PV capacity for a feeder with storage system. *Automation of Electric Power Systems,* **39**(9), 34–40.

5

Power Flow Analysis for Distribution Networks with High Photovoltaic Penetration

5.1 Introduction

Recent developments of distribution networks with high PV penetration can be attributed to phenomenal growth in integrated PV capacity. For most of the distributed PVs, the PV power is firstly consumed by local load, with the surplus power sold to the grid. This causes significant changes in distribution networks, including changes from passive to active networks, change of power flow characteristics, increases in uncertainty of load prediction, impacts on distribution of voltage and power loss, and even changes of network structure and control method. Therefore, power flow study is important and fundamental for the various analysis of those distribution networks with high PV penetration. It can evaluate operational status of distribution networks based on network structure and operation conditions, support the quantitative analysis of feasibility, reliability, and economics for system planning and operation, and lay the foundation for system stability analysis.

5.2 Power Flow Calculation for Distribution Networks with Distributed Photovoltaics

The defining characteristics of medium and low voltage distribution networks are high R/X ratio and complex network structure. For distributed PV generation, the power flow calculation models are quite different from those of traditional generators. The power flow on feeders can increase, decrease or even reverse considering PV output power. Furthermore, PV output is affected by weather and fluctuates wildly. The load profile can also change significantly, which makes power flow results harder to predict. This is the key difference in power flow calculation methods between traditional networks with a single power supply (i.e., the substation) and distribution networks with distributed PVs.

5.2.1 Comparison of Power Flow Calculation Methods for Distribution Networks

Power flow calculation methods for distribution networks are listed in Table 5.1, including the forward and backward substitution method, loop-impedance method, Newton Method, advanced fast power flow method, and so on. These power flow methods are basically divided into node-based methods and branch-based methods [2].

Grid-Integrated and Standalone Photovoltaic Distributed Generation Systems: Analysis, Design, and Control,
First Edition. Bo Zhao, Caisheng Wang and Xuesong Zhang.
© 2018 China Electric Power Press. Published 2018 by John Wiley & Sons Singapore Pte. Ltd.

Table 5.1 Comparison of typical power flow calculation methods for distribution network [1].

Method	Concept	Advantage	Disadvantage
Gauss–Seidel	Node voltage equation is solved by Gauss–Seidel iterative algorithm	Easy programming, less memory usage	Not applicable for *PV* (constant active power and constant voltage magnitude) node, poor convergence
Newton–Raphson	A nonlinear power flow equation is linearized by Taylor series expansion and a numerical solution is fixed in multiple iterations	Fast convergence, auto correction	Heavy calculation burden due to Jacobian matrix, large memory usage, high requirements for initial value, easy to reach abnormal states
Forward and backward substitution	Iteration of forward and backward process. Current is calculated based on node voltage and load in forward process, and then voltage is updated in backward process	Easy programming in radial network, fast calculation	Not applicable for *PV* node, and poor performance on looped network calculations
Loop impedance	Load is modeled as constant impedance; current is calculated based on a circuit equation formed by root node and all load nodes	Good performance on looped network calculation, simple principle	Slow calculation, large memory usage
Advanced Newton–Raphson	Based on Newton–Raphson method, approximate Jacobi matrix is formed to calculate the power flow	Good performance on looped network	Poor convergence, large memory usage
Z_{bus} Gauss	Sparse nodal admittance matrix is used in solving power flow equation	Fast convergence for system with few nodes	Poor convergence for system with more nodes

Source: Lin 2012 [5.2]. Reproduced from Shandong University of China.

Node-based methods include the node power method (NPM) and node current method (NCM). The essence of the NPM is the use of a node voltage calculation based on injection power and power balance equations. By contrast, the essence of the NCM is a node voltage calculation based on injection current and node equations. The NPM has a fast second-order convergence rate, but it also has a high requirement of initial value. On the other hand, the NCM has a low requirements of initial value, but it has a relatively slow first-order convergence rate due to a unidirectional approach. Typical algorithms include the Z_{bus}-based method, Y_{bus}-based method, improved Newton method, and so on.

Branch-based methods include the branch power method and branch current method. The essence of the branch power method is to find the solution to a set of loop equations with the loop power values and the node voltages as the variables. For the branch current method, the essence is to find the node voltages based on branch currents. Branch-based methods have the advantages of easy programming, less memory usage, fast convergence, and good numerical stability in calculations of uncomplicated distribution networks. Typical branch-based algorithms include the forward and backward substitution method, the loop-impedance method, and so on.

5.2.2 Power Flow Calculation Model for a Distributed Photovoltaics

Since the PV output is DC, an inverter is essential to convert DC to AC. There are two patterns for integrating PV system to distribution networks, depending on whether there are energy storage devices or not. One is the "nondispatchable PV system" without any energy storage devices, as shown in Figure 5.1; the other is the "dispatchable PV system" with energy storage devices, as shown in Figure 5.2. At present, owing to the advantages of convenient installation and commissioning, high reliability and integration, and low cost, nondispatchable PV systems are widely used [3].

The inverter of a distributed PV system is generally a current-controlled voltage source. To synchronize its output voltage with the grid, the inverter tracks the voltage of the distribution network by controlling the output current. Generally, the PV system is in full use, so it works at the maximum active power point. Additionally, the distributed PVs could also provide reactive power to the grid at the expense of partially reducing active power output by controlling the inverter if needed. This is conducive

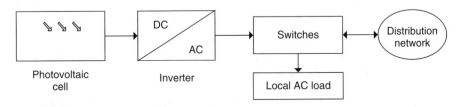

Figure 5.1 Nondispatchable grid-connected PV system.

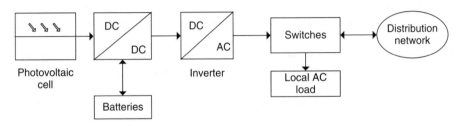

Figure 5.2 Dispatchable grid-connected PV system with energy storage.

to the stable and economic operation of the distribution network. Shigenori *et al.* [4] presented the simulation of an output-limited inverter with a current-control model and a voltage-control model. The current/active power output of the current-control model is constant like a PI node (constant power and current). Based on the voltage of the previous iteration, and the constant current and active power, the reactive power is

$$Q = \sqrt{I^2(e^2 + f^2) - P^2} \tag{5.1}$$

where P is the constant active power output of the distributed PVs, I is the constant current output of the distributed PVs; e and f are the real and the imaginary parts of the node voltage respectively.

The calculated reactive power based on the real and imaginary parts of the node voltage in the previous iteration is used in the next iteration and the node is treated as a PQ node (constant active and reactive power). For a voltage-control model, the distributed PV system is treated as a *PV* node (constant active power and voltage magnitude) like a traditional generator. If the PV is operated at its current limit, then it is treated as a PI node. In addition, a distributed PV system operated with a unity power factor could take full advantage of its active power. On this occasion, only active power is produced, so it is treated as a PQ node in the power flow calculation.

5.2.3 Power Flow Calculation Method for Distribution Network with Distributed Photovoltaics

The forward and backward substitution method is widely used in radial networks with the advantages of less memory usage as well as good convergence by avoiding the complicated matrix calculation. Especially with the characteristic of being unaffected by the R/X ratio, this method has been widely adopted in distribution networks. As described in Table 5.1, the forward and backward substitution method is not good at handling nodes of constant real power and voltage magnitude. Nevertheless, since a distributed PV system is mainly modeled as a PI or PQ node, this method shows good performance in power flow calculations for distribution networks with PVs. The process of calculation is as follows:

Step 1. Import network parameters and initialize node voltage/phase angles.
Step 2. Initialize distributed PV parameters.
Step 3. Calculate injected load current for each node.
Step 4. Calculate injected currents for the PI-type distributed PVs, and combine the injected currents of load with the current of the distributed PVs.
Step 5. Calculate the current for each branch from the end to the root node in the backward sweep.
Step 6. Calculate the voltage for each node from the voltage source in the forward sweep.
Step 7. Calculate the reactive power injected from each PI node; if the reactive power is over the limit, convert the PI node to PQ node and fix the injected current.
Step 8. Check the convergence condition of the voltage difference modulus in two contiguous iterations; go to step 9 if the convergence condition is met; go to step 3 if not.
Step 9. Output results.

5.3   Voltage Impact Analysis of Distributed Photovoltaics on Distribution Networks

5.3.1   Mathematical Model

The typical structure of a medium-voltage feeder in a radial distribution network is shown in Figure 5.3. N loads (power consumers) are distributed along this feeder. As shown in Figure 5.3, the voltage at the substation node is U_0; the voltage at node m is U_m and the corresponding load demand is $P_m + Q_m$; the impedance between load m and $m - 1$ is $R_m + jX_m$. It is assumed that the distributed PV system integrated at node n has injected power $P_v + jQ_v$.

The power flow direction pointing to the load is defined as positive for active/reactive power. The voltage drop ΔU_m is calculated as

$$\Delta U_m = \frac{\sum_{i=m}^{N} P_i R_m + \sum_{i=m}^{N} Q_i X_m}{U_{m-1}} \tag{5.2}$$

And the voltage U_m for node m is

$$U_m = U_0 - \sum_{i=1}^{m} \Delta U_i = U_0 - \sum_{i=1}^{m} \frac{\sum_{j=i}^{N} P_j R_i + \sum_{j=i}^{N} Q_j X_i}{U_{i-1}} \tag{5.3}$$

It is assumed that the impedance are distributed uniformly with resistance r and reactance x per unit distance and that l_m is the distance between $(m - 1)$th node and mth node. Thus, the voltage of the mth node is calculated as follows based on the distance between the source and the user:

$$U_m = U_0 - \sum_{i=1}^{m} \frac{\sum_{j=i}^{N} r P_j l_i + \sum_{j=i}^{N} x Q_j l_i}{U_{i-1}} \tag{5.4}$$

The load P and Q powers are positive, so the voltage drop ΔU_m is positive without a distributed PV system connected and proportional to distance l starting from the root node. Therefore, the voltage U_m decreases with increasing distance.

As described earlier, to maximize the use of the PV system, it is assumed that the PVs only generate active power in this analysis and have little influence on the reactive power loss of the feeder as no reactive power is produced by the PV system. The voltage of the

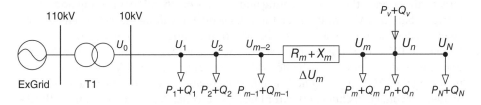

Figure 5.3 Typical structure of a medium-voltage feeder in radial distribution network.

mth node is calculated as follows with the PV system connected at the nth node when the PV node is farther to the root node than the load node ($0 < m < n$):

$$U_{mp} = U_0 - \sum_{i=1}^{m} \frac{\left(\sum_{j=i}^{N} P_j - P_v\right) rl_i + \sum_{j=i}^{N} xQ_j l_i}{U_{i-1}} \tag{5.5}$$

As can be seen from Equations 5.4 and 5.5, the voltage increases ($U_{mp} > U_m$) after installing the PV system. The improvement degree, as in the following, is dependent on various factors, including the load/PV power, the distance to PV integrating position, and the system voltage level:

$$U_{mp} - U_{(m-1)p} = -\frac{\left(\sum_{j=m}^{N} P_j - P_v\right) rl_m + \sum_{j=m}^{N} xQ_j l_m}{U_{m-1}} \tag{5.6}$$

Without considering the impact or support of reactive power, the voltage at node m is lower than the voltage of node $m - 1$ if the total load of the mth node and the nodes downstream from the mth node is larger than the PV output power $\left(\sum_{j=m}^{N} P_j - P_v > 0\right)$. The voltage drop ΔU_m becomes smaller than the situation when there is no PV generation. If their total load is smaller than the PV output power (i.e., $\sum_{j=m}^{N} P_j - P_v < 0$) then U_{mp} can be higher than $U_{(m-1)p}$ (and other upstream nodes) due to the reverse power flow caused by the PV source.

When the PV node is closer to the root node than the load node ($n < m < N$), the voltage of the mth node and the voltage drop are calculated as

$$U_{mp} = U_0 - \sum_{i=1}^{n} \frac{\left(\sum_{j=i}^{N} P_j - P_v\right) rl_i + \sum_{j=i}^{N} xQ_j l_i}{U_{i-1}} - \sum_{i=n+1}^{m} \frac{\sum_{j=i}^{N} P_j rl_i + \sum_{j=i}^{N} xQ_j l_i}{U_{i-1}} \tag{5.7}$$

and

$$U_{mp} - U_{(m-1)p} = -\frac{\sum_{j=m}^{N} P_j rl_m + \sum_{j=m}^{N} xQ_j l_m}{U_{m-1}} \tag{5.8}$$

In conclusion, assuming that the voltage at the root node stays constant, with the increase of PV capacity, there are three potential situations after a single PV system is added:

1) When the PV output power is low, as the load demand of all nodes on the downstream direction of the PV connecting node, including the PV connecting node itself, is larger than the PV output power, the voltage declines along the feeder but the voltage drop decreases compared with the situation without PV connection.
2) When the PV output power increases to the extent that it is able to cover the load demand of all the nodes on the downstream direction of the PV connecting node, including the PV connecting node itself, while unable to support the total load of the feeder, the voltage declines at the front head of the line, then increases toward the PV node and starts to decline at the downstream direction of the PV connecting node.

3) The voltage of the PV node is the highest and the voltage profile declines along both the upstream and downstream directions of the PV node when the PV output power is even larger than the total load power on the feeder.

In the case of multiple PV systems, the voltage and the voltage drop of the *m*th node are calculated as follows:

$$U_{mp} = U_0 - \sum_{i=1}^{m} \frac{\sum_{j=i}^{N}(P_j - P_{vj})rl_i + \sum_{j=i}^{N} xQ_jl_i}{U_{i-1}} \tag{5.9}$$

$$U_{mp} - U_{(m-1)p} = -\frac{\sum_{j=m}^{N}(P_j - P_{vj})rl_m + \sum_{j=m}^{N} xQ_jl_m}{U_{m-1}} \tag{5.10}$$

The voltage drop declines along the feeder if the load demand of all the nodes in the downstream direction of the PV connecting node, including the PV connecting node itself, is larger than the total PV output power (i.e., $\sum_{j=m}^{N} P_j - \sum_{j=m}^{N} P_{vj} > 0$). Otherwise, if $\sum_{j=m}^{N} P_j - \sum_{j=m}^{N} P_{vj} < 0$, the PV connecting node voltage will be even higher than the voltages of upstream nodes [5].

5.3.2 Simulation Studies

To demonstrate the impact of PV capacity and location on the voltage profile, a simulation analysis is given on a typical feeder in the following text based on the previously discussed mathematical model.

The simulation system of a 10 kV feeder with 10 electrical consumers is shown in Figure 5.4. In real operation, the voltage amplitude at the substation bus is usually kept around 1.05 p.u.; hence, in this simulation the voltage U_0 at the root node of the feeder is set as 1.05 p.u. The length of the feeder is 20 km and the distance between two adjacent consumers is 2 km. The load P_N of each node is 500 kW, resulting in a total load P_{total} of 5 MW on the feeder. The impact of reactive power is not considered in this simulation since it is assumed that the PV system generates active power only. According to the analysis in Chapter 4, if the PV output power is larger than the load demand, the surplus power can reverse the power flow, which would cause issues of protection coordination and voltage rise. Therefore, it is assumed that the PV output power is smaller than the total load demand in this simulation.

First, to explore the impact of the installation location of the PV system, a 3 MW PV system is connected through a single point as listed in Table 5.2. Since the PV output power fluctuates wildly, an extreme case that the PVs generate at their maximum

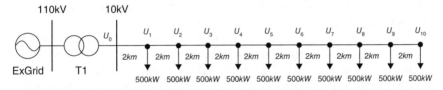

Figure 5.4 Load distributions on feeder.

Table 5.2 Location of the distributed PV system in simulation scenarios.

Connecting node	Distance to root node (km)
1	2
5	10
9	18

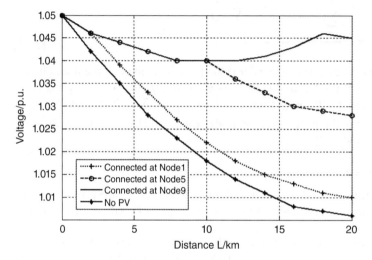

Figure 5.5 Feeder voltage profiles under different PV locations.

possible output power of 3 MW is used in the simulation study without considering the output efficiency of PV systems. The simulation results are shown in Figure 5.5. The figure shows the feeder voltage profiles when the distributed PV system is connected at node 1, node 5, and node 9, respectively. In general, the overall feeder voltage increases after the PV system is connected, and the voltage distributions are quite different for the different PV locations. The voltage still declines along the feeder when the PV system is connected at node1 and node 5, but when it is connected at node 9 the voltage of the connected node increases and becomes the local highest. This is because the PV power is larger than the local load demand and the direction of power flow is changed. On the whole, the feeder voltage is improved with the PV system integration. At a given capacity, PV systems connected at the end of the feeder raises the line voltage more effectively than when it is installed at the beginning of the feeder.

Second, to explore the impact of PV installation capacity, the PV location is set at node 1 with a 2 km distance to the root of the feeder, but the PV capacity is changed in this simulation scenario as listed in Table 5.3. The simulation results are shown in Figure 5.6, where the positive direction of power flow is defined as the power flow from upstream to downstream; the other direction is then negative.

Table 5.3 Different capacities of the distributed PV system at node 1 for different simulation scenarios.

PV capacity (MW)	0.5	1	2	3	4	5

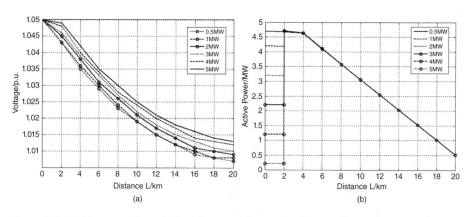

Figure 5.6 (a) Voltage profiles and (b) active power flow under different PV capacities at node 1.

Since the distributed PV power contributes to meet the load demand, the upstream power flow of the PV connection node decreases. This leads to voltage improvement and power loss reduction. The power flow downstream of the PV system has little change, but the voltage still increases as the added PV system improves the overall voltage profile along the feeder. The voltage improvement becomes more pronounced with an increasing size of PV capacity.

Next, the PV system location is changed from node 1 to node 5, 10 km away from the root of the feeder. The PV capacity is changed in this simulation scenario as listed in Table 5.3. The simulation results are shown in Figure 5.7.

As shown in Figure 5.7b, the power flow is reversed at node 5 when the capacity of the PV system increases to 4 MW. This is because the PV power is able to meet the load downstream after node 5 and the excess power flows back to supply the upstream load. Moreover, the voltage peak appears at node 5, as shown in Figure 5.7a, when the capacity

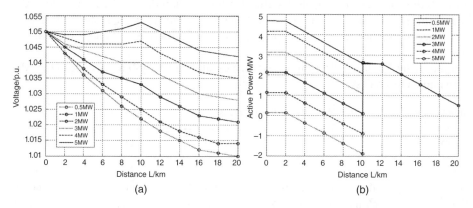

Figure 5.7 (a) Voltage profiles and (b) active power flow under different PV capacities at node 5.

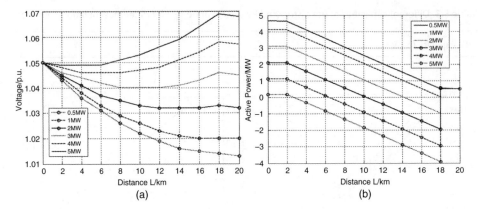

Figure 5.8 (a) Voltage profiles and (b) active power flow under different PV capacities at node 9.

of the PV system is larger than 4 MW. The voltage is even higher than that at the root of the feeder when the PV capacity at node 5 reaches 5 MW. This kind of voltage rise can cause negative impacts if the rise exceeds the acceptable range.

Similarly, Figure 5.8 shows the voltage and power flow when the PV system is located at node 9. It can be seen that, with the same PV capacity, the voltage rise is highest compared to the cases where the PV is installed at nodes 1 and 5. It is even higher than the voltage at the root of the feeder and reaches 1.07 p.u. when the PV capacity is 5 MW. This may do harm to the safe operation of the grid.

Further analysis on voltage and power flow is presented in the case study of adding PVs to multiple nodes as opposed to a single node. In this case, the structure and related parameters are the same as the aforementioned cases and the total PV capacity is 5 × 1 MW, meaning the capacity of a single PV system is 1 MW and there are five PV systems on the feeder. First, five PV systems are connected to different nodes as listed in Table 5.4.

In case 1 (see curve #1 in Figure 5.9), the PV systems are located upstream of the feeder from node 1 to node 5; in case 2 (see curve #2 in Figure 5.9), the PV systems are evenly distributed along the feeder; and in case 3 (see curve #3 in Figure 5.9), the PV systems are located downstream of the feeder from node 6 to node 10. Although the PV capacities in the three cases are all 5 MW, the voltage rise is small in case 1, while in case 3 the voltages of many nodes are even higher than that of the root node of the feeder. This is more likely to cause overvoltage issues. Since the PV systems are evenly distributed along the feeder and supply the load locally, there is no voltage rise issue in case 2.

Additionally, Table 5.5 lists three cases where all five PV systems are located at the same node in each case.

Table 5.4 PV locations under distributed installation cases.

Case ID	PV1	PV2	PV3	PV4	PV5
Case 1	Node 1	Node 2	Node 3	Node 4	Node 5
Case 2	Node 1	Node 3	Node 5	Node 7	Node 9
Case 3	Node 6	Node 7	Node 8	Node 9	Node 10

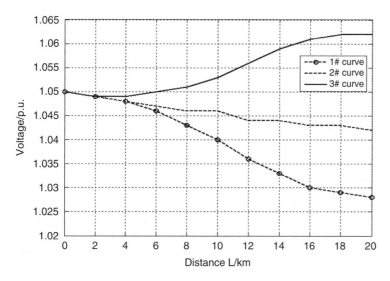

Figure 5.9 Voltage profiles for distributed installation of multiple PV systems.

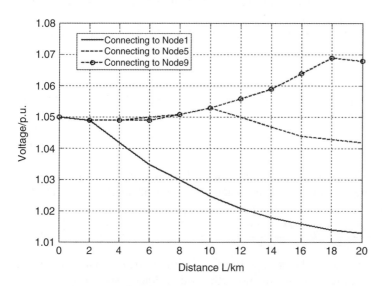

Figure 5.10 Voltage profiles for centralized installation of multiple PV systems.

Table 5.5 PV locations under centralized installation cases.

Case ID	PV1	PV2	PV3	PV4	PV5
Case 6	Node 1	Node 1	Node 1	Node 1	Node 1
Case 7	Node 5	Node 5	Node 5	Node 5	Node 5
Case 8	Node 9	Node 9	Node 9	Node 9	Node 9

As can be seen from Figures 5.9 and 5.10, the voltage profile has a larger variation range between 1.010 and 1.070 p.u. for the centralized installation scheme, as opposed to the distributed installation, which has a range between 1.025 and 1.065 p.u. Therefore, the distributed installation is a better approach for improving the voltage profile in a smoother and safer way than the centralized installation of a single large system.

5.4 Loss Analysis in Distribution Network with Distributed Photovoltaics

5.4.1 Mathematical Model

Similar to other distributed generation sources, distributed PVs can certainly impact the power flow of the distribution network and change the distribution of line losses. Current research has shown that distributed PVs contribute to reduced distribution loss as long as the integration can be reasonably designed [6, 7]. Otherwise, a random integration scheme may negatively influence the system. In this section, the distribution loss is analyzed by considering the distributed PVs. Transformer (e.g. dedicated transformer for the end user and PV step-up transformer) loss and distribution losses between the transformer and PCC are not considered.

For the typical distribution line shown in Figure 5.11, the load demand and distribution current are

$$S = P + jQ \tag{5.11}$$

where S is the apparent power of the load, P is the active power of the load, and Q is the reactive power of the load, and

$$\dot{I} = \frac{P - jQ}{\dot{V}} \tag{5.12}$$

where $\dot{I}$ and $\dot{V}$ are the load current phasor and the node voltage phasor respectively.

It is assumed that the distribution line impedance is uniformly distributed. Thus, the distribution loss is

$$\text{Loss} = I^2 R = \frac{(P^2 + Q^2)}{V^2} R = \frac{(P^2 + Q^2)}{V^2} rl \tag{5.13}$$

where r is the resistance per unit length, l is the length of the distribution line, and I and V are the magnitudes of the load current and of the node voltage respectively.

Applying Equation 5.13 to the aforementioned medium-voltage feeder of typical structure in the radial distribution network shown in Figure 5.3, the distribution loss between the $(m - 1)$th and mth load is Loss_m:

$$\text{Loss}_m = \frac{\left(\sum\limits_{i=m}^{N} P_i\right)^2 + \left(\sum\limits_{i=m}^{N} Q_i\right)^2}{V_m^2} rl_m \tag{5.14}$$

Figure 5.11 Diagram of typical distribution line.

where l_m is the distance between $(m-1)$th and mth node. The total distribution loss of the feeder is as follows:

$$\text{Loss} = \sum_{i=1}^{N} \frac{\left(\sum_{j=i}^{N} P_j\right)^2 + \left(\sum_{j=i}^{N} Q_j\right)^2}{V_i^2} rl_i \tag{5.15}$$

It is assumed that the PVs are located at the mth node, which acts as a "negative" load. Thus the final load demand at the node equals to the original load reducing P_V, which means that the power from upstream to the mth node would be reduced, while the power flowing to the downstream would keep still. Loss_p is the distribution loss after the PVs are added:

$$\text{Loss}_p = \sum_{i=1}^{m} \frac{\left(\sum_{j=i}^{N} P_j - P_V\right)^2 + \left(\sum_{j=i}^{N} Q_j\right)^2}{V_i^2} rl_i + \sum_{i=m+1}^{N} \frac{\left(\sum_{j=i}^{N} P_j\right)^2 + \left(\sum_{j=i}^{N} Q_j\right)^2}{V_i^2} rl_i \tag{5.16}$$

Equation 5.17 gives the change of distribution loss after PVs are added. To obtain a loss reduction due to the PV integration, Equation 5.18 needs to hold true.

$$\Delta\text{Loss} = \text{Loss}_p - \text{Loss} = \sum_{i=1}^{m} \frac{P_v^2 - 2P_v \times \sum_{j=i}^{N} P_j}{V_i^2} rl_i \tag{5.17}$$

$$\Delta\text{Loss} = \sum_{i=1}^{m} \frac{P_v^2 - 2P_v \times \sum_{j=i}^{N} P_j}{V_i^2} rl_i < 0 \tag{5.18}$$

Owing to positive values of v, r, and l, Equation 5.18 can be rewritten as

$$\sum_{i=1}^{m} \left(P_v^2 - 2P_v \times \sum_{j=i}^{N} P_j\right) < 0 \tag{5.19}$$

Since P_V is positive, the condition then becomes

$$mP_v - \sum_{i=1}^{m} \left(2 \times \sum_{j=i}^{N} P_j\right) < 0 \tag{5.20}$$

The condition given in Equation 5.18 can be further simplified into

$$mP_v - 2\left[mN - \frac{m(m-1)}{2}\right] P_N < 0 \tag{5.21}$$

by assuming that the load demand is P_N at each node and the distance between the two neighboring nodes is the same.

The PV output would reduce the distribution loss when it meets the condition

$$P_v < 2\left[N - \frac{(m-1)}{2}\right] P_N = \left(2 - \frac{m-1}{N}\right) P_{total} \tag{5.22}$$

where P_{total} is the total load of the feeder and $P_{total} = NP_N$ is assumed.

Based on Equation 5.21, and using the analytical method, the relationship between the PV location and distribution loss is obtained as follows:

$$f(m) = mP_v - 2mNP_N + (m^2 - m)P_N \tag{5.23}$$

The derivative of $f(m)$ can be obtained as

$$f'(m) = P_v - 2NP_N + (2m - 1)P_N \tag{5.24}$$

$f'(m)$ is the derivative of Equation 5.23. Thus, the change of distribution loss reaches its maximum when $f'(m) = 0$, which also means the distribution loss reaches its minimum value. The optimal location m_0 exists as shown in Equation 5.25 when the PV capacity meets the conditions in Equation 5.22:

$$m_0 = \left(1 - \frac{P_v}{2P_{total}}\right)N + \frac{1}{2} \tag{5.25}$$

From Equation 5.25, the location m_0 is related to the PV capacity P_v. Larger capacity comes with smaller m_0, which means the optimal location is closer to the head of the feeder. Conversely, a smaller PV capacity comes with a larger m_0, which means the optimal location is closer to the end of the feeder.

However, if the PV capacity is too large, which leads to $m_0 < 1$, the distribution loss increases along the whole feeder; if the PV capacity is too small, which leads to $m_0 > N$, so the distribution loss decreases along the whole feeder. Hence, there are three possible situations after the PV generation is added: (1) the distribution loss decreases, (2) the distribution loss initially decreases and then increases, and (3) the distribution loss increases.

5.4.2 Simulation Results

The simulation system and the related parameters are the same as those in Figure 5.4. In this case, the PVs are connected to nodes 1, 5, and 9, respectively. The range of the PV capacities is listed in Table 5.6.

The original total distribution loss is 144.7 kW without PV integration. The losses with the PV integration are shown as a parabola in Figure 5.12. As the PV capacity increases, the distribution loss decreases at first and then increases. The optimal PV capacity, which leads to the minimum distribution loss (i.e., the extreme point of the parabola), decreases with increasing distance between the PV location and the head of the feeder. For a given PV system location, the extent of distribution loss reduction is larger when the PV system is closer to the feeder end. The minimum distribution loss exists in the case that meets the condition of the PV system location and capacity defined in Equation 5.25. For example, when the PV system is located at node 9 (i.e.,

Table 5.6 Range of the PV capacities.

PV capacity (MW)	1	2	3	4	5	6	7	8
Load ratio[a]	0.2	0.4	0.6	0.8	1.0	1.2	1.4	1.6

a) Load ratio = PV capacity/Total load.

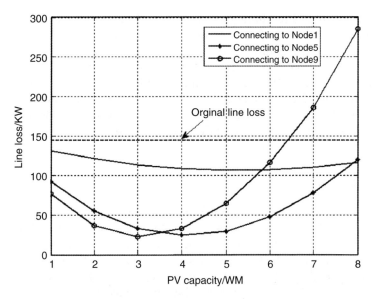

Figure 5.12 Distribution loss under different PV capacities.

$m = 9$), the optimal PV capacity is 3 MW and the PV size should be less than 6 MW according to Equation 5.22. As clearly indicated in Figure 5.12, when the PV capacity is 7 MW in Figure 5.12, the distribution loss is 180 kW, which is even greater than the case without PVs.

As shown in Figure 5.13, the optimal PV system location is closer to the head of the feeder when the PV capacity is larger. The minimum distribution loss is 36.8 kW with a 2 MW PV system and 28.5 kW with a 5 MW PV system, which are both smaller than

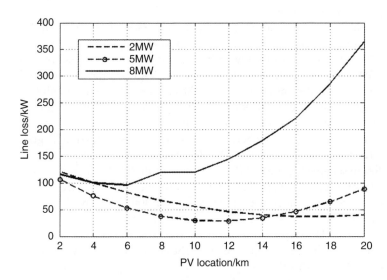

Figure 5.13 Distribution losses under different PV locations.

the case without any PV system. Thus, having an appropriate PV capacity at a suitable location can reduce distribution loss and enhance operational efficiency.

As can be seen from the foregoing discussion, the capacity and location of the PVs have great impacts on the distribution loss. The distribution loss is reduced when the condition in Equation 5.22 is met. The reduction in general is greater when the PVs are installed closer to the end of the feeder, but there is a limit for the distribution loss reduction. If the PV size is higher than the total load demand, the excess power will flow back to the grid, which can lead to an increase in the distribution loss.

5.5   Case Study

In this section, real case studies are given to demonstrate the theory and to investigate the impact of distributed PVs on voltage, power flow, and distribution loss. Based on the feeder model discussed previously, the power flow analysis cases are studied in a single feeder under different PV capacities, locations, and patterns. Furthermore, the impact on distribution operation with more PV systems integrated to the distribution network is further analyzed based on a substation-level system model. It is assumed that the power factor is unity for a PV system. PV control is not considered here.

5.5.1   Patterns for Distributed Photovoltaics Integration

There are mainly two patterns of integrating PVs: (1) connection to a medium/high-voltage network through specific transformers, and (2) connection to a low-voltage network through user distribution transformers. Generally, residential customers tend to connect PVs to low-voltage distribution networks (which are at 0.4 kV in China), while commercial and industrial customers are more likely to connect PVs to medium-voltage distribution networks. Utilizing a microgrid is another way to integrate PVs and other renewable resources. This is not only a method of integrating PVs, but also a way to improve the grid operation performance.

According to the guidelines of China's State Grid, *Typical Grid-Integration Design of Distributed Generation – The Integration Fascicle*, a typical design scheme for connecting a 1.6 MWp PV system to a 0.4 kV low-voltage network is given in Figure 5.14 (document ID: XGF380-Z-Z1 in Ref. [8]). There are two 2 MVA user transformers and four PV modules which are divided into two groups. The capacity of each PV module is 400 kWp and the integration inverter is 500 kVA. The PV system supplies power for the local load firstly, then feeds back the surplus power to the grid through the inverters and transformers. In Figure 5.14, the PCC is where the consumer accesses the public grid, and the point of interconnection (PoI) is where the PV system power is integrated.

Another typical design scheme (document ID: XGF10-Z-Z1 in Ref. [8]) for a 2.0 MWp PV connection to a 10 kV medium/high-voltage network is given in Figure 5.15. There are two user transformers, with capacities of 500 kVA and 800 kVA. Five PV modules are divided into two groups. Each group is connected to the distribution network through inverters and boosted by step-up transformers. The capacity of each PV module is 400 kWp and the corresponding inverter is 500 kVA. The capacities of the two step-up transformers are 800 kVA and 1250 kVA.

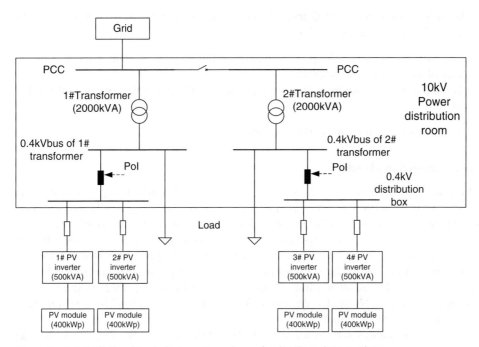

Figure 5.14 A 0.4 kV-level typical integration scheme for distributed PVs in China.

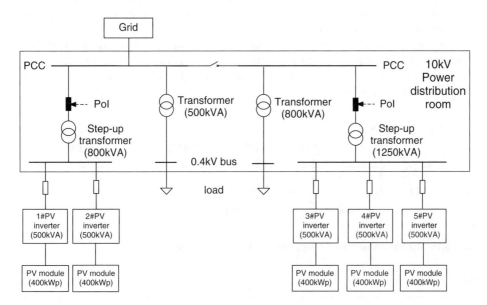

Figure 5.15 A10 kV-level typical integration scheme for distributed PVs in China.

According to Ref. [8], the PCC and PoI(Point of Interconnection) are defined as follows and used in the case study:

1) For a PV system with a step-up transformer, the PoI is located at the high-voltage side of the step-up transformer.
2) For a PV system without a step-up transformer, the PoI is the junction of the PV system.

The PCC is the join point on the grid side.

5.5.2 Analysis on a Feeder

The simulation system is an actual feeder (named SA_F1), which is part of the same 110-kV-level substation in a coastal city of East China, which was mentioned in previous chapters. SA_F1 is a radial feeder with two PV systems. As shown in Figure 5.16, one is connected to node 10kV1 through a 10 kV specific step-up transformer T10; the other is connected at the low-voltage side (i.e., node 400V2) of a common distribution transformer T11. The feeder supplies power for many electric consumers and the key nodes of the feeder are shown in the figure. Nodes 10kV1 and 10kV2 are the PCCs of the two PV systems. Nodes 10kV3 and 10kV4 are the branch nodes. There are different loads for each node or branch.

Based on the analysis of historical data, the load demand is heavy on SA_F1, which has a peak demand of 69.8% of the total allowable transmission capacity of the line. Table 5.7 lists the information of the PV systems. The PV systems represent two typical schemes with different installation capacities and capacity ratios. The capacity ratio is the ratio of PV capacity to original user transformer capacity. It reflects the PV hosting capacity of the consumer to some extent.

To simplify the study, using the same load curve without any PVs, the moment with maximum solar irradiation in the year 2014 is chosen as the best case to analyze the impact of the maximum PV output on the feeder. Table 5.8 lists the data of the load and PV system at that moment.

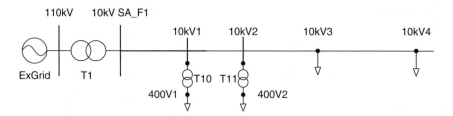

Figure 5.16 The 10 kV feeder SA_F1 supplied by a 110 kV substation.

Table 5.7 PV installation of SA_F1.

Feeder	PV system	PV capacity (MWp)	PV integration scheme	Capacity ratio
SA_F1	Automobile Company C1	2.0	10 kV level	160
	Battery Company C2	1.0	0.4 kV level	80

Table 5.8 Data of load and PV system.

	PV power (kW)	Load (kW)
Automobile Company C1	1653.6	566.9
Battery Company C2	826.8	219.6

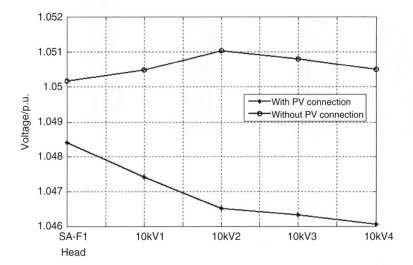

Figure 5.17 Voltage of key nodes on the feeder with/without PVs connection.

According to Table 5.8, for this case, the total PV output was 2480.4 kW and the total load was 1348.2 kW on feeder SA_F1 (including other loads). Thus, the PV output was 184% of the load, which was much larger than the load demand. The excess PV power would thus have been fed back to the feeder through the PCC, and then to the 10-kV-level bus. The power flow and voltage are changed due to the reverse power flow, as shown in Figure 5.17.

The reverse power flow due to the excess PV power causes a voltage rise on feeder SA_F1. Without PVs, the voltage decreases along the feeder from the head to the end. With the PV systems integrated, the node voltages are increased and become even higher than that of the substation with the maximum voltage at the PV node. In Figure 5.17, the voltage first increases along the feeder until node 10kV2 and then decreases after that node. The result is consistent with what was discussed in Section 5.3.

As shown in Figure 5.18, the bus voltages of both consumers C1 and C2 both increase when the PV systems are added. It is noted that the PoI voltage rise at the 10 kV level is smaller than that at the 0.4 kV level. This is mainly because at the 0.4 kV level the PV power is consumed locally on the low-voltage side of the transformer and only excess PV power flows to the feeder while integrated at the 10kV level, PV power is firstly transmitted to the feeder through the step-up transformer, then used by the local load through the user transformer. Thus, the voltage impact due to the PVs is more obvious at the 0.4 kV level.

As shown in Figure 5.18, the voltage at the PoI is higher than that of the PCC for both C1 and C2. The voltage of C2's PoI is even higher than 1.07 p.u., which is over the

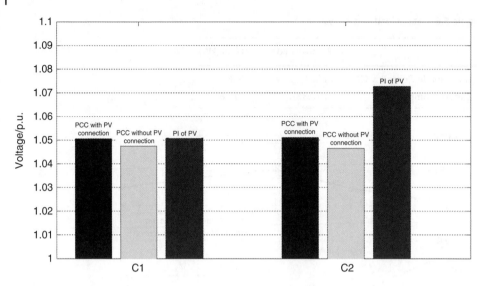

Figure 5.18 Node voltages at different locations with/without added PVs.

Table 5.9 Statistical operation data of SA_F1 with PVs in year 2014.

Index	Value
Annual PV utilization time intervals (h)	2299
Annual PV energy generation (MWh)	2607.3
Total number of time intervals (h) for PV power reversed	978
Annual maximum reversed PV power (kW)	2192.4

limit defined in China's Standard GB/T 12325-2008 (*Power Quality – Allowable Power Deviation*) [9]. Compared with that of the PCC, the voltage rise at the PoI is a much more serious issue for PV integration, especially at the 0.4 kV level.

For further analysis, based on the irradiance data of the year 2014 used in Chapter 4, the hourly data on system operations are acquired by sequential power flow simulation throughout the whole year and are listed in Table 5.9.

The annual PV energy generation reached 2607.3 MWh with total time intervals of 978 h of reverse power flow due to excess PV power. The power flow of 8760 h a year at the root of SA_F1 is shown in Figure 5.19. Positive power represents power flowing from the substation to the feeder; negative power represents power flowing from the feeder to the substation; zero power represents a balance between PV power and load.

For most of the time the power flow is from the substation to the terminal load. The reverse power flow due to PVs often occurs in spring and summer because the irradiance is strong, and the maximum reverse power appears in July. When the irradiance is insufficient in autumn and winter, the PV power is consumed by the local load demand.

As shown in Figure 5.20, the reverse power mainly ranges from 200 to 800 kW. In addition, the reverse power occurs frequently in the summer due to the higher solar irradiance. The reverse power peak is 2192.4 kW, which is due to the low level of load demand and the high level of PV power output at that moment.

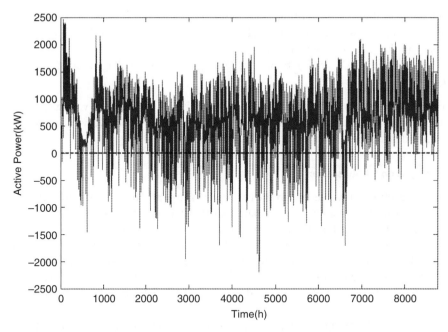

Figure 5.19 Power flow at the root of SA_F1 throughout the year of 2014.

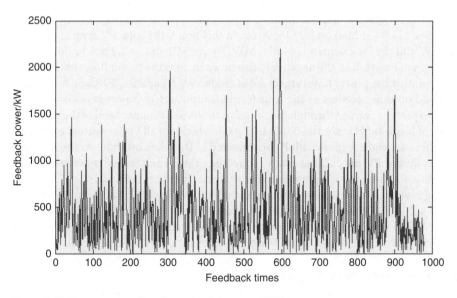

Figure 5.20 Reverse power flow throughout the year of 2014.

In summary:

1) The voltage rises as the PV power can support part of the load, and reduce the distribution loss. The extent of voltage rise is dependent on the PV capacity, PV location and integration scheme, and the load and voltage levels.

2) The PV power flows to meet other loads along the feeder if it is larger than the local load. The excess power can even reverse the power flow direction if the PV power is larger than the total downstream load.
3) Adding PVs to a 10-kV-level network has less impact on the voltage than it does on a 0.4-kV-level network.

In general, distributed PVs supply power to the local load, improve the voltage profile along the feeder, and reduce the distribution power loss. However, overvoltage has to be considered for feeders with a low level of load demand.

5.5.3 Analysis on SA Substation

Power flow direction is reversed when the PV power is larger than the total load demand. The reverse power flow may have an impact on other feeders in the same distribution network. Thus, it is important to perform a power flow analysis for the whole distribution network with distributed PVs, especially when the PV penetration level is high. This causes a series of issues for grid operation. To be consistent with the previous chapters, the SA substation distribution network is adopted to analyze the power flow with PV systems integrated. The detailed electrical features of the SA substation are listed in Table 5.10

Based on the PV development plan for this area, the relevant PV information is summarized in Table 5.11 and the statistical data is listed in Table 5.12.

The moment of the maximum irradiation in year 2014 is chosen for a case study to analyze the impact of the maximum PV power on the substation, namely the 12th hour period (11:00–12:00) on July 16, 2014. At this point, the total PV output power is 17.58 MW and the load demand is 41.86 MW. Figure 5.21 shows the power flow.

As shown in Figure 5.21, the negative active power means the power flows backward. It can be seen that the power flow is reversed at feeder SA_F1 and SA_F18, and consumed by the load on other feeders as the transformer output active power is positive. Thus, there is no power reversed through the SA substations to the upper level of the grid. The voltages of bus I–bus IV are 10.50 kV (1.05 p.u.), which are all in the normal operation range. There are eight feeders with PVs connected in the SA substation. At the moment of the maximum PV output, the range of voltages along each feeder is between 1.037 and 1.051 p.u., which is still acceptable according to China's standards.

Among seven 0.4-kV-level PV customers, the PoI voltages of five of them (i.e., PV ID 2, PV ID 4, PV ID 7, PV ID 9 and PV ID 10 – see Table 5.11) are higher than their PCC voltages. When the PV output is larger than its local load, the excess power is fed back to the grid. The extent of the voltage rise is higher if the reversed power is larger. In

Table 5.10 Electrical data for the SA substation.

Transformer capacity (Quantity × Capacity per Unit)				Separation (10 kV)	Reactive compensation device (i.e., capacitor bank)	Peak load
Quantity	Ratio	Capacity per unit (MVA)	Sum (MVA)	Quantity	(MVar)	(MW)
2	110/10	40	40 × 2 = 80	26	1.8 × 2 + 3.0 × 2	56.2

Table 5.11 Data for the distributed PV systems.

PV no.	Installation capacity (MW)	PV voltage level (kV)
1	1.5	10
2	1.0	0.4
3	1.5	10
4	1.0	0.4
5	2.0	10
6	2.0	10
7	1.0	0.4
8	2.0	10
9	1.0	0.4
10	1.0	0.4
11	2.5	0.4
12	2.0	0.4
13	3.0	10

Table 5.12 Statistical PV data for the SA substation.

Index	Substation SA
Total PV installation capacity (MW)	21.5
PV capacity at 10 kV level (MW)	12
PV capacity at 0.4 kV level (MW)	9.5
Number of PV systems at 10 kV level	6
Number of PV systems at 0.4 kV level	7

Table 5.13 Simulation results of substation SA with PVs connected.

Index	PV capacity penetration	PV energy penetration	Annual network loss ratio	Annual line loss ratio
Simulation results with PVs connection	31.291%	6.216%	2.225%	0.368%

general, the PoI overvoltage issue is more serious than the PCC overvoltage issue for 0.4-kV-level PVs; sometimes the overvoltage may even violate the limit defined in the operational standards.

The power flow over the 8760 h of the year is attained by sequential simulation studies. The results with added PVs are listed in Table 5.13 and compared with the non-PVs case in Table 5.14, which reflects the index differences with and without PVs added.

As can be seen from Table 5.13, the PVs supply 6.216% of the total load demand and effectively reduce the power needed from the grid. Owing to the reduction in the power

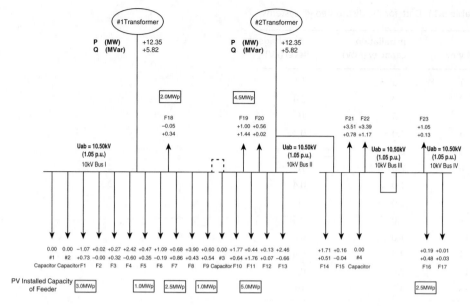

Figure 5.21 Power flow in the SA area of the maximum PV output scenario.

Table 5.14 Index comparison of the SA substation with and without PVs.

Index	Annual energy supply from the grid	Annual network loss	Annual line loss
Index trend with PVs added	↓	↑	↓

↑: index increased
↓: index decreased.

needed from the substation, the line loss decreases. However, the PV systems bring certain additional loss to the network, such as the loss in the step-up transformers and the additional PV integrating line, which can be larger than the saved power loss on the feeders. Figure 5.22 shows the detailed loss based on the simulation study of a typical day. The line loss decreases due to PV connection according to Section 5.4, but the loss of the connecting lines and transformers that integrate the PV systems to the grid is still larger. Thus, the total network loss still increases a little.

Figure 5.23 shows the peak power supply of substation SA per month. It decreases for most months when the PVs are connected. The maximum decline occurs in June, with an obvious peak-shift effect. The generated PV power is totally consumed by the load in the substation load with no power delivered to the higher level grid.

For all the 0.4 kV PoIs, the maximum and minimum voltages are 1.076 p.u. and 0.995 p.u. respectively. Some are over the voltage operation limit (1.07 p.u.). For all the PCCs, the maximum and minimum voltages are 1.054 p.u. and 1.033 p.u. respectively, which are in a reasonable operational range. Figure 5.24 shows the monthly peak voltage of 10 kV/0.4 kV nodes of the SA substation with PVs.

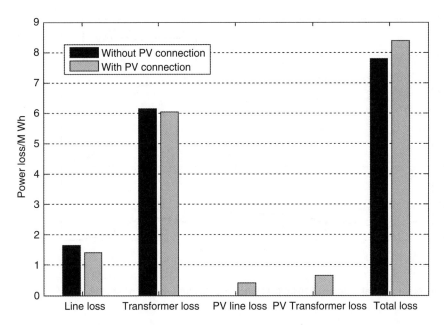

Figure 5.22 Power loss comparison of substation SA with and without PVs.

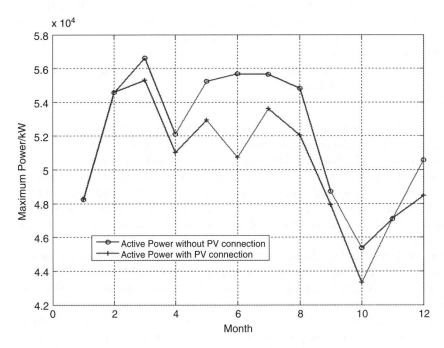

Figure 5.23 Monthly peak output power of substation SA with and without PVs.

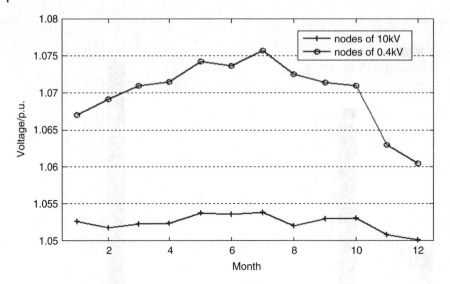

Figure 5.24 Monthly peak voltage of 10 kV/0.4 kV nodes of the SA substation with PVs.

Without PV integration, as a radial feeder with a single power source, the voltage decreases along the feeder. The 10-kV-level maximum voltage is 1.05 p.u. at the head of the line, and the 0.4-kV-level maximum voltage is less than 1.05 p.u.. With the PV systems connected, the 10-kV-level and 0.4-kV-level maximum voltages are both found at the nodes with a PV system connected. The 0.4-kV-level maximum voltage appears at the PV PoI. The voltage on the low-voltage side obviously rises, sometimes even exceeding the limit set in China's standards. The voltage on the high-voltage side is more controlled and can meet the standards.

In conclusion:

1) When the PV output power is larger than the total load of the feeders, the excess PV power flows back to the substation buses and consumed by the other feeders. The PV power can thus be dispatched to supply the load of multiple feeders.
2) Owing to the generated PV power, the node voltages rise, which might lead to operational issues in some feeders. Because both the PV power and load can have large fluctuations, the voltage is apt to violate the limit. This impact becomes more significant if the PV size increases and goes beyond the hosting capacity of local network.
3) The distribution line loss decreases with PVs added when the PV output power is less than the total load. However, the power loss of the PV system (including the step-up transformer and additional connecting lines) may be more than the reduction in the line loss. Hence, in total, the network loss may still increase. For the PVs connected at the 0.4 kV level, the PV output power can be directly consumed by customers. This is useful for reducing the network loss. However, overvoltage is a serious concern for integrating PVs at the 0.4 kV level. Thus, the PV capacity, the location, and the voltage level of integration are important factors and should be carefully and comprehensively considered in a PV integration scheme.

5.6 Summary

This chapter has presented the impact of PVs on the power flow of a distribution networks. It has also introduced various power flow calculation methods, though it is focused on the voltage variation and distribution loss caused by the PV integration. Models of voltage and distribution loss considering PVs have been established, which then used to analyze the impact of different PV capacities and locations. All the methods and models have been demonstrated via a case study of a real 110 kV substation in China. The operational data with the impact of PVs throughout a year have been given with a detailed analysis on the voltage and power flow distribution, line loss, and network losses.

The simulation results show that overvoltage is a serious issue when PV penetration is high. Although the PV penetration in the case studies in this chapter is not that high and the PV power can still be consumed by the load on the feeders in the same substation area, the voltage on the 0.4 kV side is already over the limit for most of the months in the year under study. This will only get worse when more PVs are integrated in the future. However, it is assumed that the PV power factor is unity in the simulations studies. This implies that no reactive power is generated from the PV systems. But with the development of improved PV control technology, PVs shall be able to provide reactive power support. Currently, most of the large PV systems are required to have a power factor setting in the region of 0.95 (lead–lag). The reactive power adjustment can be a potential solution to the overvoltage issue, and can further increase the PV penetration level (i.e., accommodate more PVs). Since voltage control is critical to PV integration, this is presented in Chapter 6.

References

1 Lin M. (2012) *Power Flow Algorithm for Distribution Network with Photovoltaic Forecasting*. Shandong University of China.

2 Zhang X., Liu Z., Yu E., *et al.* (1998) A comparison on power flow calculation methods for distribution network. *Power System Technology*, **4**, 45–48.

3 Liu Z. (2010) *Smart Grid Technology*. China Electric Power Press: Beijing.

4 Shigenori N., Takamu G., and Yoshikazu F. (2001) Practical equipment models for fast distribution power flow considering interconnection of distributed generators. In *2001 Power Engineering Society Summer Meeting. Conference Proceedings (Cat. No.01CH37262)*, vol. **2**. IEEE; pp. 1007–1012.

5 Xu X., Huang Y., Liu C, *et al.* (2010) Influence of distributed photovoltaic generation on voltage in distribution network and solution of voltage beyond limits. *Power System Technology*, **10**, 140–146.

6 Xie L. (2009) Grid-connected PV system modeling and impaction on voltage and power loss of the distribution network. North China Electric Power University.

7 Hoff T. (1995) The value of grid support photovoltaics in reducing distribution system losses. *IEEE Transactions on Energy Conversion*, **3**, 569–576.

8 State Grid Corporation of China. (2014) *Typical Grid-Connecting Design of Distributed Generation – The Connecting Fascicle*. China Electric Power Press: Beijing.

9 GB/T 12325-2008. (2008) *Power quality – Deviation of Supply Voltage*. Standards Press of China.

6

Voltage Control for Distribution Network with High Penetration of Photovoltaics

6.1   Introduction

As discussed in Chapter 5, the integration of high-penetration distributed PVs to the grid will bring great challenges to the safe operation of the existing distribution networks. One major problem is that the high-penetration of PVs can significantly raise the voltage level that sometimes even violate the safe operation constraint when the system load is low and the solar irradiance is high. This will directly affect consumers and the normal operation of the distribution network, as well as the service life of equipment.

Tables 6.1 and 6.2 show the acceptable deviation of supply voltage in China and in the USA respectively [1, 2].

At present, most grid-connected PV inverters operate at unity power factor (namely, $\cos \varphi = 1.0$) in order to maximize the utilization of PV generating capacity. However, during the time when the PV output is large and the system load is low, the high penetration of PVs might lead to a voltage rise in the distribution network. Voltage rise is one of the dominant constraints that limit the ability of the distribution network to accommodate more distributed PVs. To solve this problem, many voltage regulation researches have been carried out [3] and the reactive power control methods have been proposed for PV inverters and used to actively regulate the voltage [4, 5]. However, most of the current researches focus on the voltage control of PV inverters and limited to the feeder-level analysis in the distribution network. The impact of the substation-level (regional) distribution network automatic voltage control (AVC) is often neglected. With a large number of distributed PV systems connected to the distribution network, the impact of AVC on the existing voltage regulation measures of the distribution network cannot be ignored [6]. The interactions between the AVC and PV inverters are also worthy of further study. Particularly in evaluation, analysis, and optimization research, the existing AVC measures, such as the use of an on-load tap changing (OLTC) transformer and the switching of a static capacitor (SC), should be considered in order to ensure the safe and stable operation of the whole distribution network after the integration of a large number of distributed PV systems. This chapter will focus on the study of the voltage control strategy for the distribution network with AVC capabilities to achieve higher PV penetration.

Grid-Integrated and Standalone Photovoltaic Distributed Generation Systems: Analysis, Design, and Control, First Edition. Bo Zhao, Caisheng Wang and Xuesong Zhang.
© 2018 China Electric Power Press. Published 2018 by John Wiley & Sons Singapore Pte. Ltd.

Table 6.1 Acceptable deviations of supply voltage in China (GB/T 12325-2008).

	Per unit value (p.u.)	
Voltage level/voltage limit	Minimum	Maximum
220 V	0.900	1.07
10(20) kV	0.930	1.07
35 kV	0.950	1.05

Table 6.2 Acceptable deviations of supply voltage (120 V) in the USA (ANSI C84.1).

	Per unit value (p.u.)	
Voltage level/voltage limit	Minimum	Maximum
120 V	0.950	1.05
120–600 V	0.950	1.05
>600 V	0.975	1.05

6.2 Voltage Impact Analysis in the Distribution Network with Distributed Photovoltaics

The voltage impact of distributed PVs on the distribution network has been analyzed from the perspective of the power flow calculation. It has been observed that factors such as the capacity and connection location of the distributed PVs have great impacts on the voltage of the distribution network. In order to further analyze the impact mechanism of distributed PVs on the voltage of the distribution network, in this section factors that influence the voltage at the PCC, including the distribution network line parameters, PV generating capacity, initial voltage of the distribution network without the PVs, load impedance, and load power factor, will be discussed to lay the foundation for further analysis.

Under normal circumstances, when a feeder in a medium/low-voltage distribution network is short, the feeder between the PCC and the distribution network can be simplified into the series impedance shown in Figure 6.1 [7]. In Figure 6.1, U_{grid} is the voltage of the distribution network (normally, it is a positive constant), Z_{grid} is the impedance of the feeder of the distribution network, Z_{load} is the load impedance at the PCC, U_{load} is the voltage at the PCC, and I_{pv} is the output current of the PV system.

Before the distributed PV system is connected to the distribution network, the voltage at the PoI U_{load} is

$$U_{load} = \frac{Z_{load}}{Z_{load} + Z_{grid}} U_{grid} \tag{6.1}$$

Figure 6.1 Simplified circuit diagram of distributed PVs connected to distribution network.

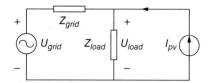

After the PV system is connected to the distribution network, the voltage at the PCC U_{load} becomes

$$U_{\text{load}} = \frac{Z_{\text{load}}}{Z_{\text{load}} + Z_{\text{grid}}}(U_{\text{grid}} + Z_{\text{grid}}I_{\text{pv}}) \tag{6.2}$$

It can be obtained from Equations 6.1 and 6.2 that the voltage rise at the PCC ΔU_{load} is

$$\Delta U_{\text{load}} = \frac{Z_{\text{load}}Z_{\text{grid}}I_{\text{pv}}}{Z_{\text{load}} + Z_{\text{grid}}} \tag{6.3}$$

When the PV system is operated at the unity power factor, the PV output current and the voltage at the PCC will have the same phase. Hence:

$$I_{\text{pv}} = kU_{\text{load}} \tag{6.4}$$

where k is a positive constant.

Since the impedance of the feeder of the distribution network is much smaller than the load impedance, substituting Equation 6.4 into Equation 6.2 and simplifying yields

$$U_{\text{load}} = \frac{U_{\text{grid}}}{1 - kZ_{\text{grid}}} \tag{6.5}$$

The feeder impedance and impedance ratio can be defined as

$$Z_{\text{grid}} = r_{\text{grid}} + jx_{\text{grid}} \tag{6.6}$$

$$m = \frac{r_{\text{grid}}}{x_{\text{grid}}} \tag{6.7}$$

Substituting Equations 6.6 and 6.7 into Equation 6.5 and the amplitude of U_{load} is obtained

$$|U_{\text{load}}| = \frac{m|U_{\text{grid}}|}{\sqrt{m^2\left(1 - kr_{\text{grid}}\right)^2 + k^2 r_{\text{grid}}^2}} \tag{6.8}$$

Combining Equations 6.4 and 6.8, the PV output power P is:

$$P = \frac{km^2|U_{\text{grid}}|^2}{m^2(1 - kr_{\text{grid}})^2 + k^2 r_{\text{grid}}^2} \tag{6.9}$$

The PV power reaches its maximum value when k takes the value shown in

$$k_{P_{\text{max}}} = \frac{m}{r_{\text{grid}}\sqrt{m^2 + 1}} \tag{6.10}$$

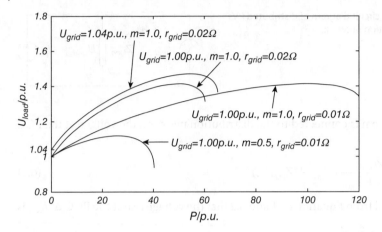

Figure 6.2 Curves of voltage variation at the PCC.

The output power of the distributed PVs keeps increasing with the increase of the value of k until the PV output power reaches its maximum.

When k takes the value shown in Equation 6.11, the voltage at the PoI reaches the maximum, which is shown in Equation 6.12:

$$k_{|U_{load}|_{max}} = \frac{m^2}{r_{grid}(m^2 + 1)} \tag{6.11}$$

$$|U_{load}|_{max} = |U_{grid}|\sqrt{m^2 + 1} \tag{6.12}$$

It can be seen from Equation 6.12 that the maximum voltage at the PCC is only related to the distribution network voltage and ratio of the feeder impedance. It can also be noted that $k_{P_{max}} > k_{|U_{load}|_{max}}$. Before the voltage at the PCC reaches the maximum $|U_{load}|_{max}$, the voltage at the PCC increases with the value of k; until the voltage at the PCC reaches the maximum, the voltage at the PCC decreases with the increase of the value of k. Equations 6.10 and 6.11 show that, before the PV power of the feeder reaches its upper limit, the output power of the distributed PVs increases with the increase of the value of k while the voltage at PCC first rises and then falls.

The aforementioned analysis demonstrates the relationship is nonlinear between the voltage at the PCC and m, U_{grid}, r_{grid}, and P. Figure 6.2 shows the relationship between the voltage at the PCC and m, U_{grid}, r_{grid}, and P.

Considering only the ascending segments of the curves shown in Figure 6.2, the following information can be obtained by comparing the four curves.

- when U_{grid}, r_{grid}, and P remain unchanged, the voltage at the PCC increases with m;
- when U_{grid}, m, and P remain unchanged, the voltage at the PCC increases with r_{grid};
- when m, r_{grid}, and P remain unchanged, the voltage at the PCC increases with U_{grid};
- when U_{grid}, m, and r_{grid} remain unchanged, the voltage at the PCC increases with p.

Generally, the distribution network load has a high power factor. When only the voltage drop caused by the load active and reactive current components flowing through the distribution network feeder is considered, the voltage phase difference between the PCC and the distribution network is small. Suppose I_r and aU_{grid} represent the load active current and voltage drop respectively, before the PV is connected to the distribution network and nI_r and bU_{grid} represent the output current of PV and the voltage at the PCC. Their relationship can be simplified as follows:

$$aU_{grid} = \frac{r_{grid}(\eta m + \sqrt{1 - \eta^2})}{\eta m} I_r \tag{6.13}$$

$$(1 - a)U_{grid} + nr_{grid}I_r = bU_{grid} \tag{6.14}$$

where a, b, and n are positive constants and η is the load power factor.

When the output power of the PV is constant, we get

$$bU_{grid}nI_r = c \tag{6.15}$$

where c is a positive constant.

Combining Equations 6.13, 6.14, and 6.15 gets that

$$b(b + a - 1) - \frac{\eta mac}{U_{grid}I_r(\eta m + \sqrt{1 - \eta^2})} = 0 \tag{6.16}$$

Taking the derivative of Equation 6.16 with η and I_r as independent variables, it can be obtained that

$$\frac{db}{d\eta} = \frac{ac\eta^{-2}}{(2b + a - 1)U_{grid}I_r\sqrt{1 - \eta^2}}\left(\frac{1}{m} + \frac{1}{m^2}\sqrt{\frac{1}{\eta^2} - 1}\right)^{-2} \tag{6.17}$$

$$\frac{db}{dI_r} = \frac{ac}{(1 - a - 2b)U_{grid}I_r^2}\left(1 + \frac{\sqrt{1 - \eta^2}}{\eta m}\right)^{-1} \tag{6.18}$$

From Equation 6.14 it can be found that

$$b + a - 1 > 0 \tag{6.19}$$

Hence

$$2b + a - 1 > 0 \tag{6.20}$$

a, n, c, U_{grid}, m, and I_r are all larger than zero and $\eta \leq 1$. From this it can be found that the right-hand sides of Equations 6.17 and 6.18 are larger and smaller than zero respectively. Therefore, when η increases and I_r decreases, b will increase accordingly. In other words, the voltage at the PCC will increase with the increase of the power factor and load impedance.

In conclusion, when the distributed PV system is connected to the distribution network, in addition to the capacity and integration location of the PV system, the distribution network line parameters, PV generating capacity, initial voltage of the distribution network without the PVs, load impedance, and load power factor will have an impact on the voltage at the PCC. As a result, relevant measures should be taken after the integration of high-penetration distributed PVs to limit the possible overvoltage [8–10]. Now the basic measures for the distribution network voltage control will be discussed in the following sections.

6.3 Voltage Control Measures

Usually, the distribution network voltage regulation measures include those taken at the feeder level and substation level. First, at the feeder level, pole-mounted capacitor banks and voltage regulators are used for the local balance of the reactive power and voltage regulation of the feeder or users. Second, the substation AVC system containing OLTC and SC is used to keep the voltage of the whole substation level within the safe operation constraints and ensure the reactive power balance of the regional distribution network. The voltage regulation and control measures taken at the feeder and the substation level can keep grid voltage within an acceptable range and minimize reactive power loss. On this basis, when a large number of distributed PVs are connected to the distribution network, the available reactive capacity of the PV inverters can be used to limit the system voltage rise so as to enable the PVs to participate in the grid operation effectively. This will make positive contributions to the grid voltage regulation [11–14]. This section will introduce the AVC system regulation at the substation level, voltage regulation measures at the feeder level, and PV inverter control locally.

6.3.1 Automatic Voltage Control System

In a distribution network, the medium-voltage or high-voltage (MV/HV) substation AVC system is usually based on the pre-developed control strategy to keep the voltage of the distribution network within the allowable operation range. In this way, it can reduce the reactive power flow between different voltage-level grids as much as possible so as to realize the local reactive power equilibrium and minimize the regional network loss. The control measures generally include the controls of OLTC and SCs, with the former used to control the transformer tap positions of the main transformer to regulate the overall voltage of the feeder; or the capacitor banks can be directly switched or switched in groups as the reactive power source. These two are discrete control measures that can not be used to realize the continuous regulation of the target voltage. The control criteria mainly include the substation low-voltage bus voltage constraint and the reactive power constraints of the main transformer.

An OLTC usually has 16 or 32 positions, each of which can increase or reduce the voltage by 1.25% or 0.625% of the rated voltage, with the maximum regulation range being ±10% of the rated voltage. It is worth noting that, theoretically, the AVC system regulate feeder voltage via remote control, but at present most of the conventional AVC systems regulate voltage according to the voltage level of the substation low-voltage bus, where the impact of the distributed PVs on the grid voltage is ignored. In essence, the AVC system is still based on the traditional radial distribution network and operates in the single-way power flow mode. This might be effective when the system has a low penetration of distributed PVs, but in the case of high penetration of distributed PVs, in order to ensure the optimal operation of the distribution network, it is necessary to change the existing control strategy of the system.

6.3.2 Feeder-Level Voltage Regulation

In addition to the substation AVC system, the distribution network feeder is usually equipped with voltage regulators, capacitor banks, and voltage regulation devices.

Take capacitor banks, for example; these include fixed capacitor banks and adjustable capacitor banks, which are usually installed on the feeder and also known as pole-mounted capacitors. Capacitor banks regulate the voltage of specific feeders or loads as a supplement to the AVC system. The control logic of capacitor banks can be generally classified into three categories: (1) sequential type, setting the time for regulation according to daily load variation; (2) voltage type, carrying out automatic regulation according to the feeder voltage level; (3) reactive power compensation, conducting effective matching according to the reactive power changes of the feeder or loads.

6.3.3 Photovoltaic Inverter

Traditionally, after the integration of distributed PVs, the PV inverters usually operate at unity power factor. Currently, the PV inverters are not required to participate in voltage regulation to maximize the active power utilization of PVs. However, technically, the PV inverters have a certain reactive power regulation capacity. In addition to outputting active power, they can provide reactive power support for the distribution network voltage regulation. In fact, in 2012, China released the national standard *Technical Requirements for Connecting PV System to Distribution Network* (GB/T 29319-2012), which is applicable to the 380 V voltage-level distribution networks with the PVs and to the newly built reconstructed and expanded PVs which are connected to the user side of the distribution network through the 10 kV (or 6 kV) voltage level. It is clearly stipulated in this standard that: (1) the power factor of the PVs should be continuously adjustable within the range of ±0.95; (2) distributed PVs should have the ability to regulate reactive output within their reactive output range, which participate in the grid voltage regulation according to the voltage level at the PCC. The regulation mode, reference voltage, voltage adjustment ratio, and other parameters of the PVs should be set by the power dispatching center.

The available reactive power capacity of the PV inverters mainly depends on the apparent power of the inverter and the active power output by the inverter. The PV inverter mainly has advantages such as fast regulation, continuous regulation of reactive power, and a positive impact on the grid voltage profile. The available reactive power capacity of the PV inverter is shown in Figure 6.3, with the vertical axis representing the active power output by the inverter, the horizontal axis representing the reactive power output by the inverter, and the bidirectional arrow indicating the inverter can output and absorb inductive reactive power, which means the power factor of the PV inverter can operate in the leading or lagging state. The capacity of the PV inverter can completely match or overmatch its installed PV capacity. Even if its capacity and the installed PV capacity are perfectly matched, the PV inverter may operate at the rated capacity (i.e., fully loaded) only for a short period of time in a day. In fact, for most of the time the inverter still has certain reactive power regulation capacity. Therefore, the inverter control strategies can be set up reasonably according to the active power output of the PVs, so as to realize optimal reactive power output and integrated coordination of the grid voltage control. Active participation of the PV inverter in distribution network voltage regulation will be the most important means of voltage regulation. The next part focuses on the PV inverter control strategies.

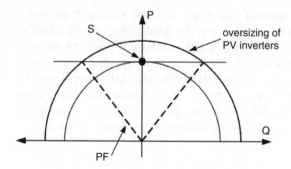

Figure 6.3 Available reactive power of PV inverter.

6.4 Photovoltaic Inverter Control Strategies

The control strategies of a PV inverter usually include unity power factor, constant power factor ($\cos \varphi \neq 1.0$), variable power factor, voltage adaptive control, and so on [15]. Except for the unity power factor control strategy, all other control strategies regulate the reactive power of the PV inverter to keep the voltage at the PCC within the allowable operation range.

6.4.1 General Control Principle

The PV modules are integrated into the power grid through the inverter, which can achieve the active and reactive power output of the PV inverter through the internal active and reactive controller (PQ controller). The control principle of the typical PQ controller is shown in Figure 6.4.

Vector control technology can be used for the active power and reactive power independent decoupling output of the PV inverter. The active power P and reactive power Q of PV inverter in the *dq* coordinate system are as follows:

$$P = V_{gd} I_d \tag{6.21}$$

$$Q = -V_{gd} I_q \tag{6.22}$$

The preset voltage reference value I_{d_ref} and current reference value I_{q_ref} are satisfied through the control of I_d and I_q, and the reference values are realized through the setting of the active power P_{ref} and reactive power Q_{ref} of the PV inverter. This kind of control strategy is known as static-state reactive power regulation. On this basis, the set value of reactive power can be dynamically adjusted based on the deviation between the preset voltage reference value and the actual measured voltage value. This deviation can be used to calculate the reactive power reference value Q_{ref} through the PI controller. The preset voltage reference value can be set as the allowable upper voltage limit at the PCC. Its control diagram is shown in Figure 6.5.

6.4.2 Constant Power Factor Control Strategy

There are two PV inverter constant power factor control strategies: the unity power factor (which is conventionally set to be 1.0) and the constant power factor ($\neq 1.0$). In the unity power factor strategy, the PV inverter operates in the MPPT mode with no reactive power regulation nor any reactive power output by the inverter. In the constant power factor strategy, the reference value of the reactive power is automatically calculated with

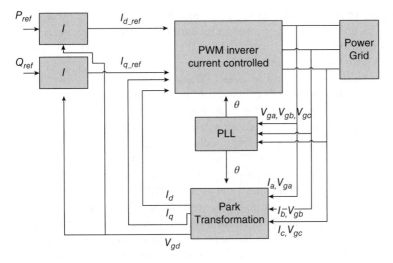

Figure 6.4 Control principle of a PV inverter.

Figure 6.5 Voltage adaptive control principle.

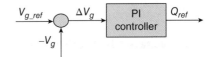

the giving power factor according to the active power output of the PV inverter to make the PV inverter absorb or output a certain reactive power; the power factor can be either leading or lagging.

6.4.3 Variable Power Factor Control Strategy

Variable power factor control of a PV inverter regulates the power factor of the inverter according to its active power. Figure 6.6 shows a set of PV inverter power factor curves, each of which is mainly determined by the slope of the curve and the power factor. There

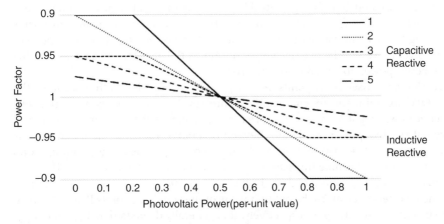

Figure 6.6 PV inverter variable power factor control curve.

are five controlling curves in the figure, each representing a typical kind of control strategy. Taking curve 3 as an example, the power factor ranges between −0.95 and +0.95. When the PV output power is in the range 0–0.2 p.u. the power factor stays at 0.95, and the inverter outputs both the active power and reactive power. As the PV output power increases, the power factor goes up and reaches 1.00 when the PV output power is 0.5 p.u. When the PV output power exceeds 0.5 p.u., the inverter outputs active power and absorbs reactive power, the power factor becomes negative, and continues to decrease with increasing PV output power until reaching −0.95.

In general, when the PV output is relatively small, the inverter will output a certain reactive power to boost the voltage at the PCC; when the PV output is relatively large, the inverter will absorb a certain reactive power to suppress the voltage rise at the PCC. In practice, the structure of the distribution network should also be taken into consideration. For example, adopting cable lines or overhead lines in a distribution network may lead to different voltage levels. Therefore, various factors should be taken into consideration comprehensively to achieve the expected control objectives through the variable power factor curve. Compared with the constant power factor control strategy, the variable power factor control strategy sets different power factors based on different PV outputs and then determines different PV inverter reactive power reference values.

6.4.4 Voltage Adaptive Control Strategy

Voltage adaptive control, also known as dynamic reactive power voltage adaptive control, is a strategy in which the PV inverter automatically regulates its reactive power output within the range of allowable capacity of the inverter according to the reference value of the voltage at the PCC. Compared with constant power factor and variable power factor control, this control strategy automatically regulates the reactive power according to the voltage at the PCC to meet the requirements of secure grid operation. It includes PV inverter Q/V droop control and P/V droop control.

6.4.4.1 Q/V Droop Control

The currently accepted Q/V droop control is also known as reactive power voltage control. Its control curve is shown in Figure 6.7, where the vertical and horizontal ordinates representing the per unit value of voltage and reactive power, with a positive value of reactive power represent the reactive power output by the inverter and a negative value for the reactive power absorbed by the inverter.

Figure 6.7 is composed of six points (A, B, V_{min}, V_{max}, C, E). The continuous line is the inverter Q/V six-point control curve and the interval $[V_{min}, V_{max}]$ constitutes the voltage reactive power regulation dead zone. In the dead zone the PV inverter does not need to carry out reactive power regulation. Beyond this range, however, the PV inverter is required to carry out reactive power regulation according to the corresponding droop curves. In order to determine this curve, it is necessary to set the parameters V_{min} and V_{max} and the voltage at points B and C. The low voltage and overvoltage protection function of the inverter determines the per unit value of the voltage at points B and C, which is set to be 0.9 p.u. and 1.1 p.u. respectively. It is generally deemed that V_{min} and V_{max} are symmetrical with respect to the system voltage reference value $V_0 = 1.0$ p.u., and

Figure 6.7 PV inverter reactive voltage control curve.

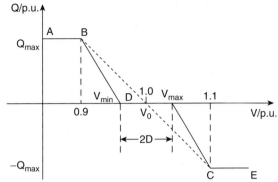

at this point the dead band $2D = 2|V_{max} - V_0|$. Therefore, in the symmetrical parameter mode, the six-point control parameter setting only requires the determination of the parameter D, while in the unsymmetrical parameter mode the parameters V_{min} and V_{max} must be set. When $V_{min} = V_{max} = V_0$, the six-point control curve changes into the four-point (A, B, C, and E) control mode, represented by the dotted line in Figure 6.7. As in this case there is no dead zone, resulting in frequent reactive power regulation of the inverter, it is seldom used in practice.

The determination of the control voltage dead zone D requires comprehensive consideration of the safe operation constraints of voltage and the frequency regulation of the PV inverter. In practice, the reactive power regulation of the inverter will be constrained by the power factor. At present, the adjustable range of the power factor of many PV inverters is ± 0.9. In this case the P–Q capacity regulation range of the inverter is shown in the shaded area in Figure 6.8. The shaded area is all the possible operation state space of the inverter. When the active power output of the inverter is P, the maximum reactive power output by the inverter is composed of the boundary line segment $\overline{S_0 S_{max}}$, the arc line segment $\overparen{S_{max} P_{max}}$, and the symmetrical fourth quadrant section.

Figure 6.8 Inverter P–Q capacity curve.

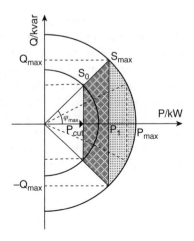

Based on the analysis of Figures 6.7 and 6.8, Equations 6.23 and 6.24 for the calculation of the inverter reactive power Q could be derived:

$$Q = \begin{cases} \frac{V}{|V|}Q_{max} & V < 0.9 \ \text{or} \ V > 1.1 \\ 0 & 1-D \leq V \leq 1+D \\ \frac{-Q_{max}}{0.1-D}(-V+1+D) & 1+D < V \leq 1.1 \\ \frac{Q_{max}}{0.1-D}(-V+1-D) & 0.9 \leq V < 1-D \end{cases} \tag{6.23}$$

$$\begin{cases} Q_{max} = P\tan(\varphi_{max}), & P \in [P_{cut}, P_1) \\ Q_{max} = \sqrt{(S_{max})^2 - (P)^2}, & P \in [P_1, P_{max}] \\ P_1 = S_{max}\cos(\varphi_{max}) \end{cases} \tag{6.24}$$

where φ_{max} is the maximum power factor angle of the PV inverter, P, and Q are the active power and reactive power output of the PV inverter respectively, S_{max} is the capacity limit of the PV inverter, Q_{max} is the maximum reactive power available to be regulated by the inverter, and P_{cut} is the cut-in power of the inverter.

6.4.4.2 P/V Droop Control

The P/V droop control strategy is more effective in a resistive low-voltage distribution network. However, it sacrifices a certain PV active power output, which will affect PV efficiency. It can also be used jointly with the Q/V droop control strategy. In the P/V droop control strategy, the reference active power output by the PV inverter is usually set based on the upper limit voltage that is allowed at the installation bus. The specific calculation formula is

$$P_i = \begin{cases} P_{max,i} - k_i(V_i - V_{crit,i}) & \forall V \geq V_{crit,i} \\ P_{max,i} & \forall V < V_{crit,i} \end{cases} \tag{6.25}$$

where i is the number of the PV inverter, k_i is the droop coefficient, V_i and $V_{crit,i}$ are the voltage and its upper limit at the PV set point, respectively, and $P_{max,i}$ is the maximum active power of the ith PV inverter.

6.4.4.3 Inverter Parameter Optimization

As the integration of PVs usually causes a voltage rise at the PCC, and low voltage scenarios are less likely to occur, so the optimization of the unsymmetrical parameters V_{min} and V_{max} will have a better voltage regulation effect than the optimization of the symmetrical parameter D. In order to set the inverter Q/V control parameters V_{min} and V_{max} and the symmetrical parameter D, a particle swarm optimization algorithm, a genetic algorithm, and other optimization algorithms can be used for the optimization and setting of all inverter parameters. In addition, these parameters are set once before the installation of the inverter. An example optimization problem is formulated in the following. The objective function and constraint conditions are shown in Equation 6.26, but the power flow equations are not listed. The objective function is to minimize the annual voltage deviation and the amount of reactive power regulation [16]:

$$f = \min \sum_{T=1}^{8760} \left(\alpha \sum_{i=1}^{N} |V_i - V_0|^2 + \beta \sum_{j=1}^{N_{\text{PV}}} |\Delta Q_j| \right)$$

$$\text{s.t.} \begin{cases} 0.9 < V_{j,\min} < V_{j,\max} < 1.1, & j = 1,2,\dots,N_{\text{PV}} \\ V_{\text{low}} < V_i < V_{\text{high}}, & i = 1,2,\dots,N \\ 0 < \Delta Q_j < Q_j^{\max}, & j = 1,2,\dots,N_{\text{PV}} \\ 0 < D_j < 0.1, & j = 1,2,\dots,N_{\text{PV}} \\ 0 < \alpha, \beta < 1, & \alpha + \beta = 1 \end{cases} \tag{6.26}$$

where N is the total bus number and N_{PV} is the total number of buses with PV installed. When $\alpha = 0$ and $\beta = 1$, the optimization target is the annual reactive power regulation. Through the analysis of reactive power regulation during all periods, the minimum annual reactive power regulation can be determined.

6.5 Modeling and Simulation

6.5.1 Approaches

After the integration of a large number of distributed PVs, in the analysis of the voltage regulation, the following factors need to be considered. (1) Solar irradiance may fluctuate dramatically in a short period of time, with the time scale being a minute or less. (2) The action time of the grid voltage regulation measures (such as OLTC tap regulation or SC switching) is usually counted in minutes or even shorter. (3) The impact of the distributed PVs on the distribution network mainly occurs in two scenarios: one is during the period of time when the PV output reaches the maximum; the other is when the PV output cannot be completely accommodated by the load resulting in reverse power flow. As a result, effective voltage control measures should be able to withstand the test in these two extreme scenarios and meet the safe operation requirements of the distribution network.

Meanwhile the requirements for the power factor of PVs are clearly stipulated in existing Chinese power grid regulations and standards in order to fully utilize the reactive power regulation capacity of the PV inverter and keep the voltage at the PCC operating within a reasonable range through the regulation of the power factor of the PV inverter. Usually in the case of large PV output, if the voltage at the PoI or at the PCC is high, the PV inverter should be controlled to absorb a certain reactive power without exceeding the limitation of the PV inverter capacity, in order to suppress the voltage rise at the PCC and improve the voltage steady-state safety margin.

This chapter is going to carry out the analysis of the optimization of the minute-level voltage control measures in the distribution network under extreme operation scenarios, and determine the optimal PV inverter control measures considering the impact of the existing substation AVC system and the safe operation constraints for the distribution network. The open distributed system simulator platform OpenDSS is employed to carry out detailed modeling of the distribution network with high PV penetration, focusing on the modeling of typical PV inverter control strategies, and conduct continuous control simulation every 5 min under extreme operation scenarios. Through the simulation analysis, we will determine appropriate PV inverter control strategies and verify the effectiveness of all control strategies; meanwhile, the effectiveness of all control strategies will be verified and the impact of high-penetration distributed PVs on the existing AVC system (including OLTC and SCs) operation will be analyzed.

6.5.2 Introduction to OpenDSS

OpenDSS (http://smartgrid.epri.com/SimulationTool.aspx) is a software package based on open source codes primarily for the simulation of the distribution network in a power system. It supports all phase/frequency domain analysis, including harmonic power flow analysis, and handles all kinds of complicated network systems or radial systems, and unsymmetrical or multiphase distribution networks. Applicable to the planning analysis of most distribution networks, it obtains solutions based on the phase/frequency domain and mainly provides results in the form of phasors. It is mainly applicable to power flow analysis, harmonic power flow analysis, and dynamic analysis.

Traditional distribution network analysis software is mainly used to study the results under the peak load conditions, without considering the temporality, intermittence, and fluctuation of different distributed PV outputs. Combining the temporal and spatial data, OpenDSS takes into consideration the changes in the location and output of the distributed PVs. It is particularly applicable to the evaluation and analysis of the planning for the integration of a large number of distributed renewable energy sources into the power system. As the chosen simulation analysis tool in the chapter, the application scope of OpenDSS mainly includes the following:

- distribution network planning and analysis;
- multiphase AC circuit analysis;
- analysis of impact of integration of distributed PVs into the distribution network;
- analysis of the changes in the load and the annual power flow of generators;
- wind turbine simulation analysis;
- transformer structure configuration analysis;
- analysis of the impact of the interference between high-order harmonics and inter-harmonics;
- probabilistic load flow study.

The main solving modes embedded in the software include interface power flow, daily/annual power flow calculation, harmonic power flow calculation, dynamic power flow calculation, and random power flow calculation.

6.5.3 Simulation Models

Using the OpenDSS simulation platform, the AVC system model and the PV model need to be developed first.

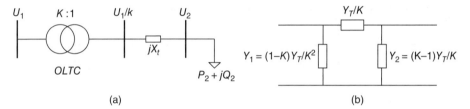

Figure 6.9 (a) Simplified and (b) equivalent circuit diagram of OLTC.

6.5.3.1 Automatic Voltage Control System

At present, the AVC measures used in substations mainly include OLTC and SCs.

6.5.3.1.1 On-Load Tap Changing Control Mode

OLTC voltage regulation is illustrated in Figure 6.9a, in which all parameters are expressed using the per unit value. K is the per unit value of the on-load voltage ratio:

$$K = \frac{K_a}{K_n} = \frac{U_h/U_{ln}}{U_{hn}/U_{ln}} = \frac{U_h}{U_{hn}} \tag{6.27}$$

where K_a is the actual on-load voltage ratio after the change of the tap position, K_n is the rated on-load voltage ratio, U_{hn} is the rated voltage of the main tap on the high-voltage side (the voltage after the change of the tap is U_h), and U_{ln} is the rated voltage on the low-voltage side. The admittances of the transformer Y_1, Y_2, and Y_T are as follows:

$$Y_1 = \frac{1-K}{K^2} Y_T, \quad Y_2 = \frac{K-1}{K} Y_T, \quad Y_T = \frac{1}{Z_T} \tag{6.28}$$

- When $K < 1$, Y_1 is inductive and Y_2 is capacitive.
- When $K = 1$, $Y_1 = Y_2 = 0$.
- When $K > 1$, Y_1 is capacitive and Y_2 is inductive.

Z_T is the short-circuit impedance of OLTC.

Figure 6.9b shows the equivalent single-line circuit model of OLTC, and the ideal on-load voltage ratio of the transformer is $K/1$.

When $K < 1$ the tap is adjusted to increase the secondary side voltage, which is equivalent to primary side parallel components, absorbing inductive reactive power from the system. The secondary side parallel components provide the system with inductive reactive power, by regulating the on-load voltage ratio of OLTC tap K to change the admittance Y_1 and Y_2, so as to change the OLTC on both sides of the system reactive power flow, thus affecting the voltage on both sides of the transformer. In order to support the regulation of the tap, the power system on the primary side must have sufficient capacity.

The control block diagram of the OLTC discrete model is shown in Figure 6.10. Δn is the step size of the tap changer, Δa is the regulation value of the on-load voltage ratio of the tap changer, and *band* is the bandwidth value of the tap regulation. T_d and T_m are the time delay elements, used to prevent unnecessary tap changes in response to transient voltage violations and to introduce the desired time delay before a tap movement As a

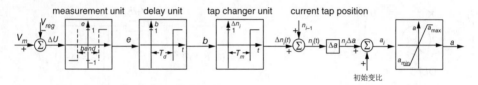

Figure 6.10 OLTC discrete control block diagram.

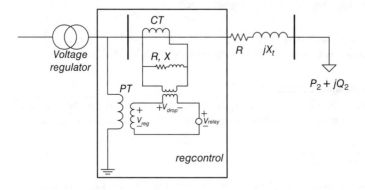

Figure 6.11 Control diagram of OLTC control structure in OpenDSS.

result, the logic of the tap change during the period of time t_k is

$$n_{k+1} = \begin{cases} n_k + \Delta n & V_m > V_{\text{base}} + \text{band}/2, \quad n_k < n^{\text{max}} \\ n_k - \Delta n & V_m < V_{\text{base}} - \text{band}/2, \quad n_k > n^{\text{min}} \\ n_k & \text{other situations} \end{cases} \tag{6.29}$$

where n^{max} and n^{min} are respectively the upper and the lower limit positions of the tap changer.

The transformer model in OpenDSS is shown in Figure 6.11. OpenDSS uses the Reg-control module to simulate the transformer control, taking into consideration of the line voltage drop compensation factor and characteristics of on-load voltage ratio of transformer. In addition, it has the function of measuring and monitoring the remote node voltage.

6.5.3.1.2 Static Capacitor Control Mode

A capacitor bank can generate inductive reactive power (i.e., consuming capacitive reactive power), which is proportional to the square of its terminal voltage. However, capacitor banks have some drawbacks. When the node voltage drops, the supplied reactive power will be reduced, leading to the further drop of the voltage level in the power system. Figure 6.12 is a simplified circuit with a shunt capacitor.

Suppose U_1 remains unchanged before and after the compensation:

$$\text{before compensation:} \quad U_1 = U_2' + \frac{PR + QX}{U_2'} \tag{6.30}$$

$$\text{after compensation:} \quad U_1 = U_{2C}' + \frac{PR + (Q - Q_C)X}{U_{2C}'} \tag{6.31}$$

Figure 6.12 Simplified circuit of shunt capacitor compensation.

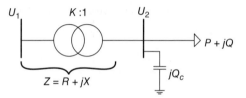

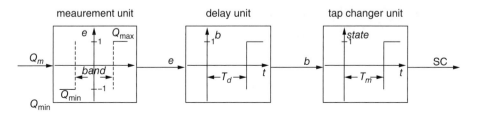

Figure 6.13 SC discrete control block diagram.

From these two equations the size of the reactive power Q_C at the specified voltage U_{2C} could be obtained:

$$Q_C = \frac{U'_{2C}}{X}\left[(U'_{2C} - U'_2) + \left(\frac{PR + QX}{U'_{2C}} - \frac{PR + QX}{U'_2}\right)\right]$$

$$\approx \frac{U'_{2C}}{X}(U'_{2C} - U'_2) = \frac{U_{2C}}{X}\left(U_{2C} - \frac{U'_{2C}}{K}\right)K^2 \qquad (6.32)$$

where U'_{2C} and U'_2 are voltage on the high-voltage side.

The block diagram of the SC discrete model is shown in Figure 6.13.

By measuring the reactive power Q_m at the high-voltage side of the main transformer, the capacitor banks would be turned on if Q_m is larger than the maximum allowable value Q_{max}; on the other hand, if it is smaller than the minimum allowable value Q_{min} then the shunt capacitors will be switched off. The capacitor bank model in OpenDSS is shown in Figure 6.14. OpenDSS uses the Capcontrol modules to simulate the capacitor switching control. In the figure, the vertical axis could represent various types of controlling variables, such as the current, voltage, reactive power, power factor, or the time, and so on. When the measured variable exceeds the ON setting value, the capacitor band is turned on, and when the measured variable decrease below the OFF setting value, the capacitor band will be switched off. Various variables can fully meet the need for simulation analysis.

The reactive power of the distribution network generally gets balanced locally. First of all, certain reactive power compensation devices are installed in the substation for a reasonable amount of compensation capability. Various voltage regulation means are then coordinated together to achieve the desired effects. In terms of the reactive voltage optimization and control of local distribution networks, the control is generally realized by combining SCs and OLTC to minimize the reactive power flow through the main transformer and to keep the voltage on the secondary side of the substation within an acceptable range to ensure a good voltage profile of all feeders.

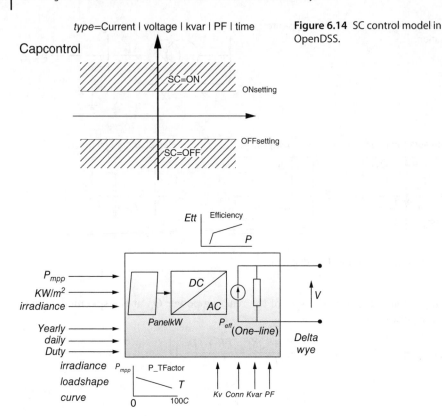

Figure 6.14 SC control model in OpenDSS.

Figure 6.15 PV system model.

6.5.3.2 Photovoltaic System Model

The PV system model is mainly composed of a PV array model, a PV inverter model, and its control strategy. The details of the modeling of the PV system in OpenDSS are given in the following subsections.

6.5.3.2.1 Photovoltaic System Structure Model

As is shown in Figure 6.15, the PV system structure model includes two parts: the PV array and the PV inverter. It is assumed that the inverter can rapidly track MPPT power, which simplifies the modeling of independent components and facilitates the analysis of the impact on the distribution network. In the model, state variables include the following: irradiance, which indicates the radiation intensity of the solar spectrum radiation intensity and can be set through the application of the loadshape coefficient; PanelkW, which represents the output power of the PV panels considering radiation intensity and temperature; P_TFactor, which represents the power versus temperature (P–T) curve; Efficiency, which represents the efficiency curve of the inverter. The output power is affected by irradiance, temperature T, and the maximum PV output P_{mpp}. Calculation modes include annual mode (Yearly), daily mode (Daily), duty mode and so on.

Reactive power is mainly determined by the fixed reactive power output or power factor value (PF), and the output reactive power can also be adjusted to regulate voltage.

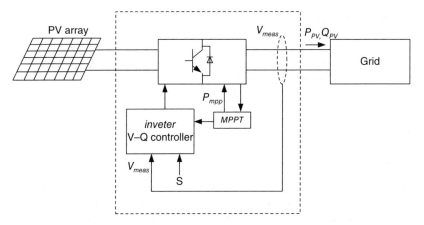

Figure 6.16 PV control block diagram.

6.5.3.2.2 Photovoltaic Inverter Control Strategy Model

First, the modeling of the constant power factor control strategy is relatively, just specify or update its power factor in the software, and 0.98 is taken as the constant power factor (lagging) in the simulation as default.

Second, in terms of the variable power factor control strategy, according to the requirements for the power factor of inverter stipulated in the Chinese national standard technical requirements for connecting a PV system to a distribution network (GB/T 29319-2012), and based on the actual situation of the regional distribution network in the following case studies, curve 4 of Figure 6.6 is used for the simulation study. Also, taking the SA substation supply region as an example, as there can be a number of cable lines in the distribution network, the capacitive charging power can help maintain the feeder voltage at an appropriate level. When the PV output is relatively low, the reactive power output of the PV inverter does not need to be regulated to support the grid voltage. As a result, some changes have been made to curve 4. When the PV output is in the range (0, 0.5 p.u.) the inverter power factor is 1.00, which means the inverter operates at unit power factor. When the PV output is within (0.5 pu, 1.0 pu], the inverter power factor is represented by Curve 4 in Fig. 6.6.

Finally, in terms of the reactive voltage control strategy, OpenDSS contains the inverter control modules and information about the voltage at the PCC. The detailed block diagram is shown in Figure 6.16, and the reactive power output of the PV system is limited by the capacity of the PV inverter. OpenDSS can be used to customize the V–Q control curve of the inverter according to the constraints on the reactive power output change rate. The typical V–Q control model is shown in Figure 6.17. According to the requirements of the supply voltage (0.93–1.07 p.u.) stipulated in the Chinese national standard power quality deviation of supply voltage, 0.01 p.u. is taken as the control margin. In other words, the actual voltage control range is 0.94–1.06 p.u.; namely, $V_1 = 0.92$ p.u., $V_2 = 0.94$ p.u., $V_3 = 1.06$ p.u., and $V_4 = 1.08$ p.u. in Figure 6.17.

It should be further explained that the curve below the horizontal axis in Figure 6.17 indicates that the actual value of the voltage at the PCC exceeds the allowable upper limit, and the inverter needs to absorb certain reactive power. On the other hand, the curve above the horizontal axis indicates the actual value of the voltage at the PCC

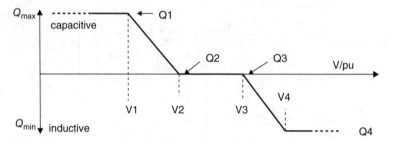

Figure 6.17 *V–Q* control curve of PV inverter.

exceeds the allowable lower limit of the control voltage and that the inverter needs to provide certain reactive power support. It should be noted that the reactive power calculated based on the curve is not the final controlled reactive power output of the PV inverter. Instead, it is the maximum value of the reactive power control in the iterative optimization calculation. As this control strategy is a complete iterative calculation process, the control goal is to meet the specified voltage control constraints and achieve the minimization of the PV reactive power regulation.

6.6 Case Study

6.6.1 Basic Data for Simulation

Again, the distribution network of the 110 kV SA substation introduced in previous chapters is adopted to carry out further analysis. The system diagram is shown in Figure 5.21. A brief introduction to the AVC system of the SA substation is given [17]. The SA substation has two main transformers, each of which has a rated capacity of 40 000 kVA, with a rated voltage of $10 \pm 8 \times 1.2\%/10.5$ kV. The transformers are equipped with OLTC with the model of SZ9 – 40 000/110 and both of them have 17 positions; the voltage range on the high-voltage side is 99–121 kV. The relationship between the transformer tap positions and the voltage on the high-voltage side is shown in Table 6.3.

Through the analysis of the main transformers in the SA substation in 2014, it is evident that in the AVC system the optimization control range of the bus voltage of the main transformers on the low-voltage side is 1.015–1.060 p.u. To be specific, when the per unit value of the bus voltage on the low-voltage side is smaller than 1.015 p.u., the tap positions of the main transformers will be moved up; when the per unit value of the bus voltage on the low-voltage side is greater than (or equal to) 1.060 p.u., the tap stalls of the main transformers will be moved down.

The SA substation has four sets of capacitor banks (#1, #2, #3, and #4). The rated capacities of #1 and #4 capacitor banks are both 1800 kVar, while the rated capacities of #2 and #3 capacitor banks are both 3000 kVar. The capacity of the single capacitor bank cannot be adjusted. The #1 and #2 capacitor banks are connected to the 10 kV bus I, while #3 and #4 capacitor banks are connected to the 10 kV bus II and bus III respectively.

In the AVC system, the annual operation of the four sets of capacitor banks in 2014 shows that when the main transformers reversely (from the low-voltage side

Table 6.3 Tap stalls of main transformer (no. 1 or no. 2) and high-voltage side voltage in SA substation.

Position	Voltage (kV)	Per unit value	Position	Voltage (kV)	Per unit value
1	121.000	1.1000	10	108.625	0.9875
2	119.625	1.0875	11	107.250	0.9750
3	118.250	1.0750	12	105.875	0.9625
4	116.875	1.0625	13	104.500	0.9500
5	115.500	1.0500	14	103.125	0.9375
6	114.125	1.0375	15	101.750	0.9250
7	112.750	1.0250	16	100.375	0.9125
8	111.375	1.0125	17	99.000	0.9000
9	110.000	1.0000			

to the high-voltage side) transmit a reactive power of greater than 1000 kVar, the capacitor banks will be switched off. In a few cases, when the power factor of the bus on the low-voltage side is too high, the capacitor banks will also be removed. The reactive power transmitted by the main transformers in the SA substation (from high-voltage side to low-voltage side) is determined based on the principle of local balance of reactive power; namely, the principle of minimizing the supplied reactive power.

Previous analysis shown in Chapter 5 demonstrates that, the periods of time when the power flow is reversed and flows back to the bus in the SA substation due to high-penetration PVs appear mainly in spring and summer. Particularly in summer, when the load is light and the PV output is high, the reverse power flow occurs more frequently. According to the aforementioned analysis, here this chapter focuses on the analysis of the two extreme scenarios in which the integration of the distributed PVs has a great impact on the distribution network. The selection of the extreme scenario days is mainly based on the solar irradiance and daily load level characteristics, as well as their dynamic characteristics. Taking 2014 as the research year, according to the meteorological data provided by the local weather bureau and the load data of this supply region in 2014, the maximum PV output appeared on July 16, 2014. However, the PV output could not be completely accommodated by the loads in this supply region on that day. The second scenario in which the period of time when the maximum reverse feeding of PV power appeared fell on May 1, 2014. Considering the time scale of the fluctuations of solar irradiance and the actual controlled time cycle, an interval of 5 min is taken as a data point and a total of 288 data points in a whole day are taken for the control simulation. The detailed information is shown in Figures 6.18 and 6.19.

See Chapter 5 for the electrical data of the substation and the detailed data of the distributed PVs integrated into the supply region. The load data, including the active power and reactive power, needs to be extracted from the power company's supervisory control and data acquisition (SCADA) system every 5 min. For the distributed PVs integrated into the distribution network at the user side, the detailed model of this user's internal

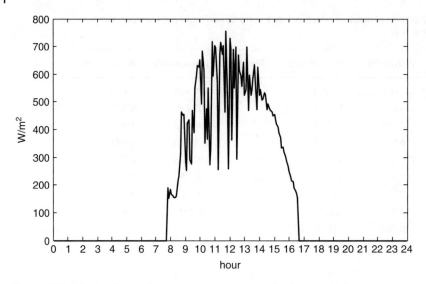

Figure 6.18 Solar irradiance change pattern on extreme day (May 1, 2014).

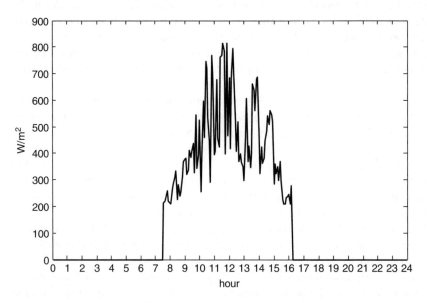

Figure 6.19 Solar irradiance change pattern on extreme day (July 16, 2014).

electrical network should be established. See relevant sections in Chapter 5 for the two typical methods for the integration of distributed PVs.

In the OpenDSS simulation platform, the substation AVC system simulation model is built first, including the tap of the OLTC and switching of SCs. Second, the simulation models are developed for the PV inverter typical control strategies, including constant power factor control strategy, variable power factor control strategy, and reactive voltage control strategy. On this basis the comprehensive voltage regulation simulation analysis will be carried out in the following.

Table 6.4 Information for distributed PVs on feeder SA-F1.

Name of PV owner	Installed PV capacity (MWp)	Voltage (kV)
Battery company C2	1.0	0.4
Automobile company C1	2.0	10

The main simulation conditions/parameters for OpenDSS are summarized as follows:

1) Power flow calculation method: fixed point iteration method.
2) Calculation mode: duty-cycle mode.
3) Simulation time step is 5 mins; every 5 min is defined as one period of time and there are a total of 288 points of time in a whole day.
4) An ideal voltage source model is used on the high-voltage side of the main transformers in the substation during each period of time (i.e., 5 min interval).
5) The load simulation model is constant power (PQ).
6) The overall efficiency of the PV inverter system is 80%.

6.6.2 Analysis of Power Flow and Voltage in Extreme Scenarios with Automatic Voltage Control

As in the 110 kV SA substation (see Figure 5.21), the existing AVC system of the substation is used for voltage regulation and control, without using the PV inverter voltage regulation strategies, feeder SA-F1 for power flow analysis on July 16, 2014 (Monday, a working day), and during the holiday of May 1, 2014 (International Labor Day, a holiday in China), are selected to evaluate the adaptability of the existing AVC system. In the simulation, the power factor of the PV inverter is set to be 1.00; namely, the PV inverter operating at MPPT, without absorbing (or outputting) reactive power. The total installed capacity of the distributed PVs on this feeder is 3 MWp. Table 6.4 shows the information of the distributed PVs on feeder SA-F1, and the integration methods are described in Section 5.5.1 in detail.

The full-day load on feeder SA-F1 on the working day of July 16, 2014 (a working day), and during the holiday of May 1, 2014 (a holiday), is shown in Figure 6.20. It can be seen that the feeder load on working days is higher during the day than at night, while the feeder load during holidays is relatively low and stable. The load level on working days is much larger than that during holidays. In the two scenarios, different impacts will be imposed on both the operation of the distribution network and the accommodation of the distributed PVs.

6.6.2.1 Working Day (July 16, 2014) Scenario

The simulation shows that on working days the tap position of the main transformers nos. 1 and 2 remain unchanged and the operations of the capacitor banks (#1, #2, #3, and #4) also remain unchanged during the whole day before and after the integration of the distributed PVs. As the load on working days is relatively high, capacitor banks #2 and #3 are operating during the whole day, bearing the base load of the reactive load of the distribution network; capacitor banks #1 and #4 are not put into operation. The changes

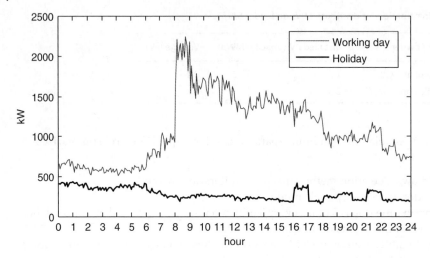

Figure 6.20 Comparison of the full-day load profile between a working day and a holiday on feeder SA-F1.

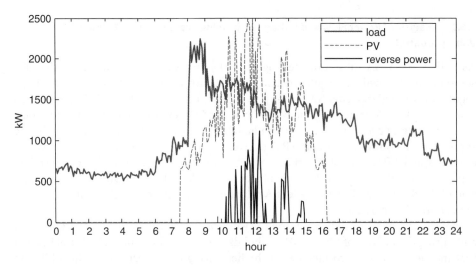

Figure 6.21 Load profile and PV output on feeder SA-F1 on a working day.

in the reactive power transmitted on the high-voltage side of nos. 1 and 2 main transformers before and after the integration of the distributed PVs are also slight because the PVs do not participate in the reactive power regulation, making a small impact on the reactive power transmitted by the distribution network.

Figure 6.21 shows the profiles of the distributed PVs and load on feeder SA-F1 during the whole day on a working day (July 16, 2014). It can be seen from Figure 6.21 that as the load on a working day is relatively high, the distributed PV output can be directly accommodated by the feeder loads for most periods of time; but at noon the output from the distributed PVs is slightly higher than the feeder load and the excess PV output flows to other feeders.

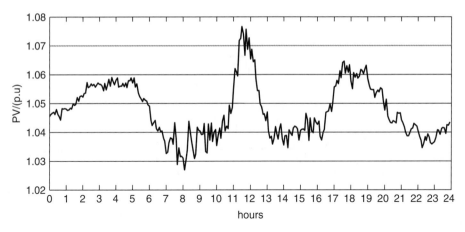

Figure 6.22 Voltage of C2 at the PCC on a working day.

Take C2 as an example. The simulation result of the C2 voltage during the whole day on a working day is shown in Figure 6.22. It can be seen from Figure 6.22 that the voltage fluctuation of C2 at the PCC is relatively large. Especially around noon, the maximum voltage of C2 exceeds 1.070 p.u., which is higher than the allowed limit in Chinese national standards. Through analysis it can be seen that the load on working days is high, especially during peak load periods when the bus voltage on the low-voltage side of the main transformers is low. At this point, in order to meet the bus voltage constraints, the AVC system usually will move up the tap position to improve the voltage level of the whole distribution network. In this case, if the PV output is high, the voltage of C2 at the PCC will become higher, exceeding the upper limit of the safe constraints of voltage and affecting the safe operation of the regional distribution network.

6.6.2.2 Holiday (May 1, 2014) Scenario

The simulation results show that under the holiday scenario the tap positions of the main transformers nos. 1 and 2 remain unchanged and the actions of the capacitor banks (#1, #2, #3, and #4) also remain unchanged during the whole day before and after the integration of the distributed PVs. Capacitor bank #3 is operating during the whole day, bearing the base load of the reactive load; capacitor bank #4 is only operating during some periods of time; capacitor banks #1 and #2 are not put into operation during the whole day; the changes in the reactive power transmitted on the high-voltage side of the main transformers nos. 1 and 2 before and after the integration of the distributed PVs are slight.

Figure 6.23 shows the profiles of the distributed PVs and the load on feeder SA-F1 during the whole day on holidays. It can be seen from Figure 6.23 that the load on feeder SA-F1 during holidays is relatively low, while the output of the distributed PVs is high; in addition to meeting the local load supply demand, the surplus PV output flow to other feeders in the regional distribution network, with the maximum daily 2 MW reverse power.

Continue to take the PV user C2 as an example. The simulation results of the voltage at the PCC during the whole day are shown in Figure 6.24. The voltage fluctuation of C2 at the PCC is relatively large. Especially at noon, the voltage of the C2 exceeds

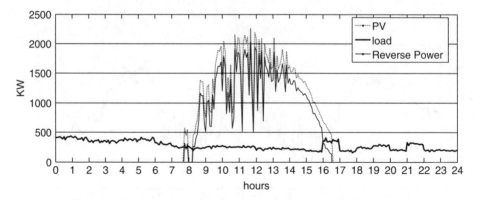

Figure 6.23 Change trends in load, PV output, and reverse power flow on feeder SA-F1 under the holiday scenario.

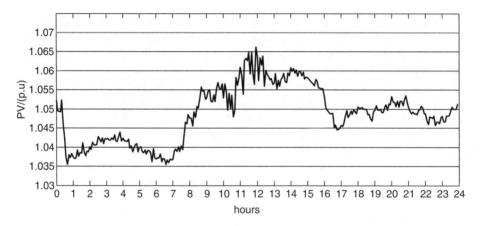

Figure 6.24 Voltage of C2 at the PCC under the holiday scenario.

1.065 p.u., approaching the upper limit of the allowable operation voltage stated in Chinese national standards. The load during the whole day is low. Compared with the scenario of working days, the tap positions of the main transformers are set low, but during periods of time when the PV output is large, due to the low load demand, a large amount of PV output is feed into the grid, resulting in a high voltage at the PCC.

Through the aforementioned analysis, in the selected extreme scenarios the control strategy of the existing AVC system, built based on the traditional distribution network, cannot completely meet the requirements for the safe operation of the distribution network after the integration of high-penetration distributed PVs. This is mainly manifested in the fact that during holidays and on working days during periods of time when the output of distributed PVs is high, the voltage of users at the PCC is high and during some periods of time the voltage of users will even exceed the voltage upper limit. The main reason that AVC did not work properly under the holiday scenario is the reverse feeding of PV output power; the main reason for this on working days is that moving up the tap positions of the main transformers through the AVC system raises the voltage level of

Table 6.5 Actions of all capacitor banks in different control strategies.

No. of main transformer	No. of capacitor bank (rated capacity)	Action times under different control strategies		
		Constant power control	Variable power control	Reactive voltage control
1	1 (1.8 MVar)	1	1	1
	2 (3.0 MVar)	0	0	0
2	3 (3.0 MVar)	0	0	0
	4 (1.8 MVar)	1	1	1

the regional distribution network and makes it easier to have overvoltage issues even if the reversed PV power is not high.

6.6.3 Participation of Photovoltaic Inverter in Voltage Regulation

Through the discussion in Section 6.6.2 it can be seen that the existing AVC system cannot completely meet the requirements for the safe operation of the regional distribution network once high-penetration PVs have been integrated into the system. Best use of the PV inverter control strategies should be made to comprehensively regulate the voltage of the distribution network. This section will analyze and compare the three control strategies – namely, PV inverter constant power factor control, variable power factor control, and reactive voltage control – in the two extreme scenarios, on working days (July 16, 2014) and during holidays (May 1, 2014), based on the existing AVC system.

6.6.3.1 Working Day (July 16, 2014) Scenario

The action times of the capacitor banks under the three control strategies are shown in Table 6.5. It can be seen from in Table 6.5 that the action times of all the capacitor banks during the whole day is the same. Owing to the high load demand on working days, capacitor banks #2 and #3 stay ON (no switching operation) during the whole day, providing reactive power for the load; both capacitor banks #1 and #4 operate once during the whole day, which is also regulated through the AVC strategy based on the load change.

Figures 6.25 and 6.26 show that, compared with the constant power factor control and variable power factor control, generally in the reactive voltage control strategy the reactive power transmitted on the high-voltage side of the main transformers nos. 1 and 2 is reduced because the voltage adaptive control strategy can automatically regulate the reactive power within the voltage control target, absorbing less reactive power. In addition, during the high PV output periods, reverse power feeding does not occur, which is mainly due to the heavy load on working days. The excess output from the distributed PVs is accommodated or transferred to other feeders. What needs to be explained is that in Figure 6.25 the reverse feeding of about 0.5 MVar occurs on no. 1 main transformer at night because a degree of overcompensation exists in the discrete control of the capacitor bank.

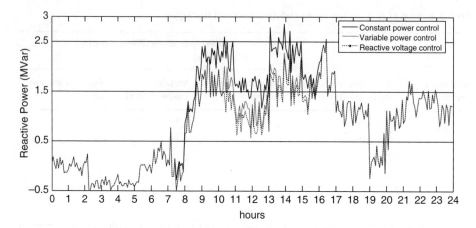

Figure 6.25 Reactive powers on the high-voltage side of main transformer no. 1 on a working day under the three control strategies.

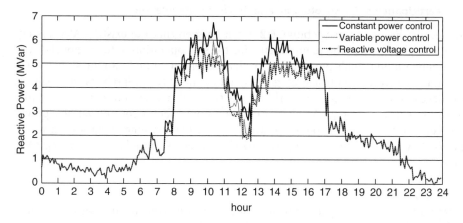

Figure 6.26 Reactive powers on the high-voltage side of main transformer no. 2 on a working day under the three control strategies.

Figures 6.27, 6.28, and 6.29 show the composition of the reactive power on the high-voltage side of the main transformer in the reactive voltage control strategy, variable power factor control strategy, and constant power factor control strategy (taking no. 2 main transformer as an example). It can be seen from the figures that the reactive power on the high-voltage side of the main transformers is composed of the reactive power absorbed by the distributed PVs, equivalent reactive load which equals to the difference between the reactive load and the reactive power output of the capacitor banks, and reactive power losses. The reactive voltage control strategy and the variable power factor control strategy have a small impact on the reactive power. However, in the constant power factor control strategy, the reactive power absorbed by the PV inverter is relatively large during the whole PV operation periods. Particularly during the peak load time, the increase of the reactive power after the integration of the distributed PVs might exceed the control range allowed by the AVC system. Compared with the reactive voltage control strategy, the variable power factor control strategy has a greater

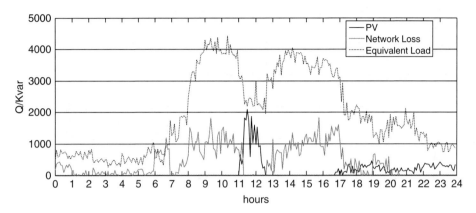

Figure 6.27 Composition of reactive power on the high-voltage side of no. 2 main transformer on a working day under the reactive voltage control strategy.

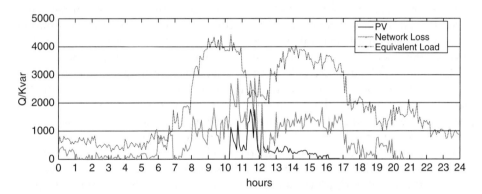

Figure 6.28 Composition of reactive power on the high-voltage side of no. 2 main transformer on a working day under the variable power factor control strategy.

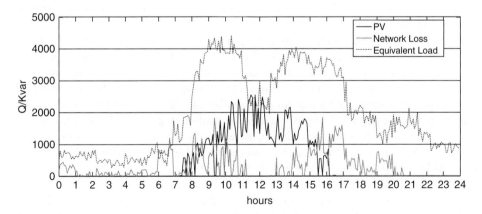

Figure 6.29 Composition of reactive power on the high-voltage side of no. 2 main transformer on a working day under the constant power factor control strategy.

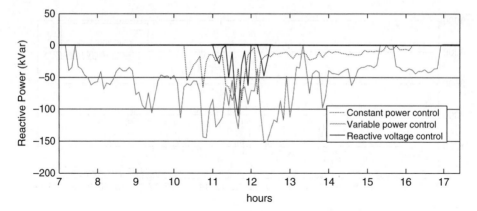

Figure 6.30 Comparison of reactive power absorbed by the PVs of C2 on a working day under the three control strategies.

impact on the reactive power on the high-voltage side of the main transformers, and the reactive power absorbed by the distributed PVs contributes about 1 MVar during the first peak load time. In general, the reactive voltage control strategy has the smallest impact on the reactive power on the high-voltage side of the main transformers, which is mainly manifested during the valley load time at noon. As during the valley load time at noon the PV output is relatively large, this causes reverse power flow at some PV nodes. This results in high voltage at the PCCs, some even exceeding the voltage control reference value. Meanwhile, the distributed PVs will start reactive power regulation, and the reactive power on the high-voltage side of the main transformers is relatively small. As the situation during the holiday scenario is similar, the analysis will not be repeated.

C2 is once again taken as an example for analysis mainly because this user integrates the PVs at the 0.4 kV level, which can make the voltage rise problem at the PCC more severe. It can be seen from Figure 6.30 that, compared with the variable power factor control strategy and the constant power factor control strategy, the reactive power absorbed by the PVs of user C2 and the reactive power regulation periods are significantly reduced on the whole, showing a better regulation performance in the reactive voltage control strategy.

Figure 6.31 shows that under the constant power factor control strategy and the reactive voltage control strategy, the voltage of the C2 at the PCC is within the set control target and meets the requirements stipulated by Chinese national standards. However, in the constant power factor control strategy, the voltage of C2 at the PCC is lower and the voltage steady-state safety margin is larger, because the PV inverter absorbs more reactive power. In the variable power factor control strategy, the voltage of C2 at the PCC is relatively high during periods of time when the PV output is high, with a maximum of 1.065 p.u., since the reactive power absorbed by the PV inverter is relatively small. Although it can meet the requirements for safe operation, the voltage steady-state safety margin at the PCC is relatively small.

Figure 6.32 shows the comparison of the reactive power absorbed by all the PVs in the regional distribution network of the SA substation under the three different control strategies. It can be seen from Figure 6.32 that the reactive power absorbed by the distributed PVs under the reactive voltage control strategy is the smallest. Compared

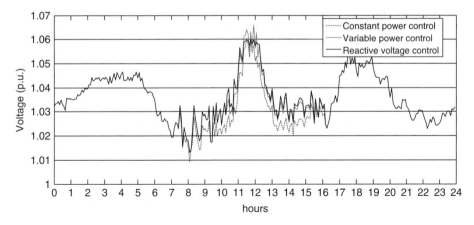

Figure 6.31 Voltage comparison of C2 at the PCC on a working day under the three control strategies.

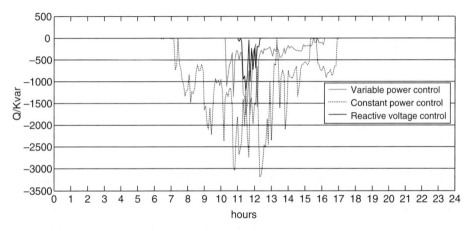

Figure 6.32 Comparison of reactive power absorbed by all PVs of substation SA on a working day under the three control strategies.

to the variable power factor control strategy, only during very few time periods is the reactive power absorbed by the distributed PVs slightly larger under the reactive voltage control strategy.

Table 6.6 shows the number of reactive power regulation periods of time and the number of reactive power regulation users during the whole day on July 16, 2014, with the three different control strategies. As the simulation time step is 5 min, the whole day is divided into a total of 288 periods of time. The number of reactive power regulation periods is the sum of all periods of time when the PVs participate in reactive power regulation; the number of reactive power regulation users is the sum of the PV users that participate in reactive power regulation during the whole day. These two indicators can be obtained through statistical analysis of the simulation results. It can be seen from Table 6.6 that both the number of reactive power regulation periods of time and the number of reactive power regulation users are significantly reduced in the reactive voltage control strategy.

Table 6.6 Number of reactive power regulation periods and number of reactive power regulation users on a working day under the three control strategies.

	Constant power factor	Variable power factor	Reactive voltage control
Number of reactive power regulation periods	129	78	20
Number of reactive power regulation users	11	11	6

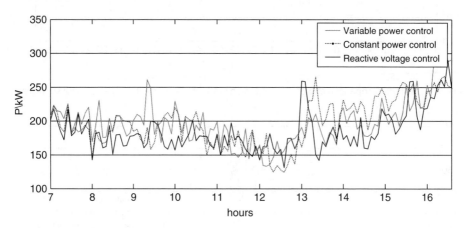

Figure 6.33 Comparison of power losses of substation SA on a working day under the three control strategies.

Figure 6.33 shows the comparison of real power losses of the SA substation under the three control strategies. It can be seen that under the reactive voltage control strategy the power losses in the SA substation are relatively lower. But the power losses during some periods of time are high because, under the reactive voltage control strategy, more reactive power needs to be absorbed during some periods of time in order to meet the preset voltage control target.

6.6.3.2 Holiday (May 1, 2014) Scenario

The action times of the capacitor banks in the whole day with the three control strategies are shown in Table 6.7. It can be seen that, compared with the constant power factor control strategy and the variable power factor control strategy, the number of the actions of capacitor bank #1 is significantly reduced in the reactive voltage control strategy. Meanwhile, capacitor bank #1 acts eight times under the constant power factor control strategy and four times under the variable power factor control strategy. Large operation times can increase the maintenance and operation cost and reduce the durability of the capacitor bank.

In order to further illustrate action times of capacitor banks in the constant power factor and variable power factor control strategies, the concept of PV output fluctuation will be introduced. To be specific, taking the PV output (in the form of per unit value) during each period of time (every 5 min) in the whole day as the reference value, compare the PV output fluctuation pattern during two continuous periods of time and define

Table 6.7 Action times of capacitor banks during the whole day (holiday scenario) under the three control strategies.

No. of main transformer	No. of capacitor bank (rated capacity)	Constant power factor	Variable power factor	Reactive voltage control
1	1 (1.8 MVar)	4	8	0
	2 (3.0 MVar)	0	0	0
2	3 (3.0 MVar)	3	3	3
	4 (1.8 MVar)	3	3	3

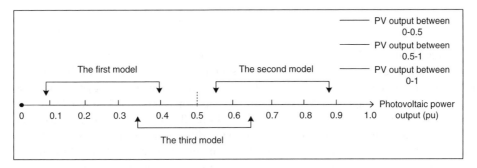

Figure 6.34 The three modes of PV output fluctuation.

three modes, which are shown in Figure 6.34. In order to facilitate analysis, the 0.5 p.u. PV output is taken as the standard for the classifying the modes. In mode 1, the PV outputs during two continuous periods of time are both less than 0.5 p.u.; in mode 2, the PV outputs during two continuous periods of time are both greater than 0.5 p.u.; in mode 3, the PV output is less than 0.5 p.u. in one of the periods and greater than 0.5 p.u. in the other. In the three modes the PV output can be increased or reduced.

Now let us go back and take C2 again as an example for analysis. Figure 6.35 shows the PV output of C2. It can be seen from it that in mode 3 the PV output (per unit value) of C2 fluctuates around 0.50 p.u. At 12:25 p.m. the PV output is about 0.70 p.u., while the PV output is suddenly reduced to 0.30 p.u. at 12:30 p.m., with the PV output variable being 0.4 MW.

According to the variable power factor control strategy, the variation of the reactive power absorbed by the PVs of C2 is about 140 kVar. However, in the constant power factor control strategy, the variation of the reactive power absorbed by the PVs of C2 is about 80 kVar. It can be seen from this that, as in mode 3, under the variable power factor control strategy, the variation of the reactive power of the PVs is larger, the reactive power fluctuation on the high-voltage side of the main transformers becomes larger, and the capacitor banks fluctuate more frequently. Moreover, further analysis showed that capacitor bank #1 is put into operation at 12:25 p.m., while the sudden reduction of the reactive power absorbed by the PV inverter reduces the reactive power transmitted on the high-voltage side of the main transformers at 12:30 p.m.; based on the principle of local balance of reactive power in the AVC system control strategy, capacitor bank

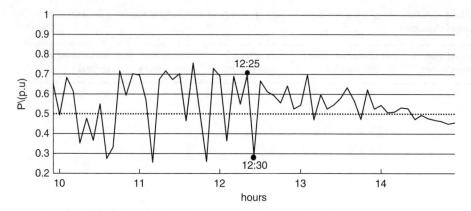

Figure 6.35 PV output fluctuation of C2.

#1 switched off. For the case in which the PV output increases from less than 0.5 p.u. to greater than 0.5 p.u., the capacitor bank is switched on. Compared with the constant power factor control strategy, the variable power factor control strategy increases the action times of the capacitor banks. It can be seen from Figures 6.34 and 6.35 that in practice there are two additional modes of PV output fluctuation. In the first mode the PV output fluctuates in the interval below 0.5 p.u., while in the second mode the PV output fluctuates in the interval above 0.5 p.u. In the first mode, compared with the constant power factor control strategy, under the variable power factor control strategy the variation of the reactive power of the PVs is obviously small (zero, according to the preset controlling strategy of the PV inverters), making a slight impact on the reactive power fluctuation on the high-voltage side of the main transformers. In the second mode, the variation of the reactive power of PV output is greater. Under the reactive voltage control strategy, the variation of the reactive power absorbed by the PVs of C2 is about 30 kVar, making a small impact on the reactive power of the distribution network. As a result, it does not lead to the switching of the capacitor banks. Therefore, compared with the constant power factor control strategy and the variable power factor control

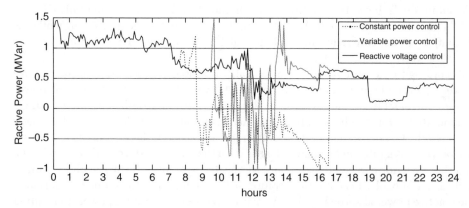

Figure 6.36 Reactive power on high-voltage side of no. 1 main transformer on holiday under the three control strategies.

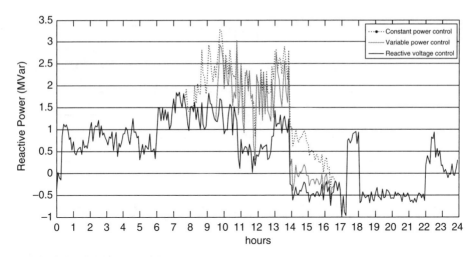

Figure 6.37 Reactive power on high-voltage side of no. 2 main transformer on a holiday under the three control strategies.

strategy, the number of action times of the capacitor banks is the smallest under the reactive voltage control strategy.

Figures 6.36 and 6.37 show that, compared with the constant power factor control strategy and the variable power factor control strategy, the reactive voltage control strategy has significantly reduced amplitude and frequency of reactive power fluctuations in the high voltage side of the main transformers. In particular, no reverse power flow occurs to no. 1 main transformer; reverse power flow occurs to no. 2 main transformer, but the reverse power flow is relatively small and within the allowable operation range. However, under the constant power factor control strategy and the variable power factor control strategy, the reserve power flow is accompanied by certain fluctuations.

Figure 6.38 shows that, compared with the constant power factor and power factor control strategies, the reactive power absorbed by the PV of C2 is significantly reduced in the reactive voltage control strategy. The reactive regulation periods of time are also greatly reduced, as is the power transmitted on the high-voltage side of the main transformers. Hence, the impact on the reactive power of the distribution network and the operation (action times) of capacitor banks after the integration of the PVs is relatively small under the reactive voltage control.

Figure 6.39 shows that, under the variable power factor control strategy, the voltage at the PCC is relatively high during some periods of time when the PV output is large, exceeding 1.060 p.u., because of the small reactive power absorbed by the PV inverter. Although it can meet the requirements for safe operation, the voltage steady-state safety margin at the PCC is small. Under the constant power factor and reactive voltage control strategies, the voltage at the PCC of C2 is within the control target range. However, during some periods of time, the voltage at the PCC of C2 is lower under the constant power factor control strategy because the PV absorbs more reactive power.

Figure 6.40 shows the comparison of reactive power absorbed by all the PVs in the regional distribution network of the SA substation with different control strategies. It can be seen from the figure that, under the reactive power voltage control strategy, the

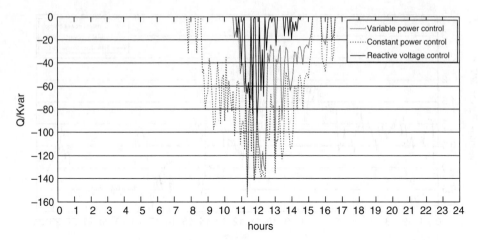

Figure 6.38 Reactive power absorbed by the PVs of C2 on a holiday under the three control strategies.

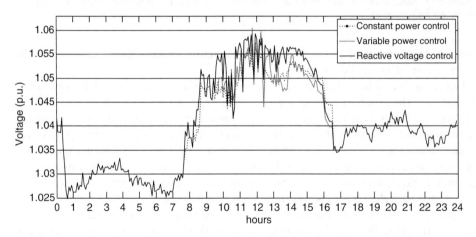

Figure 6.39 Voltage profiles at the PCC of C2 during a holiday under the three control strategies.

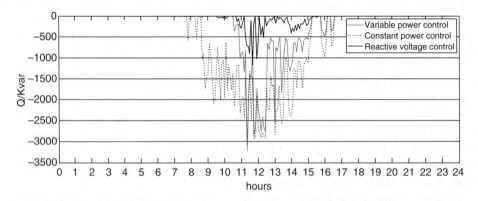

Figure 6.40 Reactive power absorbed by all the PVs in substation SA for the holiday scenario under the three control strategies.

Table 6.8 Number of PV reactive regulation periods and number of reactive regulation users for the holiday scenario under the three control strategies.

Control strategy	Constant power factor	Variable power factor	Reactive voltage control
Number of reactive regulation users	12	12	3
Number of reactive regulation periods	128	92	44

reactive power absorbed by the PVs is the smallest. The reactive voltage control strategy rarely absorbs more reactive power than the variable power factor control strategy since it carries out reactive regulation based on the preset value of the voltage at the PCC (1.060 p.u.). On the other hand, under the variable power factor control strategy the reactive regulation is carried out only based on the preset power factor (or curve). Therefore, when the tap positions of the main transformers and the operation of the capacitor banks are the same in the two control strategies, the voltage at the PCC in the reactive voltage control strategy is higher than that in the variable power factor control strategy, but within a reasonable control range.

Table 6.8 shows the number of reactive regulation periods of all the PVs and the number of reactive regulation users on May 1, 2014, under the three control strategies. It can be noted that, compared with the variable power factor control strategy and the constant power factor control strategy, both the number of reactive regulation periods and the number of reactive regulation users are significantly reduced, under the reactive voltage control strategy; thus, this strategy generally shows better regulation performance.

Figure 6.41 shows the comparison of the power losses of SA substation network for the three control strategies. It can be seen from the figure that for the reactive voltage control strategy the power losses in the SA substation are relatively low and only during some periods are the power losses higher than those under the variable power factor control strategy. This is because under the reactive voltage control strategy, when the PVs output is less than 0.50 p.u., the PV inverter will not carry out reactive regulation,

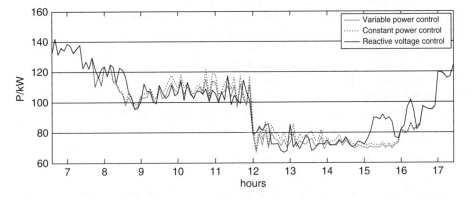

Figure 6.41 Comparison of power losses in substation SA for the holiday scenario under the three control strategies.

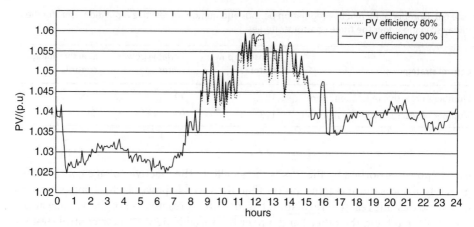

Figure 6.42 Voltage at the PCC of C2 on May 1, 2014, with different efficiencies of PVs under the reactive voltage control strategy.

reducing the reactive power transmitted from the main transformers to the PV side and reducing power losses accordingly.

The higher the efficiency of the PV inverter, the larger the active power output by the PV system under the same conditions, and thus the higher the voltage at the PCC and the greater impact on the distribution network. Figure 6.42 further shows the voltage at the PCC of C2 on May 1, 2014, in the case of different efficiencies of the PV inverter, which fully verifies the effectiveness of the PV control strategy. Under the reactive voltage control strategy, all the voltage at the PCC of this user is within the preset control target range, meeting the requirements stipulated in Chinese national standards.

The following observations can be made based on the previous analysis.

1) Under the constant power factor and variable power factor control strategies, the action times of the capacitor banks in the substation are significantly larger than those of the reactive voltage control strategies.
2) After the integration of the distributed PVs, the fluctuation of the reactive power on the high-voltage side of main transformers in the substation becomes more frequent; particularly under the constant power factor and variable power factor control strategies, during some periods of time the reactive power transmitted at the high-voltage side of main transformers exceeds the upper limit of the allowable operation range, which might affect the safe operation of the distribution network.
3) Compared with the constant power factor and variable power factor control strategies, the reactive voltage control strategy has a smaller number of periods of PV inverter reactive regulation and a smaller number of PV users of reactive regulation. Each PV inverter carries out reactive regulation based on the voltage at the PCC and the preset value. The direct control of local voltage can be achieved to meet the voltage requirements for safe operation.
4) Compared with the constant power factor and variable power factor control strategies, the reactive voltage control strategy can achieve a lower power loss because the reactive power absorbed by the distributed PVs is smaller and the reactive

power transmitted from the main transformers to the PV side is smaller as well. The reactive voltage control strategy has a small impact on the operation of the existing AVC system. Moreover, it achieves voltage control directly through the distributed PVs and it can keep the voltage at the PCC within the preset control target range.

6.7 Summary

This chapter has studied the voltage control at the substation level with high-penetration distributed PVs integrated. Firstly, the impacts on voltage due to the distributed PVs were analyzed and the three control strategies (the constant power factor control strategy, variable power factor control strategy, and the reactive voltage control strategy) and relevant modeling methods in which PV inverters participate in voltage regulation were introduced. In addition, the annual operation data from the SA substation power supply region in 2014 was taken as an example for analysis and discussion. After the detailed discussion of the extreme scenario selection criteria, the continuous system control simulation was carried out in every 5 min throughout the day on a working day (July 16, 2014) and during a holiday (May 1, 2014) based on the existing AVC system under the three PV inverter control strategies.

Considering the principle of local reactive power balancing, it has been suggested that the substation should be equipped with continuous dynamic reactive power compensation devices with a certain capacity to dynamically track the reactive power absorbed by the distributed PVs and the varying loads. This can minimize and stabilize the reactive power transmitted on the high-voltage side of the main transformers. The specific capacity of the reactive power compensative devices should be determined based on the AVC constraints, the operation constraints of the upper-level distribution networks, and the installed PV capacity.

References

1 GB/T 12325-2008. (2008) *Power quality – Deviation of Supply Voltage.* Standards Press of China.

2 ANSI C84.1-2016. (2006) *Electric power systems and equipment – voltage ratings (60 Hz).* American National Standards Institute, New York.

3 BDEW. (2008) *Technical Condition for the Connection to the Medium-Voltage Network.* German Association of Energy and Water Industries e.V.: Berlin

4 Stetz T., Yan W., and Braun M. (2010) Voltage controlling distribution systems with high level PV-penetration – improving absorption capacity for PV systems by reactive power supply. In *25th European Photovoltaic Solar Energy Conference and Exhibition/5th World Conference on Photovoltaic Energy Conversion,* September, 6–10, Valencia, Spain.

5 Demirok E., Sera D., Rodriguez P., and Teodorescu R. (2011) Enhanced local grid voltage support method for high penetration of distributed generators. In *IECON 2011 – 37th Annual Conference on IEEE Industrial Electronics Society.* IEEE; pp. 2481–2485.

6 Agalgaonkar Y.P., Pal B.C., and Jabr R.A. (2014) Distribution voltage control considering the impact of PV generation on tap changes and autonomous regulators. *IEEE Transactions on Power Systems*, **29**(1), 182–192.

7 Fan Y., Zhao B., Jiang Q., and Cao Y. (2012) Peak capacity calculation of distributed photovoltaic source with constraint of over-voltage. *Automation of Electric Power Systems*, **17**, 40–44.

8 Li Q. and Zhang J. (2015) Solutions of voltage beyond limits in distribution network with distributed photovoltaic generators. *Automation of Electric Power Systems*, **39**(22), 117–123.

9 Wang Y., Wen F., Zhao B., and Zhang X. (2016) Analysis and countermeasures of voltage violation problems by high-density distributed photovoltaics. *Proceedings of the Chinese Society of Electrical Engineering*, **5**, 1200–1206.

10 Tonkoski R., Lopes L.A.C., and El-Fouly T.H.M. (2011) Coordinated active power curtailment of grid connected PV inverters for overvoltage prevention. *IEEE Transactions on Sustainable Energy*, **2**(2), 139–147.

11 Demirok E., Casado González P., Frederiksen K.H.B., *et al.* (2011) Local reactive power control methods for overvoltage prevention of distributed solar inverters in low-voltage grids. *IEEE Journal of Photovoltaics*, **1**(2), 174–182.

12 Caldon R., Coppo M., and Turri R. (2013) Coordinated voltage control in MV and LV distribution networks with inverter-interfaced users. In *2013 IEEE Grenoble Conference*. IEEE; 1–5.

13 Degner T., Arnold G., Reimann T., *et al.* (2011) Increasing the photovoltaic-system hosting capacity of low voltage distribution networks. In *Proceedings of 21st International Conference on Electricity Distribution*; paper 1243.

14 Dall'Anese E., Dhople S.V., and Giannakis G.B. (2014) Optimal dispatch of photovoltaic inverters in residential distribution systems. *IEEE Transactions on Sustainable Energy*, **5**(2), 487–497.

15 Demirok E., Sera D., Teodorescu R., *et al.* (2009) Clustered PV inverters in LV networks: an overview of impacts and comparison of voltage control strategies. In *2009 IEEE Electrical Power & Energy Conference (EPEC)*. IEEE; pp. 1–6.

16 Xu Z., Zhao B., Ding M., *et al.* (2016) Photovoltaic hosting capacity of distribution networks and inverter parameters optimization based on node voltage sensitivity. *Proceedings of the Chinese Society of Electrical Engineering*, **6**, 1578–1587.

17 Zhou J., Lin D., and Wang Z. (2014) *The Study on Photovoltaic Hosting Control Measures for Jiaxing Photovoltaic Industrial Park*. State Grid Zhejiang Electric Power Research Institute.

7

Short-Circuit Current Analysis of Grid-Connected Distributed Photovoltaic Generation

7.1 Introduction

The growing scale and increasing penetration level of distributed PVs has a significant impact on the stability and security of distribution networks. Particularly when failure occurs in the distribution network, large numbers of distributed PV systems could be islanded, which will have serious adverse effects on the stable operation of the distribution network. As a result, it is very important to analyze PV system failures, the failure mechanisms and features, and to explore the methods for calculation and analysis of the short circuit in the distribution network with high PV penetration. The normal electrical and fault characteristics of distributed PV system failures are quite different from those of rotating power (mainly from synchronous generators) [1]. When integrated with medium- or large-scale distributed PV generation that has a low-voltage ride-through (LVRT) capability, the distribution networks become even more complex in fault analysis.

7.2 Short-Circuit Characteristic Analysis of Distributed Photovoltaic Generation

Power grid faults can be classified into symmetric faults and asymmetric faults. The asymmetric faults can be further classified into single-phase earth (i.e., single-line-to-ground (SLG) faults, two-phase short-circuit faults (i.e., line to line), and two-phase earth faults (i.e., double-line-to-ground (DLG)), while symmetric faults include three-phase short-circuit faults.

According to the National Energy Administration standard NB/T 33010-2014 [2], during the operation of a grid-connected distributed generator, when the grid voltage is too high or too low, it must respond to it. When the voltage at the grid connection point exceeds the voltage limits specified in Table 7.1, the power supply to the power grid should be cut off within a specified time.

Shown in Table 7.2, the State Grid Corporation of China corporate standard Q/GDW480-2010 contains the requirements [3] for the overcurrent capacity of grid-connected inverters for distributed PV generation.

According to the overvoltage (undervoltage) and overcurrent protection standards for distributed PV generation shown in Tables 7.1 and 7.2, for those without LVRT capability, in the case of a power grid fault the overcurrent protection and undervoltage

Grid-Integrated and Standalone Photovoltaic Distributed Generation Systems: Analysis, Design, and Control,
First Edition. Bo Zhao, Caisheng Wang and Xuesong Zhang.
© 2018 China Electric Power Press. Published 2018 by John Wiley & Sons Singapore Pte. Ltd.

Table 7.1 Voltage response requirements of distributed power supply.

$U < 50\%U_N$	The maximum opening time should be no more than 0.2 s
$50\%U_N \leq U < 85\%U_N$	The maximum opening time should be no more than 2 s
$85\%U_N \leq U < 110\%U_N$	Normal operation
$110\%U_N \leq U < 135\%U_N$	The maximum opening time should be no more than 2 s
$135\%U_N \leq U$	The maximum opening time should be no more than 0.2 s

Table 7.2 Overcurrent protection standard for distributed PV supply.

Actual output current of inverter	Requirements
$110\%I_N < I < 120\%I_N$	Continuous reliable working time should be no less than 1 min
$120\%I_N \leq I < 150\%I_N$	Continuous reliable working time should be no less than 10 s
$150\%I_N \leq I$	Immediate overcurrent protection to stop the operation

protection of the grid-connected PV inverter should be trigged within an appropriate time to respond to the fault, and the operation of the PV inverter should be stopped.

7.2.1 Short-Circuit Characteristic Analysis of Symmetric Voltage Sag of Power Grid

In the case of a voltage sag of the power grid [4], assuming that the DC-side input power is approximately constant during fault, the output power of the PVs would be reduced. The control strategy for the grid-connected PV inverter is shown in Figure 7.1.

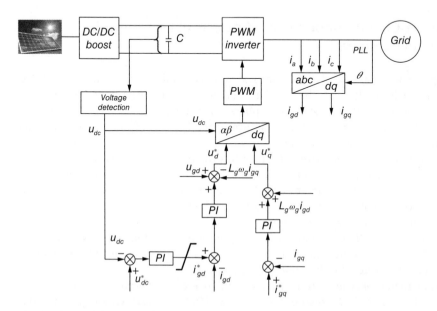

Figure 7.1 Control strategy of a PV inverter.

Figure 7.2 Equivalent circuit diagram of an inverter for short-circuit calculation.

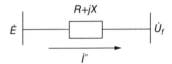

Owing to the action of the voltage outer loop PI regulator, the d-axis reference current of the grid-connected PV inverter increases; under the action of the current inner loop PI regulator, the output current increases. As a result, in the case of voltage sag, the protection will be triggered according to the actual terminal voltage and current, and the PV inverter will be opened within an appropriate time.

At the initial instant of the fault, the impulse current caused by the inverter can be calculated using the generator transient short-circuit current, of which the principle is to define the virtual internal potential of the grid-connected PV inverter $\dot{E}$, which has the typical time-varying characteristics, but can be assumed to be constant at the fault instant.

The equivalent circuit diagram of the inverter is shown in Figure 7.2 for short-circuit current calculation. $\dot{U}_f$ is the grid voltage at the PCC after the short-circuit occurs; X is the sum of the filter reactance and the reactance between the inverter and the fault point, R is the equivalent resistance of the filter between the inverter and the fault point. Therefore, the following equation can be used to calculate the initial value of the short-circuit current I'' (i.e., the subtransient fault current):

$$\dot{I}'' = \frac{\dot{E} - \dot{U}_f}{R + jX} \tag{7.1}$$

The initial value of the short-circuit current can be used to verify the fault tolerance of the equipment and provide the basis for planning and design. It should be noted that the subtransient short-circuit current of the PV inverter may be larger than the overcurrent protection value after the grid fault, but the output current of a PV generator can be restrained within a desirable limit by the controller within one control period.

7.2.2 Short-Circuit Characteristic Analysis of Asymmetrical Voltage Sag of Power Grid

When an asymmetric fault of the power grid occurs, the short-circuit current of a PV inverter can have the following features.

1) Under a stationary three-phase reference frame, the output short-circuit current of the PV inverter contains positive- and negative-sequence components. The positive-sequence current and negative-sequence current are superposed to obtain the output short-circuit current of the PV inverter. The grid-connected PV inverter activates the protection and opens the circuit within an appropriate time based on the actual terminal voltage and current of the grid-connected inverter.

The principle of generating negative-sequence currents is as follows. In the case of an asymmetric fault of the power grid, the voltage of the grid-connected PV inverter on the AC side only contains positive-sequence components, while the voltage on the grid side contains both positive- and negative-sequence components.

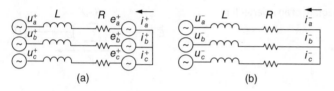

Figure 7.3 (a) Positive-sequence and (b) negative-sequence equivalent circuits of PV inverter at AC side.

The equivalent circuits of positive sequence and negative sequence are shown in Figure 7.3. The positive-sequence components of the grid voltages u_a^+, u_b^+, and u_c^+ interact with the positive-sequence components of the voltage of the grid-connected inverter on the AC side to generate the positive-sequence currents i_a^+, i_b^+, and i_c^+, while in the negative-sequence equivalent circuit, shown in Figure 7.3b, the values of the negative-sequence currents i_a^-, i_b^-, and i_c^- have a relationship with the inductance L, the resistance R of the inverter, and the negative-sequence voltages u_a^-, u_b^-, and u_c^-.

2) The positive-sequence current and negative-sequence current can be expressed in two different rotating reference frames. As shown in Equation 7.2, the electric quantity F_{gdq} in a stationary two-phase reference frame can be expressed as a function of the positive-sequence component F_{gdq+}^+ in the positive rotating reference frame and the negative-sequence component F_{gdq-}^- in the negative (reverse) rotating reference frame:

$$F_{gdq} = F_{gdq+}^+ e^{j\omega t} + F_{gdq-}^- e^{-j\omega t} \tag{7.2}$$

where the vector F represents either the grid voltages or currents. The superscripts $+$ and $-$ represent the positive- and negative-sequence component respectively, while the subscripts $+$ and $-$ represent the positive- and reverse-direction in the rotating coordinate system, respectively.

3) The output active power and reactive power of the PV inverter have double-frequency fluctuations. Transformed into the positive rotating reference frame, the physical quantity can be expressed as

$$F_{gdq+} = F_{gdq+}^+ + F_{gdq-}^- e^{-j2\omega t} \tag{7.3}$$

According to Equation 7.3, in the positive-direction synchronous rotating coordinate system, both the voltages and the currents have positive-sequence DC components and negative-sequence double-frequency AC components.

The grid voltages and the output currents of the PV inverter are transformed from the three-phase *abc* stationary coordinate system to the two-phase *dq* rotary coordinate system to obtain the *d*-axis component e_d and *q*-axis component e_q of the grid voltages, as well as the *d*-axis component i_d and the *q*-axis component i_q of the PV inverter output currents. According to Equation 7.3, when an asymmetric fault occurs, all the four components contain the DC component and the double-frequency AC component. As the PI controller cannot realize the zero static error track for the AC component, the output voltage references u_d and u_q of the inverter calculated by the PI controller also contains the double-frequency ac component. When the output voltages of the inverter are transformed to the three-phase *abc* stationary coordinate system, they contain the third-order harmonic components, as do the output currents of the inverter.

In the case of asymmetric fault of the grid voltage, the mathematical model of the three-phase grid-connected PV inverter under the positive direction and reverse direction synchronous rotating coordinate system can be expressed as

$$
\begin{cases}
E^+_{gdq+} = U^+_{gdq+} + L\dfrac{dI^+_{gdq+}}{dt} + j\omega L I^+_{gdq+} \\[2mm]
E^-_{gdq-} = U^-_{gdq-} + L\dfrac{dI^-_{gdq-}}{dt} - j\omega L I^-_{gdq-}
\end{cases}
\tag{7.4}
$$

where E and I represent the inverter output voltage and current, respectively. L is the inductance of the PV inverter, and U is the grid voltage.

The output active power p_g and reactive power q_g of the grid-connected PV inverter can be obtained as follows:

$$
\begin{cases}
p_g = 1.5\,\mathrm{Re}(S) = 1.5\,\mathrm{Re}(U_{gdq+}\hat{I}_{gdq+}) \\[2mm]
q_g = 1.5\,\mathrm{Im}(S) = 1.5\,\mathrm{Im}(U_{gdq+}\hat{I}_{gdq+})
\end{cases}
\tag{7.5}
$$

Substitute Equation 7.3 into Equation 7.5 to obtain the power model in case of the asymmetric fault:

$$
\begin{cases}
p_g = p_{g0} + p_{g\cos 2}\cos(2\omega t) + p_{g\sin 2}\sin(2\omega t) \\[2mm]
q_g = q_{g0} + q_{g\cos 2}\cos(2\omega t) + q_{g\sin 2}\sin(2\omega t)
\end{cases}
\tag{7.6}
$$

where

$$
\begin{cases}
p_{g0} = 1.5(u^+_{gd+}i^+_{gd+} + u^+_{gq+}i^+_{gq+} + u^-_{gd-}i^-_{gd-} + u^-_{gq-}i^-_{gq-}) \\[2mm]
p_{g\cos 2} = 1.5(u^+_{gd+}i^-_{gd-} + u^+_{gq+}i^-_{gq-} + u^-_{gd-}i^+_{gd+} + u^-_{gq-}i^+_{gq+}) \\[2mm]
p_{g\sin 2} = 1.5(u^+_{gd+}i^-_{gq-} - u^+_{gq+}i^-_{gd-} - u^-_{gd-}i^+_{gq+} + u^-_{gq-}i^+_{gd+})
\end{cases}
\tag{7.7}
$$

$$
\begin{cases}
q_{g0} = 1.5(u^+_{gq+}i^+_{gd+} - u^+_{gd+}i^+_{gq+} + u^-_{gq-}i^-_{gd-} - u^-_{gd-}i^-_{gq-}) \\[2mm]
q_{g\cos 2} = 1.5(u^-_{gq-}i^+_{gd+} - u^-_{gd-}i^+_{gq+} + u^+_{gq+}i^-_{gd-} - u^+_{gd+}i^-_{gq-}) \\[2mm]
q_{g\sin 2} = 1.5(u^+_{gd+}i^-_{gd-} + u^+_{gq+}i^-_{gq-} - u^-_{gd-}i^+_{gd+} - u^-_{gq-}i^+_{gq+})
\end{cases}
\tag{7.8}
$$

where p_{g0} and q_{g0} are average values of the instantaneous active power and reactive power respectively. $p_{g\cos 2}$ and $p_{g\sin 2}$ are the double-frequency fluctuation components of the instantaneous active power, while $q_{g\cos 2}$ and $q_{g\sin 2}$ are the double-frequency fluctuation components of the instantaneous reactive power. u and i are the grid voltages and currents, respectively.

7.3 Low-Voltage Ride-Through Techniques of Photovoltaic Generation

The tripping of a large-scale PV generation system from the network could lead to significant changes in the system power flow and even a massive power outage [5, 6]. At present, large and medium-size PV systems are required to have certain LVRT capabilities. As a core component in PV generation, the grid-connected PV inverter can realize the LVRT function.

The current LVRT standards are mainly for large- and medium-size PV systems of 1 MW or above. As the penetration level of PV generation increases, the impact of distributed PVs on the power system grows; hence, higher standards of LVRT capabilities of PV generation are required.

7.3.1 Review of Low-Voltage Ride-Through Standards

The PV integration standards made by many countries specify how distributed PVs should respond to abnormal grid voltage levels. The LVRT standards for distributed PVs by some countries are shown in Figure 7.4 [7, 8].

As shown in the figure, the LVRT standards are different from one country to another, with small changes in voltage drop magnitude, fault time, voltage recovery time, and final voltage magnitude.

To follow the standards, the reactive power provision is required to be adequate for voltage control [7]. The requirements for the reactive power capacity of the inverter stipulated in the Chinese standard will be introduced as an example. As shown in Figure 7.5, the horizontal axis is the amplitude of the voltage sag, which is the ratio of the voltage sag variation to the rated voltage. The vertical axis is the ratio of the reactive current to the total current in the case of different amplitudes of the voltage sag. According to the curve shown in Figure 7.5, the reactive power compensation current is determined by the degree of the voltage sag. When the voltage sag is below 10%, there is no need to provide the reactive current. When the voltage sag is higher than 10%, but lower than 80%, the PV inverter is required to increase the reactive current proportionally to the degree of voltage sags. When the voltage sag is higher than 80%, the PV inverter is required to provide a reactive current that is 105% of the rated current.

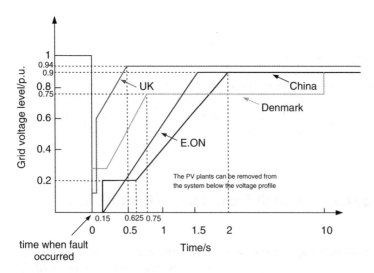

Figure 7.4 LVRT standards for PV generation made in different countries.

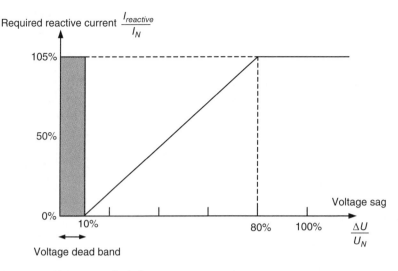

Figure 7.5 Chinese standards for reactive power support.

7.3.2 Low-Voltage Ride-Through Control Strategy for Photovoltaic Generation

The LVRT control strategy for a grid-connected PV inverter is the key to the realization of the LVRT function [9]. We will first briefly introduce a general LVRT control strategy for PV generation and then focus on the analysis of the LVRT control strategy used in this section. There are mainly four ways to achieve the LVRT function:

1) Limit the reference current of the grid-connected PV inverter. According to the control principle of the PV inverter shown in Figure 7.1, to achieve overcurrent protection of the grid-connected PV inverter it is necessary to limit the reference value of the d-axis current.

2) Add dump resistance at the DC side of the grid-connected PV inverter. In the case of voltage sag of the power grid, the input power of the grid-connected PV inverter is greater than the output power, resulting in a rise of the capacitor voltage. The dump resistance at the DC side is used to consume the excess energy in order to ensure the capacitor voltage is within the allowable range. The structure of the energy dump circuit is shown in Figure 7.6.

3) Reduce the input power of the PV cell. Adjust the operating point of the PV cell based on the voltage sag of the power grid to reduce the input power of the PV cell.

4) Change the control strategy for the grid-connected PV inverter. Change the reference value of the active and reactive currents of the inverter based on the voltage sag of the grid, in order to provide active and reactive power support for the grid voltage.

The control strategies (1) and (2) are to be analyzed. When the grid voltage is unbalanced, positive-sequence components and negative-sequence components exist

Figure 7.6 Structure of energy dump circuit on the DC side.

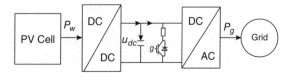

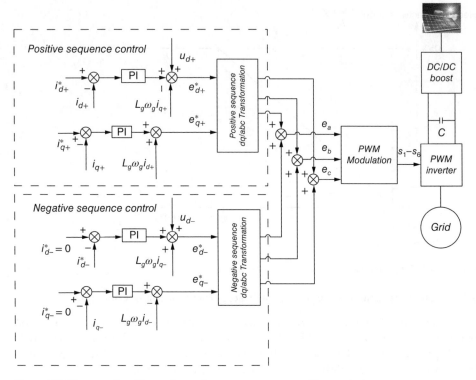

Figure 7.7 Diagram of positive and negative-sequence superposition control.

in the voltage. The positive- and negative-sequence decomposition control, as shown in Figure 7.7, can be used to eliminate the impact of negative-sequence currents [10]. In normal operation, the commands of the positive-sequence d-axis and q-axis currents are obtained from the outer voltage loop and the outer power loop, respectively, and the positive-sequence voltage command is obtained through a PI controller. When the commands of the negative-sequence d-axis and q-axis currents are zero, the negative-sequence voltage command is obtained through a PI controller and is superposed with the positive-sequence voltage command and compared with the carrier signal to obtain corresponding switching signals. The positive- and negative-sequence control method realizes the control of the negative-sequence component output to ensure that the output currents of the inverter only contain the three-phase symmetric positive-sequence currents.

Based on the control method mentioned earlier, the flow chart of LVRT control strategy for the grid-connected PV inverter is shown in Figure 7.8 [7]:

1) When it is detected that the positive-sequence voltage at the grid connection point drops below 90%, the commands for positive-sequence active and reactive current, i^*_{d+} and i^*_{q+}, are given directly.
2) In order to meet the requirements for the integration of the distributed PV generation into the medium-voltage distribution network, during the grid fault period,

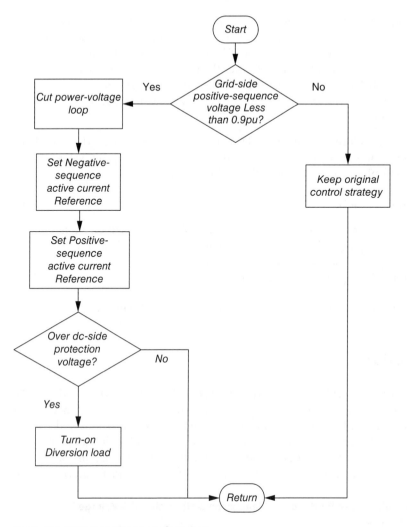

Figure 7.8 LVRT control strategy flow chart.

self-adaptive regulation is adopted on the positive-sequence reactive current command i_{q+}^*, which is given by

$$i_{q+}^* = \begin{cases} 0, & \alpha > 0.9 \\ 2(1-\alpha), & 0.4 \leq \alpha \leq 0.9 \\ 1.2, & \alpha < 0.4 \end{cases} \qquad (7.9)$$

All the values in Equation 7.9 are per unit values, and α is the amplitude of the positive-sequence voltage on the power grid with voltage sags.

3) In order to provide the maximum allowable active power without causing overcurrent of the inverter, the active current command is i_{d+}^*, which is the smaller value between i_{d0+}^* and $(i_{d\,max}^2 - i_{q+}^{*\,2})^{0.5}$, and i_{d0+}^* is the active current command of the

current inner loop of the grid-connected PV generation prior to the fault. It is worth noting that the maximum allowable positive-sequence *d*-axis current can be 1.5 times larger than the rated current of the PV inverter in this control method.

4) When the output power of the inverter after the fault occurs is less than that before the fault, the unbalanced power raises the voltage on the DC side of the PV inverter. When the voltage on the DC side exceeds a certain value, the dump circuit on the DC side kicks in.

According to the aforementioned LVRT control strategies, when the grid fault occurs, the following equation can be used to calculate the peak value of the steady-state three-phase fault current of the grid-connected PV inverter:

$$I_{abc} = \sqrt{\left(i_{d+}^*\right)^2 + \left(i_{q+}^*\right)^2} \tag{7.10}$$

The positive-sequence *q*-axis current reference is obtained based on Equation 7.9, while the positive-sequence *d*-axis current reference is obtained based on

$$i_{d+}^* = \min\left\{ i_{d0+}^*, \sqrt{\left(i_{d\,max}\right)^2 - \left[2\left(1-\alpha\right)i_{d0+}^*\right]^2} \right\} \tag{7.11}$$

where $i_{d\,max}$ is the maximum allowable positive-sequence *d*-axis current.

7.4 Simulation Studies

In the case of a power grid fault, the grid-connected PV inverter without the LVRT function kicks out, while the grid-connected PV inverter with the LVRT function can continue to keep the grid-connected operation based on the grid-connected voltage and provides certain reactive power support. In this section the different grid fault types will be investigated by carrying out electromagnetic transient simulations of a single grid-connected PV inverter with and without the LVRT function under fault conditions.

7.4.1 Fault Simulations of Photovoltaic Generation without the Low-Voltage Ride-Through Function

The topological structure diagram of simulation cases is shown in Figure 7.9 The simulation model parameters are as follows.

1) Rated PV power: 15 kWp.
2) Voltage on the DC side (i.e., DC bus voltage): 800 V.
3) The maximum rated output phase current: 33 A.
4) Fault occurrence time: 0.5 s.
5) Fault duration: 0.5 s.

When a fault occurs, the three-phase voltage of the PV inverter at the grid connection point symmetrically drops to 20%; the simulation results are shown in Figure 7.10. Without LVRT function, the protection will be activated and the switch devices will be blocked when overcurrents are detected. It can be seen from Figure 7.10 that the three-phase output current, output active and reactive power, positive sequence *d*-axis current of the PV inverter rapidly drop to zero, and eventually the PV inverter goes out of

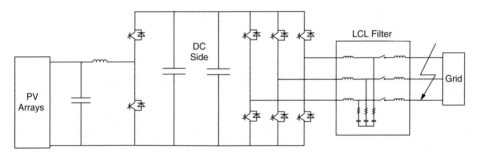

Figure 7.9 The topological structure diagram of simulation cases.

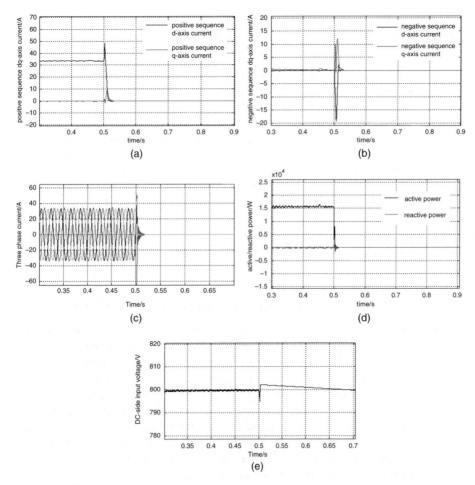

Figure 7.10 PV output waveforms in the case of three-phase grid voltage symmetrically dropping to 20%. (a) Positive-sequence *d*-axis and *q*-axis currents of inverter; (b) negative-sequence *d*-axis and *q*-axis currents of inverter; (c) three-phase output current of inverter; (d) Output active and reactive power of inverter; (e) voltage on the DC side of inverter.

operation. In other words, in the case of a short-circuit fault in the distribution network, the contribution of the PV generation without the LVRT function to the steady-state fault current is zero.

7.4.2 Fault Simulation of Photovoltaic Generation with the Low-Voltage Ride-Through Function

The simulation model parameters are the same as those given in Section 7.4.1. The control method proposed in Section 7.3.2 is chosen as the LVRT method and the amplitude limit of When fault occurs, three cases are simulated to analyze the voltage at the grid connection point: 80% voltage drop of three phases, 80% voltage drop of two phases, and 80% voltage drop of a single phase.

7.4.2.1 Case 1: 80% Three-phase Voltage Drop

According to the simulation results in Figure 7.11, the following observations can be made.

1) The reference value of the positive-sequence q-axis current of the PV inverter can be obtained from Figure 7.8 and Equation 7.9. It can be seen from Figure 7.11a that the value increases from zero to 1.2 when the smaller value between i^*_{d0+} and $(1.5^2 - i^{*}_{q+}{}^2)^{0.5}$ is chosen as the reference value of the positive-sequence d-axis current of the PV inverter. The reference values of the negative-sequence d-axis and q-axis currents of the PV inverter are zero. It can be seen from Figure 7.7 that under the action of the inner loop PI regulator for negative-sequence current, the negative sequence d-axis and q-axis currents of the PV inverter are reduced to zero within 100 ms. The steady-state three-phase short-circuit current of the PV inverter is the sum of the vectors of the positive-sequence active and reactive currents.

2) According to Figure 7.11d and e, the output active power of the PV inverter is reduced from 16 kW to 3 kW while the reactive power is increased from 0 to 4 kW.

7.4.2.2 Case 2: 80% Two-phase Voltage Drop

According to the earlier simulation results, the following remarks can be made.

1) The reference value of the positive sequence q-axis current of the PV inverter can be obtained from Figure 7.8 and Equation 7.9. It can be seen from Figure 7.12a that the current increases from zero to 1.0064, while the smaller value between i^*_{d0+} and $(1.5^2 - i^{*}_{q+}{}^2)^{0.5}$ is still taken as the reference value of the positive-sequence d-axis current of the PV inverter. The reference values of the negative-sequence d-axis and q-axis currents of the PV inverter are zero. It can be seen from Figure 7.7 that owing to the regulation of the negative-sequence current inner loop, the negative-sequence d-axis and q-axis currents of the PV inverter are reduced to zero within 200 ms. The steady-state three-phase short-circuit current of the PV inverter is the sum of the vectors of the positive-sequence active and reactive currents.

2) As the positive-sequence d-axis current decreases and the positive sequence q-axis current increases, it can be seen from Figure 7.12c that the output active power of the PV inverter is reduced and the reactive power is increased to provide the power grid with reactive power support. In the case of an asymmetric grid fault, it is different from the symmetric grid fault that the output power of the PV inverter

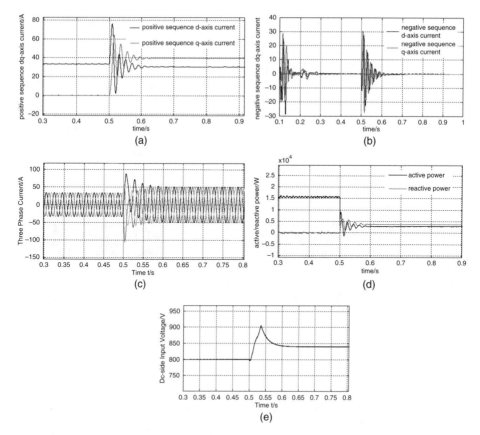

Figure 7.11 PV output waveforms in the case of symmetrical 80% drop in three-phase grid voltage: (a) positive-sequence *d*-axis and *q*-axis currents of inverter; (b) negative-sequence *d*-axis and *q*-axis currents of inverter; (c) three-phase output current of inverter; (d) output active and reactive power of inverter; (e) voltage on the DC side of inverter.

contains the double-frequency fluctuations. Based on the analysis in Section 7.2.2, when eliminating the double-frequency fluctuation in the output current is taken as the goal of control, the components of the double-frequency fluctuations in the active and reactive power cannot be removed.

7.4.2.3   Case 3: 80% Single-phase Voltage Drop

1) The reference value of the positive-sequence *q*-axis current of the PV inverter can be obtained from Figure 7.8 and Equation 7.9. It can be seen from Figure 7.13a that the value increases from zero to 0.533, while the smaller value between i^*_{d0+} and $(1.5^2 - i^{*2}_{q+})^{0.5}$ is still taken as the reference value of the positive-sequence *d*-axis current of the PV inverter. The reference values of the negative-sequence *d*-axis and *q*-axis currents of the PV inverter are zero. It can be seen from Figure 7.7 that Owing to the regulation of the negative-sequence current inner loop, the negative-sequence *d*-axis and *q*-axis currents of the PV inverter are reduced to zero within 100 ms.

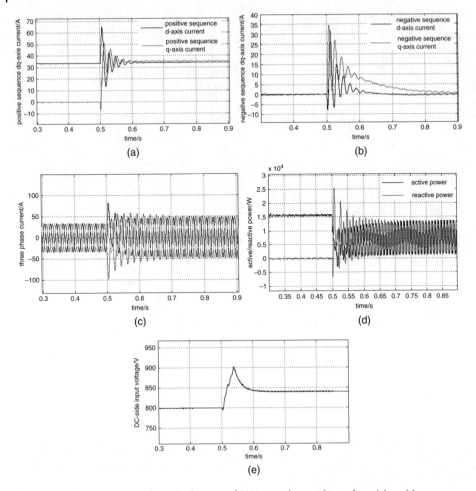

Figure 7.12 PV output waveforms in the case of 80% two-phase voltage drop: (a) positive-sequence *d*-axis and *q*-axis currents of inverter; (b) negative-sequence *d*-axis and *q*-axis currents of inverter; (c) three-phase output current of inverter; (d) active and reactive power of inverter; (e) voltage on the DC side of inverter.

The steady-state three-phase output short-circuit current vector of the PV inverter is the sum of the vectors of the positive-sequence active and reactive currents.

2) As the positive-sequence *d*-axis current decreases and the positive-sequence *q*-axis current increases, it can be seen from Figure 7.13 that the output active power of the PV inverter is reduced and the reactive power is increased to provide the power grid with reactive power support. The active power and reactive power contain the double-frequency components for the same reason as analyzed in case 2 (Section 7.4.2.2).

3) It can be seen from the simulation results that in the cases of different types of grid faults, the PV inverter can maintain the grid-connected operation, and provide certain reactive power support to achieve the LVRT control goal.

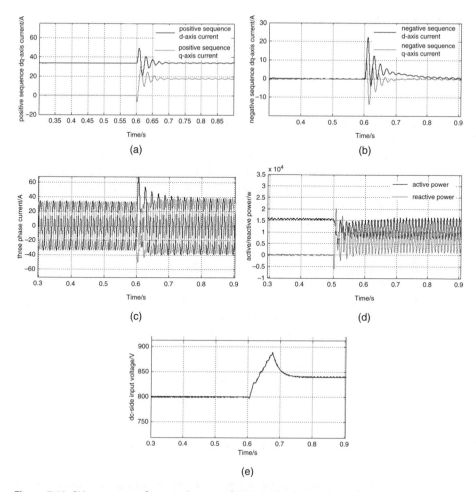

Figure 7.13 PV output waveforms in the case of 80% single-phase voltage drop: (a) positive-sequence *d*-axis and *q*-axis currents of inverter; (b) negative-sequence *d*-axis and *q*-axis currents of inverter; (c) three-phase output current of inverter; (d) active and reactive power of inverter; (e) voltage on the DC side of inverter.

7.5 Calculation Method for Short-Circuit Currents in Distribution Network with Distributed Photovoltaic Generation

The ever-increasing penetration of distributed PV generation has a great impact on the stability of a power system. In particular, the large-scale grid defection of distributed PVs will address serious adverse effects on the normal operation of the power grid [11, 12]. As a result, it is necessary to develop a method to calculated and analyze the short-circuit currents in the distribution network with high PV penetration.

7.5.1 Distribution Network Model

The traditional distribution network is a single-source power network, and the Thevenin-equivalent circuits can be used for the network analysis. When a fault occurs to the distribution network, the node equation is

$$YU = I \tag{7.12}$$

where Y is the system node admittance matrix, U is the node voltage vector, and I is the node injection current vector.

After large-scale distributed PV generation integration, the power supply of the system is changed from a single-source power supply to multiple-source power supplies. As a result, the size and phase of the fault current will change accordingly. During the fault period, the rotating distributed generators can be equivalent to the series model of the voltage source and impedance; therefore, traditional methods can be kept on. However, this traditional linear equivalent model is no longer applicable to PV systems, which are highly nonlinear and should be modeled as a voltage-controlled current source [13–15]. Therefore, new methods for the fault analysis of the distribution network with high PV penetration should be developed.

A detailed analysis of a typical distribution network with PVs is shown in Figure 7.14 In Figure 7.14, $\dot{E}_S$ is the equivalent potential of the system, and Z_S and Z_{Ln} are the equivalent impedance of the system and the impedance of line Ln, respectively. A short-circuit occurs at the end point of line L3. $\dot{U}_{SB.f}$ is the fault voltage at point SB of the system bus; $\dot{U}_{Pn.f}$ is the fault voltage at Pn, the connection point of the nth PV system PVn; $\dot{I}_{S.f}$ and $\dot{I}_{PVn.f}$ are the fault current provided by the system and PVn, respectively; $\dot{I}_{Ln.f}$ is the fault current flowing through line Ln, while $\dot{I}_{LD1.f}$ and $\dot{I}_{LD2.f}$ are the fault current flowing towards LD1 and LD2, respectively.

7.5.2 Calculation Method for Short-Circuit Currents in a Traditional Distribution Network

The calculation of short-circuit current is important in the design and operation of a power system. The short-circuit current calculation employing the electromagnetic

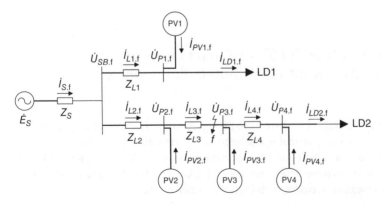

Figure 7.14 A typical distribution network containing multiple distributed PV generation.

transient model of the network elements is accurate, but the method is complex and requires a large amount of calculation, failing to meet the practical needs. At present, the calculation standards or methods for short-circuit currents mainly include the operational curve law [16], IEC standard [17], and ANSI standard [18].

7.5.2.1 Operational Curve Law

The operational curve law is a useful and simple method for calculating short-circuit currents. By the operational curve law, the direct-axis subtransient reactance x''_d of each generator is taken as its equivalent internal reactance. Ignoring the loads in the network, a simplified network is obtained, which contains only the nodes of generators, short-circuit points, and the transfer impedance between them. As per unit calculation is used to calculate all the impedances in the equivalent network according to the unified power base value, the transfer impedance between each power supply and short-circuit point is set to be the rated capacity to obtain the calculation reactance of each power supply. Then the calculated reactance is used to obtain the operational curve and the per unit value of the short-circuit current between each power supply to the short-circuit point at a certain moment. The total short-circuit current at the short-circuit point is the sum of the actual values converted from the per unit value of the short-circuit current between each power supply and the short-circuit point.

7.5.2.2 IEC Standard

In the IEC standard, all rotating machines (including generators, synchronous motors, asynchronous motors), and static inverters that are capable of power inversion operation instantaneously in the case of a short-circuit fault are treated as power supplies, e.g. the distribution network components.

The calculation of network component impedances is specified in the standard. Generally, the parameters of resistance R and reactance X are used for all network components. The short-circuit current I'' in this standard can be expressed as $I'' = (CU_n)/Z_k$, by setting an equivalent voltage source $CU_n/0.3^{0.5}$ to the short-circuit point, where U_n is the nominal voltage, Z_k is the equivalent short-circuit impedance in the short-circuit point and coefficient C is the voltage correction factor relevant to the U_n. The coefficient C can be taken as 1.05 or 1.1 when calculating the maximum short-circuit current and as 0.95 or 1 when calculating the minimum short-circuit current for the power system of 100 V $\sim$ 1000 V, and for the system higher than 1kV, respectively.

7.5.2.3 ANSI Standard

The ANSI standard for the calculation of short-circuit currents mainly provides the overcurrent reference value settings for circuit breakers. The ANSI standard simplifies the calculation of short-circuit currents through the equivalent network of an equivalent voltage source E and an equivalent series impedance (both of which take the per unit values). The voltage value E is the nominal voltage on the short-circuit bus and reflects the voltage prior to the short circuit. The ANSI standard provides a specific reference value of the equivalent reactance of all system elements [18]. The ANSI standard has a simple method for the calculation of short-circuit currents:

$$I = K\frac{E}{X} \tag{7.13}$$

where K is determined by the short-circuit point in the system and the ratio of X to R from the perspective of the short-circuit point.

7.5.3 Calculation Method for Short-Circuit Currents in a Distribution Network with Distributed Photovoltaic Generation

It can be seen from the theoretical analysis in Sections 7.2 and 7.3 that for both symmetric and asymmetric faults of the power grid, the output current on the AC side of the PV generation only contains positive-sequence current components. In addition, when voltage at the PCC collapses, the transient process of the output fault current of the PV generation is very short and negligible. In the case of a power grid fault, the fault current components can be expressed as

$$I_{\text{PV.f}\_q+} = \begin{cases} 0, & \dfrac{U^+_{\text{P.f}}}{U_{\text{P0}}} > 0.9 \\[3mm] 2\left(1 - \dfrac{U^+_{\text{P.f}}}{U_{\text{P0}}}\right) & 0.4 \le \dfrac{U^+_{\text{P.f}}}{U_{\text{P0}}} \le 0.9 \\[3mm] 1.2, & \dfrac{U^+_{\text{P.f}}}{U_{\text{P0}}} < 0.4 \end{cases} \tag{7.14}$$

$$I_{\text{PV.f}\_d+} = \min\left\{ I_{\text{PV}\_d0+}, \sqrt{\left(I_{\text{PV}\_d\max}\right)^2 - \left[2\left(1 - U^+_{\text{P.f}}/U_{\text{P0}}\right) I_{\text{PV}\_d0+}\right]^2} \right\} \tag{7.15}$$

where U_{P0} and $U^+_{\text{P.f}}$ are the rated voltage and the positive sequence component of the fault voltage at the PCC respectively, $I_{\text{PV.f}\_q+}$ is the positive-sequence reactive current actually output by the PV generation when a fault occurs, $I_{\text{PV.f}\_d+}$ is the positive-sequence actual output active current of the PV when a fault occurs, I_{PV_d0+} is the active current command of the current inner loop of the PV system before the fault occurs, and $I_{\text{PV}\_d\max}$ is the maximum allowable positive sequence active current of the PV system.

If the active component of the fault current of the PV generation is oriented toward the voltage vector at the PCC, as shown in Figure 7.15, the vector of the fault current can be expressed as

$$\dot{I}_{\text{PV.f}} = (I_{\text{PV.f}\_d+} \cos\delta + I_{\text{PV.f}\_q+} \sin\delta) + j(I_{\text{PV.f}\_d+} \sin\delta - I_{\text{PV.f}\_q+} \cos\delta) \tag{7.16}$$

where δ is the phase angle of $\dot{U}_{\text{P.f}}$.

Figure 7.15 PV fault current vector.

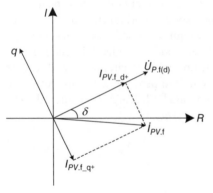

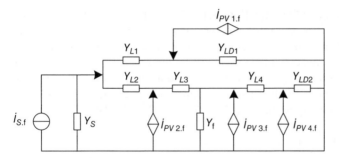

Figure 7.16 Three-phase short-circuit equivalent network in a distribution network with PVs.

It can be seen from Equations 7.15 and 7.16 that the size and phase angle of the PV fault current is determined by the fault positive-sequence voltage at the PCC. As a result, the PV generation when a fault occurs can be simplified as a current source controlled by the positive-sequence voltage at the PCC and can be expressed as

$$\dot{I}_{\mathrm{PV.f}} = f\left(\dot{U}_{\mathrm{P.f}}^{+}\right) \tag{7.17}$$

7.5.3.1 Calculation Method for Symmetric Fault Short-Circuit Currents

It is assumed that a three-phase fault occurs at point "f" and the transfer resistance is R_{f} ($1/Y_{\mathrm{f}}$ in Figure 7.15). With the voltage-controlled current source model taken into consideration, the equivalent network when a fault occurs in the distribution network is shown in Figure 7.16.

In Figure 7.16, $\dot{I}_{\mathrm{S.f}}$ and Y_{S} are the system equivalent current source and equivalent admittance, respectively; Y_{Ln} and Y_{f} are the equivalent admittance of line Ln and the transfer resistance R_{f} respectively, and Y_{LD1} and Y_{LD2} are the equivalent admittance of LD1 and LD2, respectively.

According to the aforementioned equivalent network, the fault node equation of the distribution network containing distributed PV generation can be given as

$$Y\begin{bmatrix} \dot{U}_{\mathrm{SB.f}} \\ \dot{U}_{\mathrm{P1.f}} \\ \dot{U}_{\mathrm{P2.f}} \\ \dot{U}_{\mathrm{P3.f}} \\ \dot{U}_{\mathrm{P4.f}} \end{bmatrix} = \begin{bmatrix} \dot{I}_{\mathrm{S.f}} \\ \dot{I}_{\mathrm{PV1.f}} \\ \dot{I}_{\mathrm{PV2.f}} \\ \dot{I}_{\mathrm{PV3.f}} \\ \dot{I}_{\mathrm{PV4.f}} \end{bmatrix} \tag{7.18}$$

where

$$Y = \begin{bmatrix} Y_{\mathrm{S}} + Y_{\mathrm{L1}} + Y_{\mathrm{L2}} & -Y_{\mathrm{L1}} & -Y_{\mathrm{L2}} & 0 & 0 \\ -Y_{\mathrm{L1}} & Y_{\mathrm{L1}} + Y_{\mathrm{LD1}} & 0 & 0 & 0 \\ -Y_{\mathrm{L2}} & 0 & Y_{\mathrm{L2}} + Y_{\mathrm{L3}} & -Y_{\mathrm{L3}} & 0 \\ 0 & 0 & -Y_{\mathrm{L3}} & Y_{\mathrm{L3}} + Y_{\mathrm{L4}} + Y_{\mathrm{f}} & -Y_{\mathrm{L4}} \\ 0 & 0 & 0 & -Y_{\mathrm{L4}} & Y_{\mathrm{L4}} + Y_{\mathrm{LD2}} \end{bmatrix} \tag{7.19}$$

In the node equation shown in Equation 7.18, the fault current provided by each PV generation $\dot{I}_{\mathrm{PVn.f}}$ can be expressed in Equations 7.14, 7.15, and 7.16.

Simultaneous equations, including Equations 7.14–7.16, 7.18, and 7.19, can be used to establish the equation set for the node voltages. However, owing to the coupling effect among the PV generations and the strong nonlinear relationship between the PCC voltage and the PV current, this equation set cannot be directly solved through linear transformation. In order to solve the aforementioned problems, in this section we use the iterative correction method. The iterative equations, modified equations, and convergence criteria are shown in Equations 7.20, 7.21, 7.22, 7.23, and 7.24:

$$
Y \begin{bmatrix} \dot{U}^{(k)}_{\text{SB.f}} \\ \dot{U}^{(k)}_{\text{P1.f}} \\ \dot{U}^{(k)}_{\text{P2.f}} \\ \dot{U}^{(k)}_{\text{P3.f}} \\ \dot{U}^{(k)}_{\text{P4.f}} \end{bmatrix} = \begin{bmatrix} \dot{I}_{\text{S.f}} \\ 0 \\ 0 \\ 0 \\ 0 \end{bmatrix} + \begin{bmatrix} 0 \\ \dot{I}^{(k-1)}_{\text{PV1.f}} \\ \dot{I}^{(k-1)}_{\text{PV2.f}} \\ \dot{I}^{(k-1)}_{\text{PV3.f}} \\ \dot{I}^{(k-1)}_{\text{PV4.f}} \end{bmatrix} \tag{7.20}
$$

$$
I^{(k)}_{\text{PV}n.\text{f}\_q} = \begin{cases} 0, & \dfrac{U^{(k)}_{Pn.\text{f}}}{U_{P0}} > 0.9 \\[2ex] 2\left(1 - \dfrac{U^{(k)}_{Pn.\text{f}}}{U_{P0}}\right), & 0.4 \le \dfrac{U^{(k)}_{Pn.\text{f}}}{U_{P0}} \le 0.9 \\[2ex] 1.2, & \dfrac{U^{(k)}_{Pn.\text{f}}}{U_{P0}} < 0.4 \end{cases} \tag{7.21}
$$

$$
I^{(k)}_{\text{PV}n.\text{f}\_d} = \min \left\{ I_{\text{PV}n\_d0}, \sqrt{\left(I_{\text{PV}n\_d\max}\right)^2 - \left[2\left(1 - U^{(k)}_{Pn.\text{f}}/U_{P0}\right) I_{\text{PV}n\_d0}\right]^2} \right\} \tag{7.22}
$$

$$
\dot{I}^{(k)}_{\text{PV}n.\text{f}} = \left(I^{(k)}_{\text{PV}n.\text{f}\_d} \cos \delta^{(k)}_n + I^{(k)}_{\text{PV}n.\text{f}\_q} \sin \delta^{(k)}_n\right) + j\left(I^{(k)}_{\text{PV}n.\text{f}\_d} \sin \delta^{(k)}_n - I^{(k)}_{\text{PV}n.\text{f}\_q} \cos \delta^{(k)}_n\right) \tag{7.23}
$$

$$
\max \left| \dot{U}^{(k)}_{i.\text{f}} - \dot{U}^{(k-1)}_{i.\text{f}} \right| < \varepsilon \tag{7.24}
$$

The node admittance matrix Y is a symmetric matrix and a positive definite matrix. Therefore, the aforementioned iterative method is convergent. As the maximum output current of the PV generation is 1.5 p.u., there is no need to take the PV current under normal operations as the iterative initial value. Instead, the rated current of the PV generation $\dot{I}_{\text{PV}n0}$ is used.

$$
\dot{I}^{(0)}_{\text{PV}n.\text{f}} = \dot{I}_{\text{PV}n0} \tag{7.25}
$$

7.5.3.2 Calculation Method for Asymmetric Fault Short-Circuit Currents

When an asymmetric short-circuit fault occurs to the distribution network, according to the method of symmetric components, the fault network can be decomposed into the positive-sequence network and negative-sequence network. As the PV

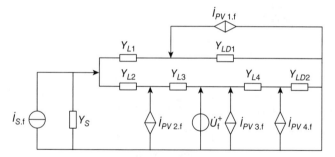

Figure 7.17 Positive-sequence network.

Figure 7.18 Negative-sequence network.

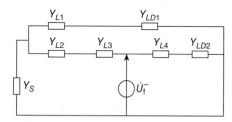

generation is equivalent to a voltage-controlled current source model that only outputs positive-sequence current and it only exists in the positive-sequence network.

Suppose a two-phase short-circuit fault occurs at point f in the distribution network shown in Figure 7.14; its positive-sequence network and negative-sequence network are shown in Figures 7.17 and 7.18 respectively.

In Figures 7.17 and 7.18, $\dot{U}_f^+$, and $\dot{U}_f^-$ are the positive-sequence equivalent voltage and negative-sequence equivalent voltage at the fault point, and

$$\dot{U}_f^+ = \dot{U}_f^- \tag{7.26}$$

Derived from Figure 7.17 and 7.18, the combined sequence network of a two-phase short-circuit fault is shown in Figure 7.19.

Comparing Figures 7.19 and 7.15, the combined-sequence network of the two-phase short circuit has the same form as the equivalent network of the three-phase short circuit. As a result, the node equation is established and the iterative method in Section 7.5.3.1 is used to obtain the positive-sequence voltage of each node in the combined-sequence network of the two-phase short circuit. The value of $\dot{U}_f^-$ is obtained according to Equation 7.26 and substitute it into the negative-sequence network to obtain the negative-sequence voltage of each node and the positive-sequence current

Figure 7.19 Combined-sequence network of a two-phase short-circuit fault.

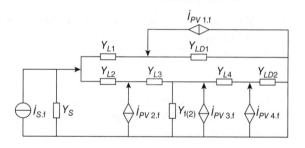

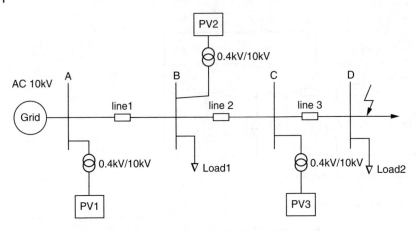

Figure 7.20 Typical distribution network with distributed PV generation.

Table 7.3 Parameters of the distribution network with PV generation.

Subsystem		Parameter name	Value
AC feeder		Line voltage (kV)	10
Transformer		Capacity (MVA)/short-circuit impedance (mH)	40/0.27
Line	line1	Line impedance Re (Ω)/Le (mH)	0.35/0.5
	line2	Line impedance Re (Ω)/Le (mH)	0.245/0.35
	line3	Line impedance Re (Ω)/Le (mH)	0.245/0.35
PV	PV1	Power (kW)	20
	PV2	Power (kW)	20
	PV3	Power (kW)	20
Load	Load1	Power (kW)	30
	Load2	Power (kW)	60

and negative-sequence current of each branch. Other asymmetric short-circuit faults can be handled in the same way as that of the two-phase short-circuit fault.

7.5.4 Fault Simulation Studies of Distribution Network with Distributed Photovoltaic Generation

A 10 kV radial distribution network low-voltage feeder is taken as an example to establish the electromagnetic transient simulation model, which is applicable to a distribution network with distributed PV generation. A typical distribution network is shown in Figure 7.20.

The simulated system includes three sources of PV generation, three lines, and two AC loads. The control strategies for PV generation with the LVRT function proposed in Section 7.3.2 are used for the control of all the PV generation. Corresponding parameters are shown in Table 7.3.

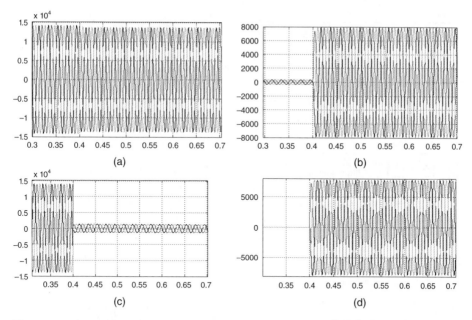

Figure 7.21 Three-phase short-circuit fault waveforms of bus D through a 0.1 Ω fault resistance without integration of distributed PV generation: (a) system voltage waveforms; (b) system current waveforms; (c) voltage waveforms at fault point; (d) current waveforms at fault point.

Table 7.4 Comparison of simulation results in case 1.

Measuring object / Measured value		System source (10 kV side)	PV1 (0.4 kV side)	PV2 (0.4 kV side)	PV3 (0.4 kV side)	Fault point (10 kV side)
Pre-fault	Three-phase output phase current i (A)	320				0
	Line voltage at connection point u (V)	14 150				13 800
Post-fault	Three-phase output phase current i (A)	7 850				7 750
	Line voltage at connection point u (V)	13 675				1 340

Case 1: As shown in Figure 7.20, there is no distributed PV generation integrated into the distribution network. A line-to-ground fault with a 0.1 Ω fault resistance occurs at bus D and the three-phase voltage drops symmetrically. The fault time is at 0.4 s and the duration time is 0.5 s. The simulation results are shown in Figure 7.21 and Table 7.4.

Case 2: As shown in Figure 7.20, the distributed PV generation with LVRT function are integrated into the distribution network. A line-to-ground fault with a 0.1 Ω fault resistance occurs at bus D and the three-phase voltage drops symmetrically. The fault happens at 0.4 s and the fault duration is 0.5 s. The simulation results are shown in Figure 7.22.

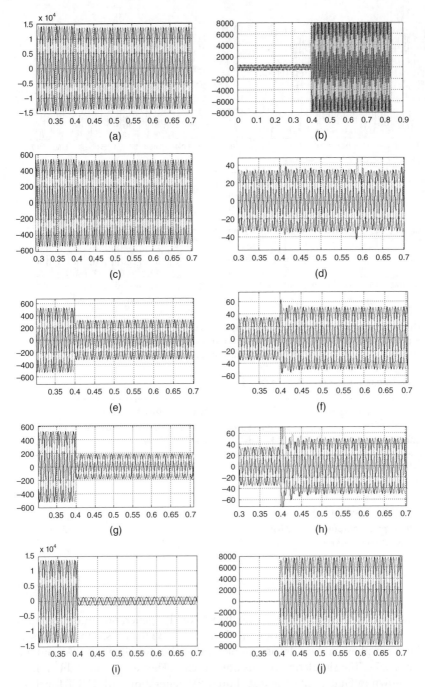

Figure 7.22 Three-phase fault waveforms of bus D through a 0.1 Ω fault resistance with integration of distributed PV generation: (a) system voltage waveforms; (b) system current waveforms; (c) voltage waveforms of PV generation 1, (d) current waveforms of PV generation 1; (e) voltage waveforms of PV generation 2; (e) current waveforms of PV generation 2; (g) voltage waveforms of PV generation 3; (h) current waveforms of PV generation 3; (i) voltage waveforms at fault point; (j) current waveforms at fault point.

Table 7.5 Comparison of simulation results in case 2.

Measured value		System source (10 kV side)	PV1 (0.4 kV side)	PV2 (0.4 kV side)	PV3 (0.4 kV side)	Fault point (10 kV side)
Pre-fault	Three-phase output phase current i (A)	450	34	34	34	0
	Line voltage at connection point u (V)	14 100	534	526	523	13 200
Post-fault	Three-phase output phase current i (A)	8 000	35	50	50	7 700
	Line voltage at connection point u (V)	13 336	528	301	189	1 500

Both voltage and current data in the table are the maximum values.

Based on the simulation waveforms shown in Figure 7.22, the voltage and current data of PV generation1, 2, and 3 before and after the fault occurs and those at the fault point are compared and shown in Table 7.5 in details.

According to the simulation results, the following conclusions can be drawn.

1) For both of the cases when a symmetric short-circuit fault occurs at bus D, the largest voltage sag happens at the fault point. In addition, the longer the distance between the PV integration point and the fault point, the smaller the voltage sag.

2) We can know from the PV LVRT control strategies proposed in Section 7.3.2 after a fault occurs, each distributed PV generation changes its control strategy according to the degree of voltage sag at the integration point, and regulates the active and reactive current and provides the power grid with reactive power support. After a fault occurs, the short-circuit current at the fault point increases rapidly and reaches the steady-state value.

3) Compared with case 1, after adding the PV integration points in the 10 kV line, the steady-state fault current at the fault point is the sum of the short-circuit currents provided by the distributed PV generation and the grid. The integration of the PV generation changes the distribution of the short-circuit fault currents.

Case 3: As is shown in Figure 7.20, the PV generation with LVRT function are integrated into the distribution network. A line-to-ground fault through a 0.1 Ω fault resistance occurs at bus B and the three-phase voltage drops symmetrically. The fault happens at 0.4 s and the simulation duration is 0.5 s. The simulation results are shown in Figure 7.23.

Based on the simulation waveforms shown in Figure 7.23, the voltage and current data of PV generation1, 2, and 3 before and after the fault occurs and those at the fault point are compared and shown in Table 7.6 in details.

According to the simulation results, the following conclusions can be drawn.

1) It can be seen from the LVRT control strategies for PV generation proposed in Section 7.3.2 that, after a fault occurs, each distributed PV generation changes its operation control strategy according to the voltage sag degree at the integration

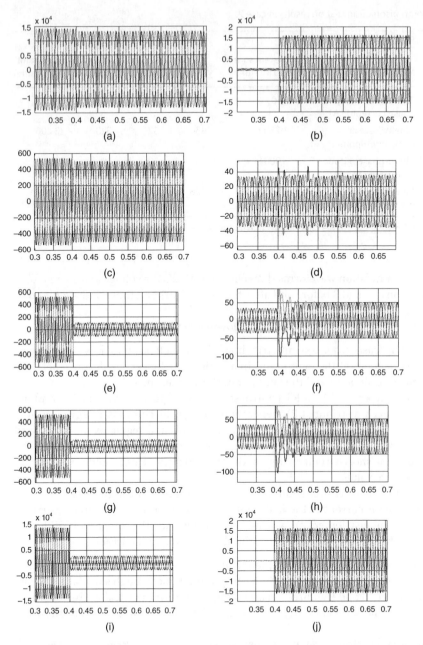

Figure 7.23 Three-phase fault waveforms of bus B through 0.1 Ω fault resistance with integration of distributed PV generation: (a) system voltage waveforms; (b) system current waveforms; (c) voltage waveforms of PV generation 1, (d) current waveforms of PV generation 1; (e) voltage waveforms of PV generation 2; (e) current waveforms of PV generation 2; (g) voltage waveforms of PV generation 3; (h) current waveforms of PV generation 3; (i) voltage waveforms at fault point; (j) current waveforms at fault point.

Table 7.6 Comparison of simulation results in case 3.

Measuring object / Measured value	System source (10 kV side)	PV1 (0.4 kV side)	PV2 (0.4 kV side)	PV3 (0.4 kV side)	Fault point (10 kV side)
Pre-fault — Three-phase output phase current i (A)	450	34	34	34	0
Line voltage at connection point u (V)	14 100	537	526	523	13 850
Post-fault — Three-phase output phase current i (A)	16 000	35	50	50	15 900
Line voltage at connection point u (V)	13 100	500	106	102	2 750

point, and regulates the active and reactive current and provides the power grid with reactive power support. After a fault occurs, the short-circuit current at the fault point increases rapidly and then drops to its steady-state value.

2) Compared with case 1, according with that in case 2, the integration of the PV generation changes the distribution of the short-circuit fault currents.
3) In this section, different faults of a distribution network with distributed PV generation are simulated by setting a three-phase line-to-ground fault at the end of the distribution network without distributed PV generation and the three-phase line-to-ground faults at different points of the distribution network with PV generation. When a fault occurs in the distribution network with PV generation, the short-circuit current at the fault point is the sum of the short-circuit currents provided by the distributed PV generation and the grid. Compared with traditional power systems, with the increasing PV penetration, the magnitude and the flow direction of the short-circuit currents in distribution networks in the case of faults will change significantly.

7.6 Summary

This chapter has introduced the short-circuit characteristics of distributed PV generation for the case of symmetric and asymmetric grid voltage sag, discussed the different LVRT standards for PV generation and the LVRT control strategies, and characterized the fault currents of PV generation. For the calculation of the short-circuit currents of a distribution network with high PV penetration, an iterative method for the calculation of distribution network fault currents was discussed, which is applicable for the calculation of the short-circuit currents in the case of symmetric and asymmetric faults of the power grid.

The short-circuit simulation results of a distribution network with distributed PV generation indicate that the grid-connected PV inverter of PV generation without a LVRT function will be shut down when faults occur, making zero contribution to the steady-state short-circuit currents. If with LVRT function, the magnitude of the active and reactive currents will be regulated so that the PV generation can continue to support the distribution network with active and reactive power. In this case the steady-state fault current at the short-circuit point is the sum of the short-circuit currents provided by the distributed PV generation and the grid.

References

1 Barker P.P., de Mello R.W. (2000) Determining the impact of distributed generation on power systems. I. Radial distribution systems. In *2000 Power Engineering Society Summer Meeting (Cat. No.00CH37134)*. IEEE; 1645–1656.

2 NB/T 33010-2014. (2014) *Operation and control specification for distributed resources interconnected with power grid*. China Electric Power Press.

3 Q/GDW 480-2010. (2010) *Technical rule for distributed resources connected to power grid*. China Electric Power Press.

4 Lopez J., Sanchis P., Roboam X., and Marroyo L. (2007) Dynamic behavior of the doubly fed induction generator during three-phase voltage dips. *IEEE Transactions on Energy Conversion*, **22**(3), 709–717.

5 Katiraei F., Iravani M.R., and Lehn P.W. (2005) Micro-grid autonomous operation during and subsequent to islanding process. *IEEE Transactions on Power Delivery*, **20**(1), 248–257.

6 Girgis A. and Brahma S. (2001) Effect of distributed generation on protective device coordination in distribution system. In *LESCOPE 01. 2001 Large Engineering Systems Conference on Power Engineering. Conference Proceedings. Theme: Powering Beyond 2001 (Cat. No.01ex490)*. IEEE PES; pp. 115–119.

7 Kong X., Zhang Z., Yin S., et al. (2013) Study on fault current characteristics and fault analysis method of power grid with inverter interfaced distributed generation. *Proceedings of the Chinese Society of Electrical Engineering*, **33**(34), 65–74.

8 Jayakrishnan R. and Sruthy V. Fault ride through augmentation of microgrid. In *2015 IEEE International Conference on Technological Advancements in Power and Energy (TAP Energy)*. IEEE; 357–361.

9 Baran M.E. and Ismail E.-M. (2005) Fault analysis on distribution feeders with distributed generators. *IEEE Transactions on Power Systems*, **20**(4), 1757–1764.

10 Song H.S. and Nam K. (1999) Dual current control scheme for PWM converter under unbalanced input voltage conditions. *IEEE Transactions on Industrial Electronics*, **46**(5), 953–959.

11 Nimpitiwan N., Heydt G.T., Ayyanar R., and Suryanarayanan S. (2007) Fault current contribution from synchronous machine and inverter based distributed generators. *IEEE Transactions on Power Delivery*, **22**(1), 634–641.

12 Walling R.A., Saint R., Dugan R.C., et al. (2008) Summary of distributed resources impact on power delivery systems. *IEEE Transactions on Power Delivery*, **23**(3), 1636–1644.

13 Baran M.E. and El-Markaby I. (2005) Fault analysis on distribution feeders with distributed generators. *IEEE Transactions on Power Systems*, **20**(4), 1757–1764.

14 Plet C.A., Brucoli M., McDonald J.D.F., and Green T.C. (2011) Fault models of inverter-interfaced distributed generators: experimental verification and application to fault analysis. In *2011 IEEE Power and Energy Society General Meeting*. IEEE; pp. 1–8.

15 Boutsika T.N. and Papathanassiou S.A. (2008) Short-circuit calculations in networks with distributed generation. *Electric Power Systems Research*, **78**(7), 1181–1191.

16 Li G.-Q. (1995) *Power System Transient Analysis*, 2nd edn. China Electric Power Press: Beijing.

17 Zhou Y.-S. (1994) Calculating the short circuit current of a three-phase alternating current system with IEC method. *Technology of Nitrogen Fertilizers*, **32**(3), 8–13.

18 ANSI C37.5-1969 (1969) *Methods for determining values of a sinusoidal current wave, a normal-frequency recovery voltage, and a guide for calculation of fault currents for application of AC high-voltage circuit breakers rated on a total current basis*. American National Standards Institute: New York.

8

Power Quality in Distribution Networks with Distributed Photovoltaic Generation

8.1 Introduction

The distributed PV integration would cause harmonics pollutions in the following aspects:

1) The grid-connected PV inverter based on power electronics conversion technology injects harmonic currents into the grid, which will impact the grid equipment, equipment and other power consumption equipment.
2) Owing to varying weather conditions (solar irradiance, temperature, etc.), the output of PV generation is highly random and in particular, when the equivalent grid capacity at the integrating spot is relatively small, large PV output fluctuations will cause voltage deviations, fluctuations, and flickers.
3) With a low-voltage PV inverter integrated, the impedance of the filter at the output of the inverter changes the impedance of the low-voltage side of the grid, and under certain circumstances this can lead to harmonic resonance [1].
4) With the increasing number of family (household) PV generation, some improperly single-phase integration of a small-scale PV generation results in unbalanced three-phase currents in the grid and the overload of neutral current in the low-voltage grid.
5) PV inverters without isolation can inject DC currents directly into the power grid, which has adverse effects, such as saturation of transformers, reactors and substation grounding corrosion.

With the growing capacity of distributed grid-connected PV generation, analyzing and evaluating the impact of PV generation on the power quality is significant to ensuring the safe and stable operation of the distribution network.

8.2 Power Quality Standards and Applications

The distributed PV integration-related technical standard system has been basically established. The power quality requirements in these standards are based on Chinese and international standards, but the Chinese and international standards differ in calculating and evaluating power quality limit values. Therefore, the evaluation of a

Grid-Integrated and Standalone Photovoltaic Distributed Generation Systems: Analysis, Design, and Control,
First Edition. Bo Zhao, Caisheng Wang and Xuesong Zhang.
© 2018 China Electric Power Press. Published 2018 by John Wiley & Sons Singapore Pte. Ltd.

grid-connected PV generation project for the distribution network asks for clearly defined applicable standards and the power quality evaluation methods.

8.2.1 Power Quality Standards for Grid-Connected Photovoltaic Generation

International organizations such as IEEE and IEC started the development of standards for grid-connected PV generation in 2000. The requirements of power quality for grid-connected PV generation are stipulated by standards IEEE Std 929-2000 and IEC 61727-2004. The Standardization Administration of the People's Republic of China (SAC) specifies the power quality requirements for grid-connected PV generation in GB/T 19939-2005 (Technical requirements for grid integration of PV system) and GB/T 20046-2006 (Photovoltaic (PV) systems – characteristics of the utility interface). The core international and Chinese standards for the power quality of grid-connected PV generation are:

- IEEE Std 929-2000   IEEE recommended practice for utility interface of photovoltaic (PV) systems
- IEEE Std 1547-2003   IEEE standard for interconnecting distributed resources with electric power systems
- IEC 61727-2004   Photovoltaic (PV) systems – characteristics of the utility interface
- GB/T 19964-2012   Technical requirements for connecting photovoltaic power station to power system
- GB/T29319-2012   Technical requirements for connecting photovoltaic power system to distribution network
- GB/T 19939-2005   Technical requirements for grid integration of PV system.

8.2.2 Power Quality Requirements Stipulated in Standards for Grid-Connected Photovoltaic Generation

In China, the requirements of power quality under technical standards for grid-connected PV generation are divided into two perspectives.

Firstly, power quality standards for the public supply network based on GB/T 19964-2012 and GB/T 29319-2012 are directly used for PV generation. In these standards, the PV generation is treated equally with ordinary power users. For example, GB/T 19964-2012 states: "in the grid integration of PV generation, the harmonic current injected into the grid should meet the requirements specified in GB/T 14549 (Quality of electric energy supply – harmonics in public supply network)." The actual application of the standards, however, can be flexible and adaptive. For instance, when the short-circuit capacity of the power grid is large there can be less restrictions on the PV power quality; on the other hand, when the short-circuit capacity of the power grid is small, there are more stringent restrictions on the power quality of the PV generation.

Secondly, power quality restrictions required in GB/T 19939-2005, IEC 61727-2004, and IEEE Std 1547-2003 are from the perspective of the equipment specifications, and in these standards the limit values are only related to the PV capacity. When the capacity is fixed, the limit values specified in the standards are fixed no matter where the PV generation is installed. The standards set up requirements for PV generation, which must meet product standards. But there is no consideration for grid integration. For example, in terms of voltage fluctuation and flicker, GB/T 20046-2006 uses IEC 61000-3-3 and stipulates the test method for the output flicker of the power consumption equipment

Table 8.1 Brief comparison of power quality provisions in China and international PV integration standards.

Standard	GB/T 19964-2012	GB/T 29319-2012	GB/T 19939-2005	IEC 61727-2004	IEEE 1547-2003
Standard application scope	connected into grid at 35 kV level; 10 kV connected into public supply network	connected into the user's internal network at 10 kV level; connected into grid at 380 V level	Low voltage grid connection	Low voltage grid connection; the PV capacity should be no more than 10 kVA	Capacity: no more than 10 MVA (mainly medium voltage connection)
Voltage deviation	GB/T 12325	GB/T 12325	GB/T 12325	n/a	n/a
Voltage fluctuation DC and flicker P_{lt}/P_{st}	GB/T 12326	GB/T 12326	No requirements	IEC 61000-3-3 IEC 61000-3-5	IEC 61000-3-7
Harmonic	Harmonic current: GB/T 14549; inter-harmonic: GB/T 24337	Harmonic current: GB/T 14549; inter-harmonic:GB/T 24337	HRI_2 to HRI_{33}: 4%, 2%, 1.5%, 0.6%; $THD_i\% \leq 5\%I_N$	HRI_2 to HRI_{33}: 4%, 2%, 1.5%, 0.6%; $THD_i\% \leq 5\%I_N$	$TDD_i\% \leq 5\%$
Voltage unbalance factor	GB/T 15543	GB/T 15543	GB/T 15543	n/a	n/a
DC injection component	DC component $\leq$0.5% of the AC current rating	n/a	DC component $\leq$1% of the AC current rating	DC component $\leq$1% of the AC current rating	DC component $\leq$0.5% of the AC current rating

THD: total harmonic distortion; TDD: total demand distortion.

when the consistency testing circuit conditions are met. Standards GB/T 19939-2005, GB/T 20046-2006, and IEEE Std 1547-2003 have more applications than other similar standards because they are relatively fair and easy in implementation with a clear technical guidance. A comparison of the major technical standards with different power quality requirements is given in Table 8.1, and the relative application methods are recommended as well.

8.2.2.1 Voltage Deviation

The GB/T 19964-2012 and GB/T 29319-2012 standards in China require that after the integration of the PV system the voltage deviation at the PCC should meet the requirements of GB/T 12325 (Power quality – deviation of supply voltage) while the PV system is integrated to the power grid.

It is stipulated in GB/T 12325-2008 that the three-phase supply voltage deviation of 20 kV and below should be within ±7% of the nominal voltage, and the allowable voltage deviation of the 220 V single-phase voltage should between +7% and −10% of the nominal voltage.

The IEC 61727-2004 and IEEE Std 1547-2003 standards are mainly for medium- and small-capacity PV generation. IEC 61727-2004 is applicable to PV generation with a capacity of no greater than 10 kW, while IEEE Std 1547-2003 is applicable to PV generation with a capacity of no greater than 10 MW. These two standards do not stipulate the voltage deviation for the PCC, but when it comes to the protection of the power grid and the PV generation, both of the standards require the PV generation to respond to the abnormal power grid voltage.

As the line resistance of the distribution network cannot be ignored, both the active and reactive power outputs of PV generation affect the power grid voltage. Both GB/T 19964-2012 and GB/T 29319-2012 require that PV generation should have power factor regulation ability. However, owing to the PV capacity constraints, the power factor regulation ability is limited and it is very difficult to only rely on the PV generation carry out the voltage regulation. Hence, for GB/T 19964-2012 and B/T 29319-2012 it is suggested that the voltage deviation should be used as a reference and whether the PV generation provides the corresponding reactive power support according to its generation capacity should be checked. The PV system may only be responsible for the portion of the voltage regulation responsibility it should take.

8.2.2.2 Voltage Fluctuation and Flicker

It is stipulated in Chinese standards GB/T 19964-2012 and GB/T 29319-2012 that, with the PV integration, voltage fluctuation and flicker of the PCC should meet the requirements defined in GB/T 12326-2008 (Power quality – voltage fluctuation and flicker). It is stipulated in GB/T 12326-2008 that the applicable change frequency of the voltage should meet the requirement of $1 < r \le 10$ (r is the number of changes per hour) for PV integration at a voltage of 35kV and below. In the continuous operation or switching of the PV generation the voltage change limit value $d\%$ should be no greater than 3%. It is required that flicker of the PCC in the distribution network caused by PV generation should be no greater than the value set by the distribution network operator, which can be calculated through the evaluation process.

The IEC 61727-2004 standard is mainly for low-voltage grid-connected PV systems. For a PV system with a rated current no greater than 16 A, it is required in the standard that it should meet the following provisions that are also delineated in IEC 61000-3-3: (1) short-term flicker (minutes) $P_{st} \le 1.0$ and long-term flicker (hours) $P_{lt} \le 0.65$; (2) steady-state relative voltage change $d_c \le 3\%$ and maximum relative voltage change $d_{max} \le 4\%$; (3) during the voltage changing period, the duration of d_c greater than 3% should be no longer than 200 ms. It is required to meet the provisions of IEC 61000-3-5, For a PV system with a rated current greater than 16 A, which mainly makes adaptive extensions to IEC 61000-3-3 and stipulates new standard test circuit parameters.

In the requirements of IEEE Std 1547-2003, the grid-connected medium- and high-voltage PV generation should not lead to voltage flicker, and its evaluation method is based on the IEC 61000-3-7 standard. The core of the standard is the stipulation of the three-level assessment method for evaluating a user's voltage fluctuation and flicker, the user limit distribution method, and a simplified method for the prediction of the severe degree of flicker.

Both Chinese and international standards have certain application conditions for voltage fluctuation and flicker. It is suggested that in actual applications the GB 17625.2 (GB/Z 17625.3) standard should be used for the low-voltage PV inverter equipment test to obtain the equipment consistency testing results. The GB/Z 17625.5 (IEC 61000-3-7) standard should be used in the evaluation before the grid integration of the PV generation. In the evaluation process, attention should be paid to the difference between the calculated limit value and the value stipulated in GB/T 12326-2008. For example, the calculation of severe degree of flicker requires the limit value of P_{st}, which, however, is not stipulated in the GB/T 12326-2008 standard. After the grid integration of the PV generation, the GB/T 12326-2008 standard should be used as a reference for the site testing evaluation.

8.2.2.3 Voltage Unbalance Factor
It is stated in Chinese standards GB/T 19964-2012 and GB/T 29319-2012 that after grid connection of the PV generation the voltage unbalance factor should meet the requirements of GB/T 15543, in which it is stipulated that, in normal operation of the grid, the negative-sequence voltage unbalance factor $\varepsilon_u\%$ should be no greater than 2% or 4% in the short term during normal system operations.

International standards such as IEC 61727-2004 and IEEE Std 1547-2003 have no requirements for the voltage unbalance factor. In site applications, the following attention should be paid to the following two aspects: (1) In the PV grid integration design plan, phase sequence rotation should be considered to reduce injection of unbalanced currents and avoid neutral line overload. (2) The GB/T 15543-2008 standard only stipulates the voltage, but the grid enterprises should stipulate pertinent current unbalance or negative-sequence and zero-sequence current indicators from the perspective of the planning.

8.2.2.4 DC Injection
It is delineated in GB/T 19964-2012 standard that the DC current component fed to the power grid by grid-connected PVs should be no greater than 0.5% of the rated AC current. In the GB/T 19939-2005 standard the limit value of the DC current component should not be greater than 1% of the AC current rating. However, the DC current component is not stipulated in the GB/T 29319-2012 standard.

According to IEEE Std 1547-2003, the isolated transformer should be installed at the output terminal of the grid-connected inverter to limit the DC component injected into the inverter and the DC component fed to the power grid should not be greater than 0.5% of the rated AC current.

8.2.2.5 Current Harmonics
In the GB/T 19964-2012 and GB/T 29319-2012 standards, PV generation is regarded as an ordinary electrical load and it is stipulated that the harmonic injection current of the PCC into which the PV system is integrated should meet the requirements of the GB/T 14549 standard. The following items should be noticed in the standard:

1) The standard only stipulates current harmonics to the 25th order, which cannot fully reflect the harmonic characteristics of fast turn-off electronic devices.
2) The current harmonic limit values are a function of three parameters: the minimum short-circuit capacity, the capacity of the power supply equipment, and the contracted capacity. In other words, according to the principle of capacity distribution, PV generation and other harmonic sources integrated into the grid share the total harmonic current injection, and the limit values of the current harmonics are not only related to the installed capacity, but also to the short-circuit capacity and power supply capacity of the grid.
3) The current THD limit is not stipulated.

Table 8.2 Limit values of PV current harmonics stipulated in GB/T200426-2006 and IEEE Std 1547.

Harmonic order	Limit value of harmonic current (%) $\times I_N$
$h < 11$	$<4.0\% \times I_N$
$11 \leq h < 17$	$<2.0\% \times I_N$
$17 \leq h < 23$	$<1.5\% \times I_N N$
$23 \leq h < 35$	$<0.6\% \times I_N$
$35 \leq h^{a)}$	$<0.3\% \times I_N$

The limit values of even harmonics in this range should be 25% less than the limit values of the lower odd harmonics.
I_N is the rated current of the PV generation referred to the voltage level at the power quality assessment point.
As voltage distortions can cause more serious current distortions, the test of harmonics is very complicated. The injected harmonic currents should not include any current harmonics caused by voltage harmonics before PV generation connection.
a) Provisions added to IEEE Std 1547-2003.

GB/T 19939-2005 and IEC 61727-2004 utilize the current harmonic evaluation method given in IEEE Std 929 as a reference. They stipulate that HRI_2 to HRI_{33} should be classified into four categories – namely, 4%, 2%, 1.5%, 0.6% (see Table 8.2) – and that at rated output the current THD of the inverter should be no greater than 5%; that is, $THD_i\% \leq 5\%$. The IEEE Std 1547 standard is stricter than the IEEE Std 929 standard because it stipulates that the HRI_h ($h > 33$) should be less than 0.3%. Meanwhile, in IEEE Std 1547 the limit value of current THD ($THD_i\%$) is changed to total demand distortion ($TDD_i\%$), which is required to be no greater than 5%. The difference between $THD_i\%$ and $TDD_i\%$ is the base value, but their evaluation indicators are consistent with each other.

The calculation formulas of relevant indicators are as follows:

- harmonic ratio for hth current

$$HRI_h\ (\%) = \frac{I_h}{I_1} \times 100 \tag{8.1}$$

- total harmonic current

$$THD_i\ (A) = \sqrt{\sum_{h=2}^{40} I_h^2} \tag{8.2}$$

- THD of current

$$THD_i\ (\%) = \frac{1}{I_1}\sqrt{\sum_{h=2}^{40} I_h^2} \tag{8.3}$$

- total demand distortion for current

$$TDD_i\ (\%) = \frac{1}{I_L}\sqrt{\sum_{h=2}^{40} I_h^2} \tag{8.4}$$

where I_1 is the fundamental current and I_L is the demand current of load, which is equivalent to the larger one between the rated load current and the maximum load current recorded in the past year (15 min or 30 min average).

By comparing these two standard perspectives, it can be found that the Chinese standards for the technical conditions of PV connection (such as GB/T 19964-2012, GB/T 29319-2012, and GB/T 19939-2005) and IEC 61727-2004 are different in the provisions on the limit value of harmonic current. As a result, in the PV project assessment, the application of standards should be clearly stated according to the application needs.

The GB/T 19964-2012 standard, from the perspective of the power user integration assessment, sets the limit values of PV current harmonics while considering the harmonic values allowed by local grid conditions. According to this standard, when the local grid is strong, it could accept larger harmonic currents. On the other hand, when the local grid is weak, it sets up a more stringent limits PV current harmonics. Therefore, this standard takes the actual hosting capacity of the local power grid into consideration.

The GB/T 20046-2006 and IEEE Std 1547 standards focus on the technical level of equipment. The limit values of the current harmonic injection set in the standards are only related to the installed capacity, which makes them more applicable in actual applications. According to the characteristics of PV generation and based on the insulated gate bipolar transistor devices that are widely used in PV inverters, the standards delineate the current harmonic requirements. The requirements are stricter than those for the ordinary thyristor converter equipment by requiring $\text{THD}_i\% \leq 5\%$. Two provisions of IEEE Std 1547 on current harmonics can be used as a reference: (1) HRI_h ($h > 35$); (2) $\text{TDD}_i\% \leq 5\%$. However, there is a disadvantage of the standard in that it stipulates the limit value only from the perspective of the equipment. When the grid is weak, the power grid company should pay more attention to the voltage harmonics of the public supply network.

In the following we give examples to compare the limit values of the current harmonics obtained from different perspectives.

The analysis is carried out based on the assumptions that a 6 MW PV system is integrated into a 10 kV bus of a substation in a public supply network and the substation transformer has a capacity of 50 MVA (110 kV/10 kV). Two scenarios are considered:

1) A short-circuit capacity of a 10 kV bus of 100 MVA (a small/weak power grid).
2) A short-circuit capacity of a 10 kV bus of 400 MVA (a large/strong power grid).

Table 8.3 shows the calculated limit values of the harmonic current. The harmonic current calculation method stipulated in GB/T 14549-1993 is used [2], and the last column shows the limit values of the harmonic current given in GB/T 20046-2006. Figure 8.1 shows a comparison of the first 25 harmonic currents (in amperes) given in Table 8.3, where the horizontal coordinate is the harmonic order and the longitudinal coordinate is the limit value of the harmonic current.

The following analysis is carried out based on the assumptions that a 300 kW PV generation is integrated into a 380 V bus of a substation of the public supply network. Two scenarios are considered:

Table 8.3 Comparison of limit values of current harmonics of a 6 MW PV generation integrated into a 10 kV power grid.

Harmonic order	Harmonic current (A)		
			GB/T 20046
	Scenario 1	Scenario 2	IEEE Std 1547
2	9.01	36.03	3.46
3	2.91	11.64	13.86
4	4.50	18.01	3.46
5	3.42	13.67	13.86
6	2.94	11.78	3.46
7	3.30	13.20	13.86
8	2.22	8.87	3.46
9	2.36	9.42	13.86
10	1.77	7.07	3.46
11	2.86	11.45	6.93
12	1.49	5.96	1.73
13	2.59	10.35	6.93
14	1.28	5.13	1.73
15	1.42	5.68	6.93
16	1.11	4.43	1.73
17	2.08	8.31	5.20
18	0.97	3.88	1.30
19	1.87	7.48	5.20
20	0.90	3.60	1.30
21	1.00	4.02	5.20
22	0.80	3.19	1.30
23	1.56	6.24	2.08
24	0.73	2.91	0.52
25	1.42	5.68	2.08
$26 \leq I_h <$ odd number of 35	n/a	n/a	2.08
$26 \leq I_h <$ even number of 34	n/a	n/a	0.52
$35 \leq I_h$ (odd number)	n/a	n/a	1.04
$35 \leq I_h$ (even number)	n/a	n/a	0.26

1) A distribution transformer with a capacity of 300 kVA and a short-circuit capacity of the 380 V bus of 4.878 MVA.
2) A distribution transformer with a capacity of 900 kVA and a short-circuit capacity of the 380 V bus of 13.95 MVA.

Table 8.4 shows the calculated limit values of the current harmonics using the calculation method in GB/T 14549-1993, and the limit values of the current harmonics given in GB/T 20046-2006 are listed in the last column. Figure 8.2 shows a comparison of the first 25 harmonic currents (in amperes) given in Table 8.4. It can be found that

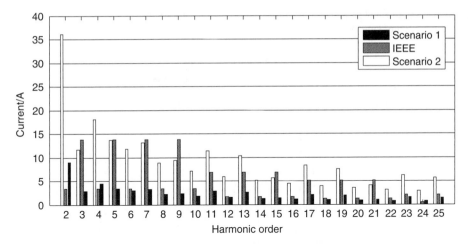

Figure 8.1 Comparison of limit values of harmonic current of 6 MW PV generation integrated into a 10kV power grid.

for the current harmonic limit values of the grid-connected PV generation the $TDD_i\%$ is required to be no greater than 5% (IEEE Std 1547) or $THD_i\%$ is no greater than 5% (GB/T 20046). The limit value of each harmonic current should be accepted by the utility company involved and the PV installation company (or the PV owner) based on the actual situation.

The applicability of $THD_i\%$ and $TDD_i\%$ in testing and evaluating the total current harmonic is now compared. In Figure 8.3, $TDD_i\%$ is the measured value at the outlet of an inverter from 05:00 to 18:30, while the demand current of load I_L is the rated current of the PV inverter. Figure 8.4 shows the $THD_i\%$ trend in the same test. It can be seen from this that the total harmonic for current has two characteristics: (1) On sunny days, the output current harmonic increases with the increase of the output power of the PV inverter and reaches a maximum around 12:00. The maximum point in the $TDD_i\%$ curve (Figure 8.3) happens at the same moment when the current harmonic reaches a maximum. (2) When the PV inverter output power is small, the fundamental current is small, while the harmonic current is not proportional to the small fundamental current. As a result, the the THDi% is a saddle-shaped curve (see Figure 8.4). The maximum value of $THD_i\%$ (which is equal to 7.9% at 06:40) cannot reflect the maximum harmonic current of the inverter. The output power of the PV inverter is required in order to select the test data needed. In the GB/T 20046-2006 standard, the $THD_i\%$ should be tested and evaluated at the rated output of the inverter according to the following procedure.

When the output of the inverter reaches the rated power, record the $THD_i\%$ at this moment and compare it with 5%. The test condition (power level) should be recorded at the same time. Owing to the varying solar irradiance and temperature, it is difficult to obtain the required test results at the rated output power. Based on the comparison, when the output of the inverter ranges from 80% to 100%, the total current harmonics, including $THD_i\%$ and $TDD_i\%$, change slightly and the test results can be used for evaluation. Therefore, using $TDD_i\%$ is more useful than $THD_i\%$. While the $TDD_i\%$ profile is recorded, the maximum value of TDDi% can be obtained and compared with 5%. The procedure is simple and effective.

Table 8.4 Comparison of limit values of current harmonics of a 300 kW PV generation integrated into a 380 V power grid.

Harmonic order	Harmonic current (A)		
	Scenario 1	Scenario 2	GB/T 20046 IEEE Std 1547
2	38.05	62.82	4.56
3	30.24	31.86	18.23
4	19.02	31.41	4.56
5	30.24	34.62	18.23
6	12.68	20.94	4.56
7	21.46	28.00	18.23
8	9.27	15.30	4.56
9	10.24	16.91	18.23
10	7.80	12.89	4.56
11	13.66	21.22	9.12
12	6.34	10.47	2.28
13	11.71	18.78	9.12
14	5.37	8.86	2.28
15	5.85	9.66	9.12
16	4.73	7.81	2.28
17	8.78	14.50	6.84
18	4.20	6.93	1.71
19	7.80	12.89	6.84
20	3.80	6.28	1.71
21	4.34	7.17	6.84
22	3.46	5.72	1.71
23	6.83	11.28	2.73
24	3.17	5.24	0.68
25	5.85	9.66	2.73
$26 \leq I_h <$ odd number of 35	n/a	n/a	2.73
$26 \leq I_h <$ even number of 34	n/a	n/a	0.68
$35 \leq I_h$ (odd number)	n/as	n/a	1.37
$35 \leq I_h$ (even number)	n/a	n/a	0.34

8.2.2.6 Voltage Harmonics

In China, power companies manage voltage harmonics according to GB/T 14549 (Table 8.5). However, there can be issues in granting harmonic permission to power users. Traditionally the harmonic source is simulated and integrated into the power grid. If the simulation results show the voltage harmonic is smaller than the required level (Table 8.5), the user is then granted permission for grid connection. However, the problem with this method lies in the fact that the grid voltage distortion is evaluated by

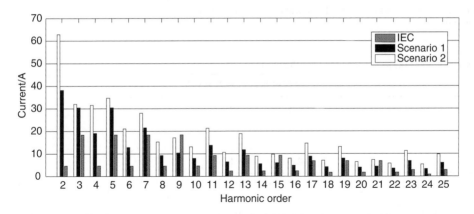

Figure 8.2 Comparison of limit values of current harmonics of 300 kW PV generation integrated into a 380 V power grid.

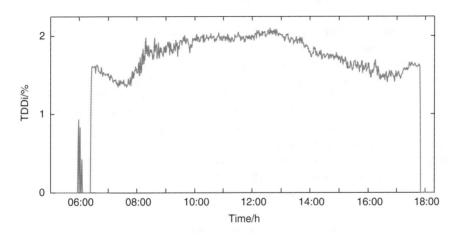

Figure 8.3 Total demand distortion of PV inverter output current harmonics.

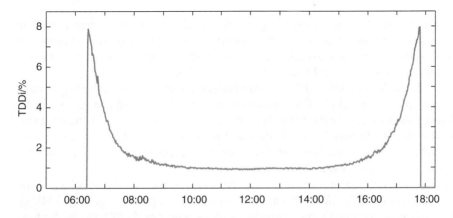

Figure 8.4 THD of PV inverter output current.

Table 8.5 Limit values of voltage harmonics of a public supply network.

Nominal system voltage (kV)	THD_u (%)	HRU_h (%)	
		Odd order	Even order
0.38	5.0	4.0	2.0
6	4	3.2	1.6
10			
35	3	2.1	1.2

the single harmonic source user under test. If each individual user passes the test, the combination of all the users' loads may cause voltage harmonics over the limit. In other words, the first user(s) may use up all the allowable room for harmonics, which is not fair to the other users.

A PV system shall not infringe the harmonic injection rights of other power users by using up all the harmonic injection capacity. A utility company can come up with its own allowable harmonic capacity distribution scheme to guarantee users to fairly share the harmonic injection capacity. Hence, the allowable grid voltage distortion caused by the integration of a single PV generation should be smaller than the given limit values listed in Table 8.5. When the background harmonic of the grid is small and the users are insensitive to harmonics, the share (i.e., the acceptable limit values) can be increased appropriately. On the other hand, in order to prevent the voltage harmonics due to a large number of harmonic source users in the region from exceeding the values given in Table 8.5, the PV harmonic injection should be strictly limited.

8.3 Evaluation and Analysis of Voltage Fluctuation and Flicker for Grid-Connected Photovoltaic Generation

The output power of a distributed PV generation is affected by solar irradiance. The solar irradiance variation can cause voltage fluctuations and flickers. The sensitivity of flicker also depends on the capacity of the power grid. Flicker occurs more frequently in low-voltage systems and at the far end of a long feeder.

In the GB 17625.2 and GB/Z 17625.3 standards, the methods for evaluating voltage fluctuation and flicker include direct measurement, simulation, and analysis methods. The direct measurement method stipulates the standard test circuit and the limit values of all indicators in detail in a scientific and objective way and then calculates the specific values at the integration point. In this section a practical evaluation method for integrating PV generation to a 10 kV bus is proposed.

In the case study, the parameters of the distribution network include the voltage at the integration point (10 kV), the short-circuit capacity of the equivalent grid (245 MVA), the angle of the internal equivalent impedance of the power grid (70°), the feeder load (3 MVA), and the power factor (0.9).

The scenarios of integrating 2MW, 4MW, 6MW, and 8MW PV generation to a 10 kV bus are studied. The rate of the PV power change frequency $r < 10$ times/min, and the PV generation operates at unity power factor. The maximum active power fluctuation of the PV generation is among 30% and 50% of the PV output.

8.3.1 Evaluation Process

According to the three-level assessment process for fluctuating load stipulated in the Chinese standard GB/T 12326-2008 as reference, the flicker caused by the PV generation is handled according to the user load size, the percentage of the contracted power capacity over the total power supply capacity, and the situation of the PCC in the power system.

8.3.1.1 First-Level Provisions

If the provisions of this level are met, the PV generation is allowed to be integrated to the grid. For the low- and medium-voltage users, the limit values of level 1 are shown in Table 8.6.

8.3.1.2 Second-Level Provisions

The value of long-term flicker value caused by the fluctuating load alone must be less than the flicker limit value of the load user.

According to the ratio of contracted power capacity S_i ($S_i = P_i/cos\varphi_i$) to the total power supply capacity, each user determines the flicker limit while considering factors such as the impact of the flicker transfer from the upper level to the lower level (the flicker transfer from the lower level to the upper level is generally ignored). The method for calculating the flicker limit of a single user is as follows.

First, calculate G_{PltMV}, the total limit of the flicker generated by all the loads integrated into PCC:

$$G_{PltMV} = \sqrt[3]{L_{PltMV}^3 - T_{HM}^3 L_{PltHV}^3} \tag{8.5}$$

where L_{PltMV} is the planning value of P_{lt} (the long-term flicker in the medium-voltage power supply system), L_{PltHV} is the limit of P_{lt} (the long-term flicker at the upper voltage level), and T_{HM} is the coefficient of the flicker transfer from the upper voltage level to the lower one (recommended to be 0.8); the flicker transfer from the extra-high voltage system to the lower voltage level system is not considered.

Table 8.6 First-level limit values of low- and medium-voltage users.

r (power change/min)	$k = (\Delta S/S_{sc})_{max}$ (%)
$r < 10$	0.4
$10 \leq r \leq 200$	0.2
$r > 200$	0.1

ΔS is the change of apparent power; S_{sc} is the short-circuit capacity of the PCC.

It is noted that the equation for calculating the short-term flicker severity P_{st} is removed from the limit provisions in the GB/T 12326-2008 standard.

Then, the limit value E_{Plti} of an individual user flicker interference P_{lt} is obtained as

$$E_{Plti} = G_{PltMV} \sqrt[3]{\frac{S_i}{S_t} \frac{1}{F}} \qquad (8.6)$$

where S_i is the contracted apparent power of the ith PV user, S_t is the total power supply of transformer in the power supply region, and F is the simultaneity factor of the fluctuating load, with a typical value of 0.2–0.3 (but $S_i/F \leq S_t$).

8.3.1.3 Third-Level Provisions

For individual users with fluctuating loads who fail to meet the second-level provisions and exceed the flicker limit after treatment, the limit can be loosened appropriately according to the actual flicker of the PCC and the prediction of future grid development. However, the flicker of the PCC has to meet its limit provisions.

8.3.2 Calculation

For the PV integration scenarios considered, the calculation procedures are as follows.

8.3.2.1 The First-Level Evaluation for Photovoltaic Integration

According to the first-level provisions in the evaluation process of GB/T 12326, the maximum fluctuating power is calculated thus:

$$\Delta S_{max} = S_{sc} \times 0.4\% = 245.9 \times 0.004 = 0.98 \text{ MW}$$

For the grid equivalent circuit, the maximum PV integration capacity to meet the first level conditions for the voltage fluctuation and flicker can be calculated as $0.98/0.5 = 1.96$ MW if it is assumed that the maximum PV power fluctuation due to the change of illumination is 50% of the PV output.

Therefore, when the PV power in the power supply region is less than 2 MW, which generally meets the simple integration conditions. There is then no need to carry out further evaluation of voltage fluctuation and flicker. However, if the PV capacity exceeds 2 MW, we need to carry out the second-level evaluation.

8.3.2.2 The Second-Level Evaluation

Equation 8.5 can be used to calculate G, which is the total value of the flicker generated by all the loads of the PCC of 10 kV bus. G represents the short-term flicker severity P_{st}:
$G = (L_p^3 - T^3 L_H^3)^{1/3} = 0.87$.

According to Equation 8.6, the coincidence factor $F = 0.2$. As $S_i/F \leq S_t$, it is assumed that the maximum share of the flicker indicators of the PV generation is 10%; namely, $S_i/S_t = 10\%$. The maximum PV flicker limit is

$$E_i = G \sqrt[3]{\frac{S_i}{S_t} \frac{1}{F}} = 0.787 \times 0.7937 = 0.625$$

As the impact of PV generation on the grid voltage fluctuation is mainly manifested in short-term severity, $P_{st\_PV} = 0.625$.

Figure 8.5 Equivalent circuit for PV voltage fluctuation and flicker evaluation.

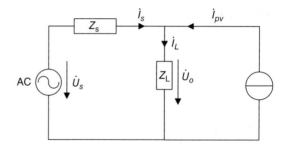

Figure 8.6 Phasor diagram.

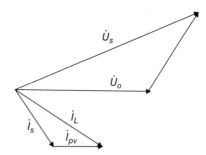

First of all, the voltage changes at the PCC caused by the PV fluctuations are carried out.

In Figure 8.5, $\dot{U}_s$ represents the equivalent grid voltage. Z_s is the equivalent impedance of the grid calculated based on the short-circuit impedance ($Z_s = R_s + jX_s = 0.1388 + j \times 0.3815$), Z_L represents the load impedance at the 10 kV bus PCC ($Z_L = R_L + jX_L = 3.6 + j \times 1.7435$), $\dot{I}_s$ represents the current injected into the PCC from the equivalent grid, $\dot{I}_{PV}$ represents the current injected into the PCC from the PV generation, and $\dot{I}_L$ represents the load current, where $\dot{I}_L = \dot{I}_s + \dot{I}_{PV}$.

According to the circuit principle, $\dot{U}_o$ is calculated using Equation 8.7, which is the voltage at the PCC. When the PV output changes and the PV current reduces by k times (k ranges from 0.3 to 0.5, and let it be 0.5), the voltage at the PCC is calculated in Equation 8.8. $\dot{U}_o$ is assigned to be the reference direction and $\dot{I}_{PV}$ is the same as the reference direction (operating at unity power factor). Therefore, $\dot{I}_{PV}$ and $\dot{U}_o$ can be expressed using scalars. The voltage change at the PCC is calculated according to Equation 8.9, as shown in Figure 8.6.

$$\dot{U}_o = \frac{\frac{\dot{U}_s}{Z_s} + \dot{I}_{PV}}{\frac{1}{Z_s} + \frac{1}{Z_L}} \tag{8.7}$$

$$\dot{U}'_o = \frac{\frac{\dot{U}_s}{Z_s} + \dot{I}'_{PV}}{\frac{1}{Z_s} + \frac{1}{Z_L}} = \frac{\frac{\dot{U}_s}{Z_s} + (1-k)\dot{I}_{PV}}{\frac{1}{Z_s} + \frac{1}{Z_L}} \tag{8.8}$$

$$\Delta \dot{U}_o = \dot{U}_o - \dot{U}' = \frac{1}{\frac{1}{Z_s} + \frac{1}{Z_L}} k\dot{I}_{PV} \tag{8.9}$$

where $k=0.5$ and

$$\frac{1}{\frac{1}{Z_s} + \frac{1}{Z_L}} = 0.1522 + j \times 0.3456 = 0.3776\angle 66.229° \; \Omega$$

$$d\% = \frac{|\Delta \ddot{U}_o|}{U_{10 \; kV}} \times 100 \tag{8.10}$$

The results in Table 8.7 show that the voltage changes caused by the power fluctuations of the PV generation (2–10 MW) at the PCC of the simulation grid are smaller than the limit values stipulated in the GB/T 12326 standard.

Secondly, the flicker evaluation is then carried out. According to the curve analysis method stipulated in the GB/T 12326 and GB/Z 17625.5 standards, which is shown in Equation 8.11, it is assumed that the PV generation fluctuates twice every 10 min (power reduction and power recovery due to clouds). The value of d_{Lim} can be obtained from the standard GB/T 12326 as $d_{Lim} = 2$:

$$P_{st} = \frac{d\%}{d_{Lim}} \tag{8.11}$$

The value of d_{Lim} can be obtained from GB/T 12326 standard as $d_{Lim} = 2$ and $d\%$ is the voltage change calculated in Table 8.8.

It can be seen from the comparison of the flicker assessment calculation results that when the grid-connected PV capacity exceeds 6 MW the flicker will exceed the limit value stipulated in the flicker standard.

Table 8.7 Voltage changes at the PCC.

Grid-connected PV capacity (MW)	Voltage change $d\%$	National standard limit (%)
2	0.378	3
4	0.755	3
6	1.133	3
8	1.510	3
10	1.888	3

Table 8.8 Voltage flicker at the PCC calculated using the curve analysis method.

Grid-connected PV capacity(MW)	Voltage change $d\%$	P_{st}
2	0.378	0.189
4	0.755	0.378
6	1.133	0.566
8	1.510	0.755
10	1.888	0.944

8.4   Harmonic Analysis for Grid-Connected Photovoltaic Generation

With the PV integration comes harmonics. The limit allocation of voltage distortion and harmonic emission should be paid special attention to [3]. This chapter will analyze the harmonic voltage distortion caused by PV harmonic currents and the impact on the distribution network in the region and focus on establishing the relationship between the capacity and PV integration location and the harmonic voltage distortion.

The GB/T 14549 standard provides the method for allocating the limit values of harmonic voltage in the public supply network and the limit values of harmonic currents of ordinary power users. In the GB/T 19939-2005 standard it is required that the THD for PV current should be no greater than 5%.

8.4.1   Fundamentals of Harmonic Analysis

8.4.1.1   Harmonic Simulation Platform

The calculation of the harmonic power flows in a regional distribution network is very complicated. The computer-program-based calculation method is well known and utilized. There are two methods for the harmonic evaluation using computer programs: the time-domain and the frequency-domain analysis methods.

The time-domain analysis method for harmonic calculation is to accurately simulate the voltage and current waveforms through time-domain simulation and obtain the harmonic simulation values through the Fourier transform. Its advantage is that the fundamental power flow and harmonic power flows can be realized at the same time, but its disadvantage is that the distribution network element time-domain model cannot accurately simulate the skin effect and other characteristics of the harmonics, and the error is quite large.

In this section, the frequency-domain analysis method is discussed. A frequency-domain-analysis-based harmonic power flow calculation method mainly includes four aspects: harmonic power flow calculation method, harmonic source modeling, modeling of the harmonic impedance matrix of the distribution network and the load. The harmonic power flow calculation process is required to complete the following functions:

- carry out the fundamental power flow calculation to obtain the equivalent impedance model of each load and the harmonic current source spectrum $I(h)$ in the typical distribution network operation mode and operation scenario;
- establish $Y(h)_{grid}$, the harmonic admittance matrix of the distribution network and load for each order of harmonic;
- solve the harmonic power flow when the harmonic current source spectrum $I(h)$ is known;
- calculate the node voltage, harmonic spectrum of currents, and THD of current;
- compare/analyze the results data, prepare reports; issue an alarm as needed.

Hence, establishing the harmonic calculation simulation platform is the first step of harmonic power flow calculation. The simulation platform includes the distributed PV harmonic source model, distribution network model, and the load model. In a harmonic analysis, the distribution network and load harmonic impedance models have a great impact on the harmonic power flow calculation results. The models are the basis for solving the harmonic currents and analyzing the harmonic distortion. Harmonic impedance definitions for the load model, the power network model, and the load reactive power compensation equipment model will have a significant impact on the harmonic impedance matrix. As a result, an accurate harmonic impedance model is the key to obtaining credible calculation results.

8.4.1.1.1 Harmonic Impedance Equivalent Model of Network Components

A power network model contains the models of power source(s), transformers, transmission lines, power capacitors, reactors, and so on. It is not accurate to do simulation studies just using a simple $R + jX$ model. As the harmonics have more significant skin effect and positive/negative/zero-sequence harmonics have different propagation laws, it is necessary to develop accurate harmonic models for power networks and components. In other words, modeling those components according to the harmonic order and sequence is essential to harmonic simulation. Many achievements have been made in the research of the modeling. Hence, well-established harmonics models and mature power quality analysis tools are directly used here [4].

8.4.1.1.2 Equivalent Load Model

Load models in harmonic power flow study can be classified into three categories: constant impedance load model, induction motor load model, and a more realistic comprehensive load aggregation model.

For the constant impedance load model, the impedance parameters can be expressed in series or parallel form. For example, the series harmonic impedance model of Shojaie and Mokhtari [5] is expressed as $Z_s(h) = R_s + jhX_s$, and the parallel load model of Hui *et al.* [6] can be expressed as $Y_s(h) = 1/R_s - jh/X_s$ when harmonics exist. For analyzing three-phase loads, the load connection modes should include star connection and delta connection. Different connection modes lead to different harmonic impedance matrices and then different model expressions in the harmonic sequence component calculation. As the analytic expression for the harmonic impedance model of the induction motor load has been well discussed in literature, readers are referred to Yoshida *et al.* [7] for more details on the induction motor load model.

Owing to diverse load types and changing load operation conditions in the medium- and low-voltage distribution networks, a comprehensive aggregation load model has been developed. In the comprehensive aggregation load model, a centralized simulation method is generally used to provide the mathematical expressions of the harmonic impedance or the circuit models. However, it is difficult to develop a general comprehensive aggregation load model. In practical cases, the model is developed via field measurements incorporated with a model matching method. A model set consisting of various forms of series and parallel structures of R, L, C, and transformers can be developed first. In the model set, $(R_1 + L_1) + (R_2 \| L_2)$ and $(R_1 + L_1) \| (R_2 + L_2)$ are typical structures. Meanwhile, the parallel branch models can be developed to specifically model reactive power compensators at the user side. Second, professionals will choose the best models from the set based on the onsite tests and their experiences [8].

8.4.1.1.3 Harmonic Sequence Component Power Flow

In a frequency-domain analysis method for calculating the harmonics, the nth harmonic is decomposed into a positive-sequence component, negative-sequence component, and zero-sequence component. The positive-sequence impedance model, negative-sequence impedance model, and zero-sequence impedance model of the distribution network for this order of harmonic are built and then the positive-sequence harmonic power flow, negative-sequence harmonic power flow, and zero-sequence harmonic power flow of this harmonic are calculated.

The simplified calculation of a three-phase load is assumed to be symmetrical. According to the simplification of the harmonic order, the harmonics are classified into positive-sequence harmonics ($h = 3n + 1$), negative sequence harmonics ($h = 3n - 1$), and zero-sequence harmonics ($h = 3n$). A distribution network sequence impedance model (positive/negative/zero) is then constructed for the corresponding harmonic order. Finally, different impedance models based on the harmonic order are selected for harmonic power flow calculation. For example, in order to calculate the power flow of the second harmonic to the 25th harmonic, we need to calculate the power flow of all eight negative-sequence harmonics (2nd, 5th,…, 23rd harmonics), the power flow of all eight zero-sequence harmonics (3rd, 6th,…, 24th harmonics), and the power flow of all eight negative-sequence harmonics (4th, 7th,…, 25th harmonics).

By solving the aforementioned harmonic network equations, the voltage and current of each harmonic at each node and the distribution of each harmonic in the system is obtained. In addition to calculating the mode harmonic voltage and branch harmonic current, calculation of the THD for voltage THD_u is also required.

8.4.1.2 Photovoltaic Harmonic Model

PVs are connected to the grid via inverters, which can produce a wide spectrum of harmonics. In this section, a two-dimensional Fourier series analysis method is used to analyze the harmonics generated by an inverter in the pulse-width modulation (PWM) process.

In Figure 8.7, $x(t)$ represents the time variable of the carrier function and $x(t) = \omega_c t + \theta_c$, in which the radian carrier frequency is $\omega_c = 2\pi/T_c$, with T_c being the carrier cycle and θ_c is any phase shift angle of the carrier waveform; $y(t)$ represents the time variable of the modulation wave and $y(t) = \omega_o t + \theta_o$, in which the modulation wave radian requency is $\omega_o = 2\pi/T_o$, with T_o being the modulation wave cycle (the fundamental frequency in China is 50 Hz) and θ_o is any phase shift angle of the modulation waveform. The function $f(t) = f[x(t), y(t)]$ represents the output voltage of each phase bridge.

Figure 8.7 A PWM strategy.

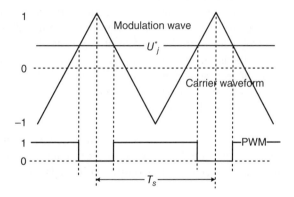

If the upper bridge switch is turned on, the output voltage of the corresponding bridge arm is $f(t) = U_{dc}$; if the lower arm switch is turned on, the output voltage of the corresponding arm is $f(t) = 0$ for an arm in a single-bridge inverter. The following two-dimensional Fourier series analysis can be obtained based on the typical one-dimensional Fourier series analysis:

$$
f(x,y) = \underbrace{\frac{A_{00}}{2}}_{①} + \underbrace{\sum_{n=1}^{\infty}(A_{0n}\cos ny + B_{0n}\sin ny)}_{②} + \underbrace{\sum_{n=1}^{\infty}(A_{m0}\cos mx + B_{m0}\sin mx)}_{③}
$$
$$
+ \underbrace{\sum_{m=1}^{\infty}\sum_{n=1}^{\infty}[A_{mn}\cos(mx+ny) + B_{mn}\sin(mx+ny)]}_{④} \tag{8.12}
$$

where

$$
A_{mn} = \frac{1}{2\pi^2}\int_{-\pi}^{\pi}\int_{-\pi}^{\pi} f(x,y)\cos(mx+ny)\,dx\,dy \tag{8.13}
$$

$$
B_{mn} = \frac{1}{2\pi^2}\int_{-\pi}^{\pi}\int_{-\pi}^{\pi} f(x,y)\sin(mx+ny)\,dx\,dy \tag{8.14}
$$

Or it can be expressed in the form of a complex number:

$$
C_{mn} = A_{mn} + jB_{mn} = \frac{1}{2\pi^2}\int_{-\pi}^{\pi}\int_{-\pi}^{\pi} f(x,y)e^{j(mx+ny)}\,dx\,dy \tag{8.15}
$$

In Equation 8.12, the main definitions are as follows:

- ① is the DC offset;
- ② are the fundamental component and baseband harmonics, corresponding to the low-order harmonic components;
- ③ are the carrier harmonics, corresponding to the components of high-frequency harmonics, whose frequencies are integral multiples of the carrier frequency;
- ④ are the sideband harmonics, the result of the interaction between the modulation wave and the carrier.

The waveform decomposition effect of the two-dimensional Fourier series analysis is equivalent to that of a traditional Fourier series analysis. However, by using the two-dimensional Fourier series analysis the harmonic components generated by both the carrier and the modulation waves is obtained clearly. In the design of the inverter control algorithm, the control of low-order harmonic components represented by ② is the focus; in the design of the filter circuit, the suppression of the high-frequency components represented by ③ is the focus. This method can be used to complete the design of the inverter control circuit.

When a single-phase inverter adopts the triangular modulation strategy, the expression of the triangular carrier is $z_1 = x/\pi$. Meanwhile, the expression of the modulation wave is $z_2 = M\cos y = M\cos\omega_0 t$, in which M is the modulation index, the amplitude ratio of the modulation wave to the carrier. When the modulation wave z_2 is greater than the carrier z_1, the output voltage is $2U_{dc}$; when the modulation wave z_2 is

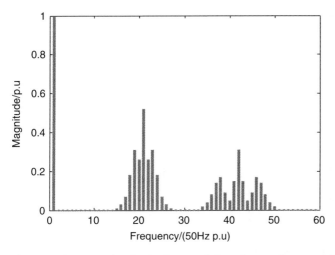

Figure 8.8 Harmonic distribution in case of triangular waveform modulation in a single-phase inverter.

smaller than the carrier z_1, the output voltage is zero. This is shown in the following equation:

$$\begin{cases} f(x,y):0 \rightarrow 2\pi & x = 2p - 1 \\ f(x,y):2\pi \rightarrow 0 & x = 2p\pi + \pi M \cos \omega_0 t \end{cases} \tag{8.16}$$

The formula for the calculating the coefficients in the two-dimensional Fourier series analysis of the natural sampling PWM in a single-phase system is

$$A_{mn} + jB_{mn} = \frac{1}{2\pi^2} \int_{-\pi}^{\pi} \int_{-\pi}^{\pi M \cos y} 2V_{dc} e^{j(mx+ny)} \, dx \, dy \tag{8.17}$$

The modulation index M is set to 0.9. When the carrier ratio is 21, Equation 8.12 can be used to obtain the harmonic distribution when a triangular waveform modulation is used for a single-phase inverter, which is shown in Figure 8.8.

In Figure 8.8, the horizontal axis represents the harmonics order and the vertical axis represents the harmonic contents. It can be seen from Figure 8.8 that the harmonics generated in the triangular waveform modulation process are composed of the low-frequency components of the fundamental wave and the baseband harmonics distributed around the carrier and the integral multiples of the carrier. If the symmetrical modulation strategy is used, the calculation function of the two-dimensional Fourier series analysis can be changed into

$$A_{mn} + jB_{mn} = \frac{1}{2\pi^2} \int_{-\pi}^{\pi} \int_{-(\pi/2)(1+M \cos y)}^{(\pi/2)(1+M \cos y)} 2V_{dc} e^{j(mx+ny)} \, dx \, dy \tag{8.18}$$

The distribution of the output harmonics of the inverter obtained is shown in Figure 8.9. It can be seen from the comparison between Figures 8.8 and 8.9 that after the symmetrical modulation strategy is used, the inverter output harmonics will be significantly reduced. Compared with the triangular waveform modulation, the output waveform of the inverter is greatly improved.

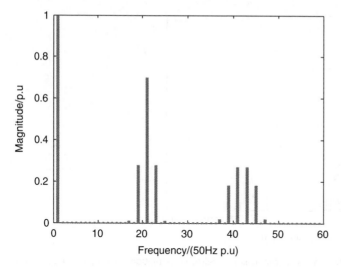

Figure 8.9 Harmonic distribution of a symmetrical modulation in a single-phase inverter.

For three-phase inverters, in order to obtain the harmonic in the output voltages, the output voltage harmonic distributions of individual phases are calculated first and the harmonic content of the phase voltage can be obtained by subtracting the corresponding two phases. When the space vector PWM (SVPWM) strategy is used, under the rated conditions, we take the A-phase as an example and express its modulation wave as follows:

$$
u_A = \begin{cases}
M \cos\left(\theta - \frac{\pi}{6}\right), & 0 \le \theta < \frac{\pi}{3}, \pi \le \theta < \frac{4\pi}{3} \\
\sqrt{3}M \cos\theta, & \frac{\pi}{3} \le \theta < \frac{2\pi}{3}, \frac{4\pi}{3} \le \theta < \frac{5\pi}{3} \\
M \cos\left(\theta + \frac{\pi}{6}\right), & \frac{2\pi}{3} \le \theta < \pi, \frac{5\pi}{3} \le \theta < 2\pi
\end{cases}
\tag{8.19}
$$

The two-dimensional Fourier series analysis for natural sampling space vector modulation is given as follows:

$$
A_{mn} + jB_{mn} = \frac{1}{2\pi^2} \sum_{i=1}^{6} \int_{y_l(i)}^{y_u(i)} \int_{x_l(i)}^{x_u(i)} 2V_{dc} e^{j(mx+ny)} \, dx \, dy
\tag{8.20}
$$

Since the modulation wave of the SVPWM is a piecewise function, the upper and lower limits are listed in Table 8.9.

The frequency modulation index of the system is set to be 21 and the amplitude modulation index is set to be 0.9. The Fourier analysis of the output voltage of the inverter is then carried out. The harmonic distribution is shown in Figure 8.10. It can be seen that when SVPWM is used, the immediate output voltage of the inverter contains the low-order sideband harmonic components. In addition, in order to optimize the distrubtion of the phase-to-phase harmonic components of the three-phase inverter, the ratio of the carrier frequency (i.e., the frequency modulation index) of the three-phase inverter to the fundamental reference frequency needs to be an odd number and a multiple of 3.

Table 8.9 Piecewise representation of SVPWM process.

L	$y_l(i)$	$y_u(i)$	$x_l(i)$	$x_u(i)$
1	$2\pi/3$	π	$-\frac{\pi}{2}\left[1+\sqrt{3}M\cos\left(y+\frac{\pi}{6}\right)\right]$	$\frac{\pi}{2}\left[1+\frac{\sqrt{3}}{2}M\cos\left(y+\frac{\pi}{6}\right)\right]$
2	$\pi/3$	$2\pi/3$	$-\frac{\pi}{2}\left(1+\frac{3}{2}M\cos y\right)$	$\frac{\pi}{2}\left(1+\frac{3}{2}M\cos y\right)$
3	0	$\pi/3$	$-\frac{\pi}{2}\left[1+\frac{\sqrt{3}}{2}M\cos\left(y-\frac{\pi}{6}\right)\right]$	$\frac{\pi}{2}\left[1+\frac{\sqrt{3}}{2}M\cos\left(y-\frac{\pi}{6}\right)\right]$
4	$-\pi/3$	0	$-\frac{\pi}{2}\left[1+\frac{\sqrt{3}}{2}M\cos\left(y+\frac{\pi}{6}\right)\right]$	$\frac{\pi}{2}\left[1+\frac{\sqrt{3}}{2}M\cos\left(y+\frac{\pi}{6}\right)\right]$
5	$-2\pi/3$	$-\pi/3$	$-\frac{\pi}{2}\left(1+\frac{3}{2}M\cos y\right)$	$\frac{\pi}{2}\left(1+\frac{3}{2}M\cos y\right)$
6	π	$-2\pi/3$	$-\frac{\pi}{2}\left[1+\frac{\sqrt{3}}{2}M\cos\left(y-\frac{\pi}{6}\right)\right]$	$\frac{\pi}{2}\left[1+\frac{\sqrt{3}}{2}M\cos\left(y-\frac{\pi}{6}\right)\right]$

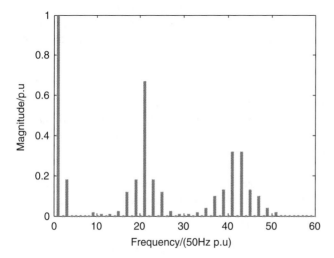

Figure 8.10 Harmonic distribution in case of three-phase SVPWM.

Based on the theoretical analysis, it can be concluded that the harmonic current of a distributed PV inverter is affected by many factors, including the carrier frequency, the modulation frequency, the modulation scheme, the output filter parameters, the harmonic voltage of the converter at the integration point, the output harmonic impedance, and the inverter output power. Without loss of generality, this can be summarized as follows:

$$I^k = g_k(U^1, U^3, U^5, \dots, U^H, C^1, C^2, \dots C^m), \quad k = 1, 3, 5, \dots, H \tag{8.21}$$

where I^k is the kth harmonic current generated by the harmonic source, $U^1, U^3, U^5, \dots$, U^H are the fundamental and harmonic components of the voltage at the harmonic source node, $C^1, C^2, \dots, C^m$ are the control parameters of the harmonic source, and H is the highest harmonic order needed to be considered. If all the control parameters C^1, $C^2, \dots, C^m$ can be obtained, each harmonic current generated by the harmonic source can then be accurately calculated. As it is very difficult to get all the control parameters, this model is too complicated to be useful.

Under stable conditions, it can be assumed that all the control parameters of the harmonic source C^1, C^2,..., C^m remain constant, and therefore, Equation 8.21 can be rewritten as

$$I^k = g_k(U^1, U^3, U^5, \dots, U^H), \quad k = 1, 3, 5, \dots, H \tag{8.22}$$

The harmonic source model can be expressed in many forms, and the constant current source model, the Norton model, and the simplified model based on least-squares approximation are frequently used.

In the constant current source model [9], it can be assumed that the harmonic current generated by the harmonic source depends on the fundamental voltage at the node. Then, Equation 8.22 can be simplified as

$$I^k = g_k(U^1), \quad k = 1, 3, 5, \dots, H \tag{8.23}$$

If the fundamental voltage remains constant during the study period, the determined harmonic injection current of the harmonic source remains constant. In this case the harmonic source is considered as the combination current sources of individual harmonics:

$$I^k = I_0^k, \quad k = 1, 3, 5, \dots, H \tag{8.24}$$

As in the actual power grid the $THD_u\%$ for voltage at each node is normally smaller than the stipulated limit, the harmonic voltage is much smaller than the fundamental voltage. Accordingly, the harmonic voltage has little effect on the harmonic injection current and can be neglected when the calculation precision is not required to be very accurate.

In the Norton model [109], the nth harmonic current generated by the harmonic source is expressed as the function of the fundamental voltage and the harmonic voltage:

$$I^{(k)} = g_k(U^{(1)}, U^{(k)}), \quad k = 1, 2, 3, \dots, H \tag{8.25}$$

The harmonic voltage and harmonic current at the output of the harmonic source can be obtained through actual tests. Meanwhile, the complicated relationship between each harmonic current and voltage can be established based on a least-squares approximation method [11]. To obtain the model parameters, multiple tests are needed to run the least-squares method.

The main purpose of this part is to analyze the impact of harmonic current injection during the stable period of time under adverse conditions. To be typical, the method of equal scaling of the actually measured parameters according to the harmonic current frequency spectrum of the PV inverter is used to establish the constant current source model. This is the most widely used method in actual projects and in most harmonic analysis software programs.

8.4.2 Harmonic Analysis of Photovoltaic Generation Connected to a Typical Feeder

In this section a typical feeder is taken as an example to explain the impacts of grid-connected PVs on the harmonics of the distribution network. The focus is laid on the introduction of different impacts at different connection locations, PV capacities, and integration modes (centralized connection or distributed connection). The typical feeder parameters given in Figure 5.3 are used in the study. The typical feeder structure, distribution network equipment parameters, load parameters, and load harmonic model are all predefined.

The short-circuit capacity of the 110 kV bus at the 110 kV substation is set to be 1500 MVA and the rated capacity of the 110 kV transformer is 50 MVA. The short-circuit impedance is set to be 17%. The LJ-240 conductor is used for the power line of the typical feeder. In this section, it is assumed that 10 loads are evenly connected into the typical feeder with each load of 500 kVA at a power factor of 0.95. The comprehensive load model of the $(R_1 + L_1) + (R_2 \parallel L_2)$ structure is utilized. Each of the two parts in the series accounts for 50%. In the model, no reactive power compensator is installed at the load side.

8.4.2.1 Harmonics Analysis of Centralized Photovoltaic Connection

According to the PV integration plan shown in Table 5.2 in Section 5.3.2, it is assumed that the PV system, with a total capacity of 3 MW and full power output, is connected to nodes 1, 5, and 9 of the typical feeder, respectively. The PV system is also assumed to be at its rated output power for the harmonic analysis. The impacts of different connection locations on the feeder harmonic voltage are analyzed. The simulation results are shown in Figure 8.11. In Figure 8.11, the comparison of the $THD_u\%$ on the feeder is made among PV nodes 1, 5, and 9. In all cases, the overall $THD_u\%$ on the feeder rises after the PV integration. When no harmonic suppression devices are connected to the feeder, the $THD_u\%$ at the end of the feeder is greater than that at the head of the feeder.

For the centralized connection, the PV generation connected at the end of the feeder has a greater impact on $THDu\%$ than that at the head of the feeder. When the 3 MW distributed PV generation is connected to node 9, the maximum $THD_u\%$ rise occurs at node 10 with $THD_u\%$ of 0.47%. According to Chinese national standard GB/T 14549-1993, the maximum limit value of the $THD_u\%$ of a 10 kV network is 4%.

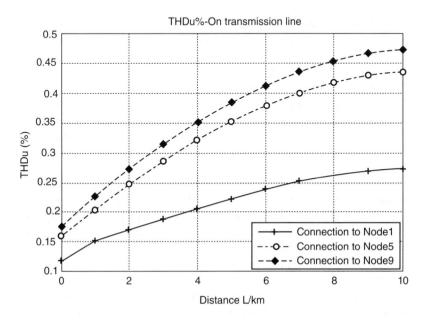

Figure 8.11 Distribution of $THD_u\%$ on feeder after the connection of PVs to different nodes.

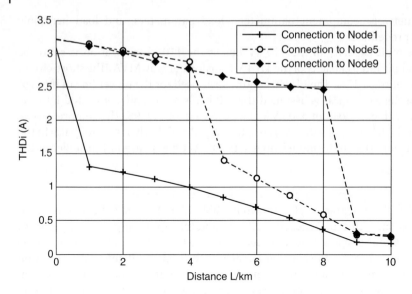

Figure 8.12 Distribution of THD_i on the feeder at different PV nodes.

Hence, in this case, the PV uses up about 11.7% of the allowable harmonic distortion. Similar procedures can be developed to analyze the 5th, 7th, 11th, and other orders of harmonics.

Through the aforementioned analysis, several conclusions can be drawn. The connection of the distributed PVs affects the harmonic voltage level of the entire distribution network. The THD_u% rise caused by the connection of the distributed PV generation is more significant at the end of the distribution line than that at the head of the distribution line. Therefore, without considering the harmonic suppression measures, it is highly recommended that PV generation should be connected to the head of the feeder.

Figure 8.12 shows the distribution of current harmonics (THD_i) at three different PV connection locations. It is found that, at different locations, the current harmonics on the typical feeder vary and the PV injection harmonic currents mainly flow toward the locations where the harmonic impedance is low. Therefore, the current harmonics flowing into the head of the feeder are greater than that coming into the end in the example. In addition, for a given capacity, if the PV generation is connected to the end of the feeder, most of its harmonic currents flow through the whole distribution line, resulting in more harmonic losses.

According to the PV connection plan shown in Section 5.3.2, the distributed PV generation is connected to node 1 on the typical feeder. The impact of different capacities of grid-connected PV generation on the distribution network voltage without changing other operation conditions is analyzed. The capacity of the grid-connected PV generation is the same as that shown in Table 5.3, and the PV system output is at its rated power. The calculation results of THD_u (percent) and THD_i (amperes) are shown in Figures 8.13 and 8.14 respectively.

Figure 8.13 shows that the THD_u% increases as the PV capacity grows when connected at node 1. When a 5 MW PV generation is connected to node 1, the THD_u% at node 10 (the end of a typical feeder) reaches 0.4%. By observing the distribution of current

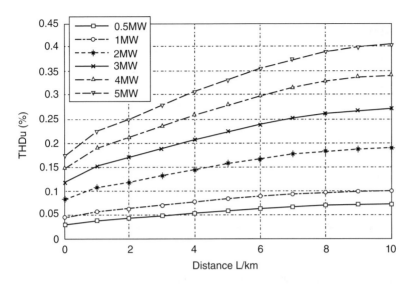

Figure 8.13 Distribution of THD$_u$% under different sizes of PV generation at node 1.

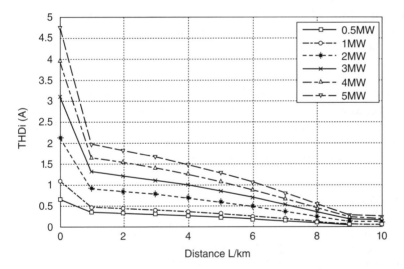

Figure 8.14 Distribution of THD$_i$ under different sizes of PV generation to node 1.

harmonics shown in Figure 8.14, there are more current harmonics flowing toward the head of the feeder. The current harmonics flowing into the head of the feeder are almost twice the current harmonics flowing into the end of the feeder.

If the aforementioned conditions remain the same and the PV generation is now connected to node 5, which is 5 km away from the head of the feeder, the calculation results of THD$_u$% are shown in Figure 8.15. It can be seen from the analysis of the harmonic power flow distribution that the PV integration raises the voltage harmonic level of the whole feeder. Compared with those results when the PV generation is connected to node 1, the THD$_u$% rise is higher in this case. It can be seen from Figure 8.15 that when the

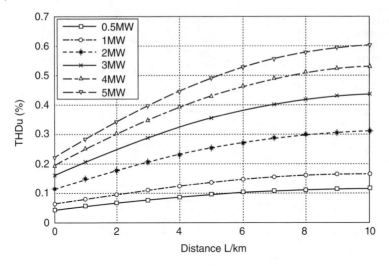

Figure 8.15 Distribution of THD_u% under different sizes of PV generation at node 5.

4 MW PV generation is connected to node 5 the THD_u% at the location 4 km from the head of the feeder is calculated to be 0.4%, and the THD_u% at the end of the feeder is 0.53%. When the 5 MW PV generation is connected to node 5, the THD_u% at the location 3 km from the head of the feeder is calculated to be 0.4%, and the THD_u% at the end of the feeder is 0.6%. It is clear that, when the capacity of the grid-connected PV generation exceeds 4 MW, the voltage harmonics at the middle section of the feeder will exceed 10% of the standard limit value stipulated by the Chinese national standards.

When the PV generation is connected to node 9, instead of node 5, the THD_u% profile along the feeder obtained by simulation is shown in Figure 8.16, where it can be seen that THD_u% still monotonically increases in this case. However, compared with the case of connection to node 5, the THD_u% rise slows down. For example, when the 5 MW

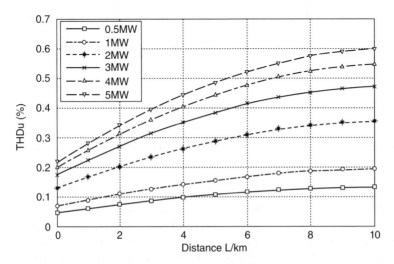

Figure 8.16 Distribution of THD_u% under different sizes of PV generation at node 9.

PV generation is connected to node 9, the THD$_u$% at the location 3 km from the head of the feeder is calculated to be 0.4% and the THD$_u$% at the end of the feeder is 0.6%. Obviously, the line impedance plays a significant role in determining the path that the harmonic currents flow through.

Generally, the connection of PV generation raises the harmonic voltage level of the feeder. Without considering the harmonic suppression equipment in the feeder, the THD$_u$% along the whole feeder increases. The connected PV generation has a much greater impact on harmonics on the end of the feeder than on the head.

8.4.2.2 Harmonics Analysis of Photovoltaic Connection in a Distributed Way

After analyzing the single-node (centralized) connection of PV generation the distribution of voltage harmonic for the multiple-node connection of PV generation is discussed. The distribution network structure parameters and load data are the same as those in the single-node connection. There are five 1 MW PV systems with a total capacity of 5 MW connected to the feeder. The distributed connection locations are shown in Table 5.4 and the simulation results are shown in Figure 8.17.

Curve 1 represents the distributed connection of five PV systems to the first half section of the typical feeder, with the five PV systems distributed from node 1 to node 5. Curve 2 is the result when the five PV systems are connected evenly to the whole feeder, such as nodes 1, 3, 5, 7, and 9. The result of curve 3 is for the situation that the PV systems are connected to the second half section of the feeder, namely node 6 to node 10. Figure 8.17 shows the distribution trend of THD$_u$%, which generally increases from the head to the end of the feeder for all the three PV connection schemes. The difference between the three schemes is that the THD$_u$% rise caused by the connection of the PV generation to the first half section of the feeder is greater than that in the other two cases. By comparing curves 2 and 3, there is only a slight difference in THD$_u$% between the PV connection evenly distributed to the whole feeder and the distributed connection to the second half section of the feeder.

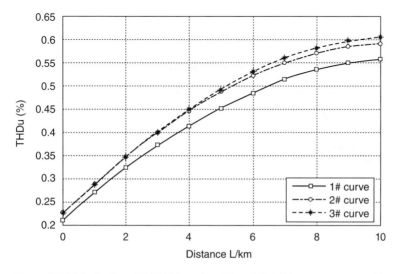

Figure 8.17 Distribution of THD$_u$% introduced by a 5 MW PV system connected in a distributed way.

8.4.3 Analysis of Practical Cases

In this section the connection of PV generation to the 110 kV SA substation in City J is taken as an example to analyze the voltage harmonics. The PV generation is connected to a total of five substation section I buses: SA_F1, SA_F5, SA_F7, SA_F9 and SA_F18. As discussed in Section 5.5.3, two PV systems are connected to SA_F1, with a total capacity of 3 MW. The capacity of the PV generation connected to each bus is shown in Table 8.10.

In the evaluation process, the annual power flow in practical cases is analyzed, which has been completed in Chapter 5. Through the analysis it is found that the most serious phenomenon of reverse power flow occurred at the substation 10 kV bus at 13:00 on May 1, 2014. Obviously the PV harmonic injection at this point might cause the most serious voltage distortion. Now the impact of the PV connection at this moment on the distribution network feeder harmonics will be analyzed.

Figure 8.18 shows the distribution of $THD_u\%$ along SA_F1. It can be seen that, without considering the exiting load harmonics of SA_F1, the integration of the PV systems of the vehicle company and the battery company to bus SA_F1 raises the harmonic voltage level of the bus. At the end of bus SA_F1, $THD_u\%$ reaches 0.36%, which alone satisfies the limit value of the harmonic voltage in the Chinese standards. However, when many PV

Table 8.10 Grid-connected PV capacities in practical cases.

Feeder	Grid-connected PV capacity(MW)		Total PV capacity(MW)
SA_F1	Vehicle company C1	2.0	3.0
	Battery company C2	1.0	
SA_F5	Science and technology company C3	1.0	1.0
SA_F7	Chemical company C4	2.5	2.5
SA_F9	Clothing company C5	1.0	1.0
SA_F18	Industrial company C6	2.0	2.0

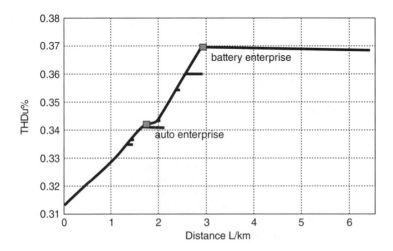

Figure 8.18 Distribution of $THD_u\%$ along bus SA_F1 in the test substation.

systems are connected to the distribution network at the same time, the total generated harmonics may violate the harmonic distortion requirements.

As the distribution network with high PV peneration is different from the connection of large PV stations, it should be analyzed from the perspective of the regional grid comprehensively. The comprehensive evaluation includes the impact evaluation of PV systems connected to the distribution network on harmonic levels. One important aspect is to judge and evaluate each allowable harmonic indicator allocated to the PV systems. Furthermore, in real applications, overloading issues of harmonic suppression equipment at the user side and the grid side (e.g., passive power filter) should be evaluated and avoided.

8.5 Summary

This chapter has discussed the impact of the grid-connected distributed PV generation on power quality. Chinese standards and other international standards on PV integration harmonic requirements are compared and discussed. The differences in power quality terms and the methods for analyzing the impacts due to distributed PV generation on the power quality of the distribution network are given. Meanwhile, examples are given to analyze the impacts caused by distributed PV generation on the current and voltage harmonics on a typical feeder model and an actual 110kV substation regional distribution network.

It can be seen from the analysis results that grid-connected PV generation raises the voltage harmonics of the whole feeder and the network, not just the integration points without considering the harmonic suppression equipment in the feeder/network. For an feeder, distributed PVs generate greater impacts on the end of the feeder than that on the head of the feeder. The voltage harmonic distortion of the distribution network caused by one individual medium- and small-size PV system alone may not be significant. However, the resultant harmonics generated from multiple harmonic sources can be very severe and violate the standards. Hence, a comprehensive harmonic analysis and evaluation is necessary for distribution networks with high PV penetration.

References

1 Yang C., Liu K., and Wang D. (2009) Harmonic resonance circuit's modeling and simulation. In *2009 Asia-Pacific Power and Energy Engineering Conference*. IEEE; pp. 1–5.

2 GB/T 14549-1993. *Quality of Electric Energy Supply – Harmonics in Public Supply Network*. China Zhijian Publishing House.

3 Bhattacharyya S., Cobben S., Ribeiro P., *et al.* (2012) Harmonic emission limits and responsibilities at a point of connection. *IET Generation, Transmission & Distribution*, **3**, 256–264.

4 Wakileh, G.J. (2001) *Power Systems Harmonics: Fundamentals, Analysis, and Filter Design*. Springer: Berlin.

5 Shojaie M. and Mokhtari H. (2014) A method for determination of harmonics responsibilities at the point of common coupling using data correlation analysis. *IET Generation, Transmission & Distribution*, **1**, 142–150.

6 Hui J., Yang H., Lin S., and Ye M. (2010) Assessing utility harmonic impedance based on the covariance characteristic of random vectors. *IEEE Transactions on Power Delivery*, **3**, 1778–1786.

7 Yoshida Y., Suzuki H., Fujiwara K., and Ishihara Y. (2016) Basic study on island-ing detection method for PV systems by harmonic impedance detection in case of including induction motor in load system. *Electrical Engineering in Japan*, **3**, 20–34.

8 Burch R., Chang G., Hatziadoniu C., *et al.* (2003) Impact of aggregate linear load modeling on harmonic analysis: a comparison of common practice and analytical models. *IEEE Transactions on Power Delivery*, **2**, 625–630.

9 Atkinson-Hope G. and Stemmet W. (2006) Assessing harmonic current source mod-elling and power definitions in balanced and unbalanced networks. *International Journal of Energy Technology & Policy*, **4**(1–2), 85–102.

10 Almeida, C.F.M. and Kagan N. (2010) Harmonic coupled Norton equivalent model for modeling harmonic-producing loads. In *Proceedings of 14th International Con-ference on Harmonics and Quality of Power – ICHQP 2010*. IEEE; pp. 1–9.

11 Zhang W. and Yang H. (2004) A method for assessing harmonic emission level based on binary linear regression. *Proceedings of the Chinese Society of Electrical Engineering*, **6**, 50–53.

9

Techniques for Mitigating Impacts of High-Penetration Photovoltaics

9.1 Introduction

As introduced in previous chapters, due to the ever-increasing amount of grid-connected PVs in distribution networks, the PV penetration will increase accordingly. When the capacity of the PVs reaches a certain level, a series of problems, such as reversal tidal current, overvoltage, short-circuit current, and power quality issues, may not only restrict the accommodation of PVs in the distribution network, but also seriously impair the normal operation of the distribution network. However, with the rapid development of active distribution networks and the maturation of different kinds of enabling technologies, the PV hosting capacity of the distribution network will be effectively increased. Various technologies can be used to avoid or eliminate the potential detrimental impacts of PVs and to increase the economic efficiency of PVs. In this chapter, it will focus on the application of energy storage technologies, demand response (DR), and an advanced strategy for cluster partition control for integrating high-penetration PVs into the distribution network.

9.2 Energy Storage Technology

Energy storage technologies can be used to convert electrical energy into other forms of energy that can be released in the form of electrical energy when needed. Together with certain energy storage systems, unschedulable PVs can be operated as schedulable generating units to achieve grid-connected operation with utility, such as supply power to the power company, peak shaving, emergency support and other services when necessary. Energy storage is also a necessary and critical part for standalone PV systems, which will be further discussed in detail in Chapter 10. Energy storage also enables grid-connected PVs to be more self-sustainable when they are islanded to run autonomously. As a quick summary, the energy storage system can help stabilize the PV output, improve voltage quality, shift peak loads, and increase the system reliability and resilience. It is one of the most effective approaches for eliminating the adverse impacts caused by PV integration and improving the PV hosting capacity.

Grid-Integrated and Standalone Photovoltaic Distributed Generation Systems: Analysis, Design, and Control, First Edition. Bo Zhao, Caisheng Wang and Xuesong Zhang.
© 2018 China Electric Power Press. Published 2018 by John Wiley & Sons Singapore Pte. Ltd.

9.2.1 Classification of Energy Storage Technologies

From the energy conversion point of view, the electrical energy storage technologies can be categorized into mechanical energy storage, electromagnetic energy storage, phase-change energy storage, and chemical energy storage, as shown in Figure 9.1 [1–3].

9.2.1.1 Mechanical Energy Storage

Pumped hydro storage power stations are still the kind of energy storage technique that occupies the largest storage capacity. With reservoirs both in the upper and lower stream, the pumped hydro storage equipment can switch between the generator state and motor state. Water is pumped to store energy at load valley periods, and discharged to generate power at peak load periods. Characterized by long service life, stable energy conversion efficiency, and small impact on the environment, it is the most mature large-scale energy storage technology in the electrical power system.

In compressed air energy storage, energy is stored by converting the electrical power to compressed air in a high-pressure sealed air storage cabinet; when discharging energy, high-pressure air is released to drive a gas turbine to generate power. It is characterized by a high safety factor, large capacity, and long service life. However, its application is limited by its low energy density and geographical limitations.

A flywheel is a kind of energy storage where the electrical energy is stored in the form of kinetic energy in a rapidly rotating flywheel. Mainly composed of a flywheel, a bearing, and a motor system, the flywheel storage has advantages such as high power density, wide operation temperature range, and no limitation on the charging and discharging times; but it also has disadvantages, such as low energy density and high cost of system safety maintenance.

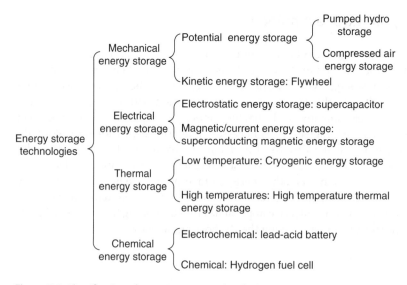

Figure 9.1 Classification of energy storage technologies.

9.2.1.2 Electromagnetic Energy Storage

The supercapacitor is a device that achieves energy storage by forming an electric double layer on the electrode surface through an applied electric field. Both the charging and discharging process are physical processes, and it has advantages such as long service life, high power density, providing high discharging current, and rapid charge.

The superconducting magnetic energy storage technology enables the energy storage device to use a superconducting coil to store the electrical energy in the form of magnetic energy. The magnetic energy is supplied to load in the form of electrical energy when needed. It has advantages such as long service life, quick response, and high energy storage efficiency. However, it suffers from low specific energy and energy density, and relatively high construction cost.

9.2.1.3 Phase-Change Energy Storage

The phase-change material is a kind of material with special functions and with the phase change in the gas-liquid process or solid-liquid process, it can form a constant temperature or an approximate constant temperature. In a phase-change process a large amount of energy is released or absorbed. The phase-change energy storage device achieves energy storage or release through the phase-change process of the phase-change material and has advantages such as small size, high energy density, flexible design, and easy utilization and management.

Cryogenic energy storage is a developing technology by using off-peak power or renewable energy sources to generate cryogenic fluid, which can then be used in a cryogenic heat engine to generate electricity. On the other hand, high-temperature phase change materials can be used for high-temperature thermal energy storage. High-temperature thermal energy storage is mainly applied to industrial thermal energy storage systems and high-temperature solar thermal energy storage systems. Typical applications include the use of water loss and water absorption of salt hydrates to complete industrial waste heat recovery, the thermal energy storage system in solar thermal power generation, and so on.

9.2.1.4 Chemical Energy Storage

Electrochemical energy storage is an energy storage technology that can achieve the interconversion of chemical energy and electrical energy through electrochemical reactions. There are many types of electrochemical energy storage, including lead–acid battery, Li-ion battery, flow cell, sodium–sulfur cell, and so on. In recent years, rapid development and great breakthroughs have been made in its safety, energy conversion efficiency, and economy. In addition, it has advantages such as easy utilization, little environmental pollution, no geographical restrictions, no restriction of Carnot cycle, high conversion efficiency, high specific energy, and high specific power.

The technical maturity of different kinds of energy storage technologies is summarized and shown in Figure 9.2 [4].

9.2.2 Electrochemical Energy Storage

At present, the energy storage technology widely used in the power system is the electrochemical energy storage approach, including lead–acid battery, Li-ion battery, and flow cell. Three electrochemical energy storage technologies will be briefly introduced [5–7].

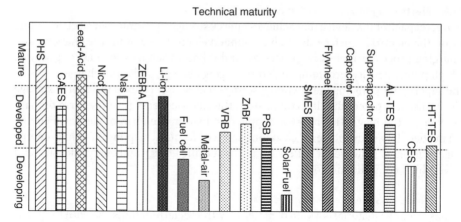

Figure 9.2 Technical maturity of energy storage technologies.

9.2.2.1 Lead–Acid Battery

In a lead–acid battery, the anode is made of a loose-porous finely-crystallized lead diox-ide, which is usually mahogany. The cathode consists of spongy metal lead, which is usually grey. The materials used at the anode and cathode to store energy are collec-tively called active material. An ordinary lead–acid battery uses pure dilute sulfuric acid as the electrolyte, which is mainly used to take part in the chemical reaction on the elec-trode plate, conduct ions, and reduce the temperature of the battery in the reaction. The formulas of the charging and discharging chemical reactions of a typical lead–acid battery are as follows:

$$\text{Reaction at anode}: \quad PbO_2 + 3H^+ + HSO_4^- + 2e^- \underset{\text{Charge}}{\overset{\text{Discharge}}{\rightleftarrows}} PbSO_4 + 2H_2O$$

$$\text{Reaction at cathode}: \quad Pb + HSO_4^- \underset{\text{Charge}}{\overset{\text{Discharge}}{\rightleftarrows}} PbSO_4 + H^+ + 2e^-$$

$$\text{Overall reaction}: \quad PbO_2 + Pb + 2H_2SO_4 \underset{\text{Charge}}{\overset{\text{Discharge}}{\rightleftarrows}} 2PbSO_4 + 2H_2O$$

It can be seen from these chemical reaction equations that, in the discharging process of a lead–acid battery, both the anode active material(lead dioxide), and the cathode active material(metallic lead), react with sulfuric acid electrolyte to generate lead sulfate, which is called the "double sulfation reaction" in the discharging process. Immediately after the discharging of the battery, the resultant lead sulfate transformed from the anode active material and cathode active material is now of very high activity. In the charging process of the battery, the loose and fine lead sulfate at the anode and cathode will be turned back into lead dioxide and lead metal under the action of the external charging current; the battery is then in the fully charged state once again.

The ordinary lead–acid battery energy storage system has been used for many years in power plants and substations as backup power, and has played an extremely impor-tant role in maintaining the safe, stable, and reliable operation of the power system. However, owing to the low energy density, low power density, long charging time, short

Table 9.1 Lead-carbon battery energy storage power station projects.

Location	System scale	Function	Year
Dongfushan, Zhejiang, China	0.2 MW/1 MW h	Frequency modulation and voltage regulation, stabilization of renewable energy output, etc.	2011
Albuquerque, New Mexico, USA	0.5 MW/2.8 MW h	Peak shaving, stabilization of PV power, etc.	2011
Goldsmith, Texas, USA	36 MW/24 MW h	Peak shaving, stabilization of wind power, etc.	2012
Luxi Island, Zhejiang Province, China	2 MW/4 MW h	Frequency modulation and voltage regulation, stabilization of renewable energy output, etc.	2014

cycle life, and high self-discharge rate of an ordinary lead–acid battery, and the possible environmental pollution it may cause, it has been used less and less in spite of its low cost. At present, some improved lead–acid batteries, such as the lead–carbon battery, are attracting more interest. Compared with the lead–acid battery, the specific power of a lead-carbon battery is substantially increased and the cycle life is also significantly improved in the case of shallow depth of discharge. Some typical projects and applications of lead–carbon batteries in distributed generation and microgrids are given in Table 9.1.

9.2.2.2 Lithium-Ion Battery

An Li-ion battery cell is mainly composed of an anode, a cathode, the electrolyte, the separator, the external connection and packaging components. The anode of the Li-ion battery has a high potential and is usually intercalated between lithium and transition metal oxide or polyanionic compound such as $LiMO_x$ and $LiMPO_4$ (where M is a transition metal, which can be one or more of Co, Ni, Mn, Fe, V, and other elements). $LiCoO_2$, $LiFePO_4$, $LiMn_2O_4$, are the common anode materials. The cathode materials of the Li-ion battery are usually carbon materials, such as graphite and non-graphite carbon. Non-graphite carbon includes soft carbon and hard carbon. Soft carbon is easily graphitized (e.g., mesocarbon microbeads), while hard carbon is usually high molecular polymer pyrolytic carbon. In addition, researchers have also worked on some new cathode materials, such as nanocrystalline transition metal oxides.

As a secondary battery, when the Li-ion battery is in the charging process, the lithium ions deintercalate from the anode inside the battery and pass through the separator via the electrolyte, and then intercalate into the cathode. The electrons transfer from the external circuit to the cathode outside the battery. In the discharging process, inside the battery, lithium ions deintercalate from the cathode, pass through the separator, and then intercalate back into the anode. The electrons again go from the external circuit to the anode. In the charging and discharging process, it is the "lithium ion" rather than "lithium" that migrates in batteries. As a result, the battery is called an "Li-ion battery."

The equations of the chemical reaction of an Li-ion battery in the charging and discharging process are as follows:

$$\text{Reaction at anode}: \quad \text{LiMO}_m \underset{\text{Discharge}}{\overset{\text{Charge}}{\rightleftharpoons}} \text{Li}_{1-x}\text{MO}_m + x\text{Li}^+ + xe^-$$

$$\text{Reaction at cathode}: \quad C_n + x\text{Li}^+ + xe^- \underset{\text{Discharge}}{\overset{\text{Charge}}{\rightleftharpoons}} \text{Li}_x C_n$$

$$\text{Over reaction}: \quad \text{LiMO}_m + C_n \underset{\text{Discharge}}{\overset{\text{Charge}}{\rightleftharpoons}} \text{Li}_{1-x}\text{MO}_m + \text{Li}_x C_n$$

Compared with other conventional batteries, the Li-ion battery has advantages such as high specific energy, high rated voltage, high power density, high power tolerance, and low self-discharge rate. Its specific energy (200 W h/kg) is about five times that of the lead–acid battery and its energy storage efficiency exceeds 90%. The working voltage of the monomer Li-ion battery is 3.7 V or 3.2 V, and its cycle life reaches 3000 times or even 5000 times in a shallow charge and discharge mode. However, the Li-ion battery has poor overcharging/charging tolerance performance, complicated protection circuits, and high cost. As a commercialized energy storage technology, the Li-ion battery now is generally used in integrating renewable energy, in microgrids and in improving the power quality of power system. Table 9.2 lists some Li-ion battery energy storage projects.

9.2.2.3 Flow Cell

The flow cell, or redox flow cell system, was first designed with financial aid provided by NASA. In 1974, L.H. Thaller made it public and applied for a patent. In contrast to traditional batteries, where the active material is contained in a solid anode or cathode, the active material in liquid form of a flow cell acts as both the electrode material and the electrolyte solution. The active material can be dissolved into the solutions in two liquid storage tanks and a pump circulates the solution to flow through the flow cell in each tank. The reduction and oxidation reactions occur at the electrode on each side of the ion-exchange membrane.

The two electrodes of the flow cell are composed of two flow stacks with different potentials. In the charging process, on one side of the ion-exchange membrane, the active material of the high potential stack is oxidized to the high valence state from the

Table 9.2 Li-ion battery energy storage power station projects.

Location	Rated output power (MW)	Function	Year
Zhangbei County, Hebei, China	20	Smoothing wind power, peak load shifting, etc.	2012
Kula, Hawaii, USA	11	Smoothing wind power, improving power quality, etc.	2013
West Virginia, USA	32	Peak load shifting, smoothing wind power output, etc.	2014
Nanji Island, Zhejiang, China	2	Smoothing renewable energy power output, frequency modulation, and voltage regulation	2014

low valence state at the anode. On the other side, the active material of the low potential stack is reduced to the low valence state from the high valence state at the cathode. In the discharging process, these two processes are reversed. The anode pair of the vanadium redox flow battery is $VO^{2+}/VO_2{}^+$, while the cathode pair is V^{2+}/V^{3+}. In the charging and discharging process, the electrode reactions are as follows:

$$\text{Reaction at the anode}: \quad VO^{2+} + H_2O \underset{\text{Discharge}}{\overset{\text{Charge}}{\rightleftharpoons}} VO_2{}^+ + 2H^+ + e^-$$

$$\text{Reaction at the cathode}: \quad V^{3+} + e^- \underset{\text{Discharge}}{\overset{\text{Charge}}{\rightleftharpoons}} V^{2+}$$

$$\text{Overall reaction}: \quad VO^{2+} + H_2O + V^{3+} \underset{\text{Discharge}}{\overset{\text{Charge}}{\rightleftharpoons}} VO_2{}^+ + 2H^+ + V^{2+}$$

The output power of the vanadium redox flow battery depends on the size and quantity of the stack, and its storage capacity depends on the capacity and concentration of the electrolyte. As a result, the scale design of a flow cell is very flexible. Increasing the stack area and quantity can enlarge the output power; meanwhile, the energy storage capacity can be increased via the electrolyte volume. The system can be operated fully automatically. It is pollution-free, of easy-maintaince, and has low operation cost. However, it also has some disadvantages, such as excessive volume, high requirements for ambient temperature, complicated structure, and so on. Some typical vanadium redox flow battery projects are shown in Table 9.3.

9.2.3 Electrochemical Energy Storage Model

At present, electrochemical energy storage is mostly widely used. Here will introduce some commonly used mathematical models for electrochemical energy storage batteries (such as lead–acid battery and Li-ion battery). This will also lay the foundations for the subsequent simulation analysis.

9.2.3.1 Mathematical Model
See Fossati *et al.* [8], Bortolini *et al.* [9], Chen *et al.* [10], and Zheng *et al.* [11] for the detailed models.

9.2.3.1.1 State of Charge Model
The charging and discharging processes can be represented by Equations 9.1 and 9.2 respectively. The charging process is

$$S_{oc}(t) = (1 - \delta)S_{oc}(t - 1) + \frac{P_{ch}(t)\Delta t \eta_{ch}}{C_N} \tag{9.1}$$

Table 9.3 Vanadium redox flow battery energy storage power station projects.

Location	Energy storage system scale	Function	Year
King Island, Australia	0.2 MW/0.8 MW h	Hybrid wind–diesel-storage system	2003
Hokkaido, Japan	4 MW/6 MW h	Smoothing the output of wind farms	2005
Faku County, Liaoning Province, China	5 MW/10 MW h	Tracking planned power generation, smoothing output, improving the power grid's renewable energy power generation acceptance capacity	2013

And the discharging process is

$$S_{oc}(t) = (1 - \delta)S_{oc}(t - 1) - \frac{P_{dis}(t)\Delta t}{C_N \eta_{dis}} \tag{9.2}$$

In Equations 9.1 and 9.2, $S_{oc}(t)$ is the state of charge (SOC) of electrochemical energy storage at the moment t, $S_{oc}(t - 1)$ is the SOC of electrochemical energy storage at the moment $t - 1$, δ is the self-discharge rate, $P_{ch}(t)$ and $P_{dis}(t)$ are respectively the charging power and discharging power of electrochemical energy storage at the moment t, Δt is the time interval (e.g., 1 h), η_{ch} and η_{dis} are respectively the charging and discharging efficiency of electrochemical energy storage, and C_N is the rated capacity of electrochemical energy storage.

9.2.3.1.2 Capacity Model

If the self-discharge rate of an electrochemical energy storage is neglected, then the following formula can be used to express the residual capacity of the electrochemical energy storage at time t:

$$C(t) = C(t - 1) + \Delta t P_{ch}(t)\eta_{ch} - \frac{\Delta t P_{dis}(t)}{\eta_{dis}} \tag{9.3}$$

where $C(t - 1)$ is the residual capacity at the moment $t - 1$.

9.2.3.1.3 Constrained Conditions

To ensure healthy and safe operation, the charge and discharge of electrochemical energy storage need to be carefully managed to meet certain requirements. The main requirements include the SOC constraint, capacity constraint, charging rate and discharging rate, charging and discharging current constraint, and charging and discharging power constraint:

$$S_{OC,min} \leq S_{OC} \leq S_{OC,max} \tag{9.4}$$

$$C_{min} \leq C \leq C_{max} \tag{9.5}$$

$$\begin{cases} r_{ch} \leq r_{ch,max} \\ r_{dis} \leq r_{dis,max} \end{cases} \tag{9.6}$$

$$0 \leq |I_{bat}| \leq I_{max} \tag{9.7}$$

$$\begin{cases} 0 \leq P_{ch} \leq P_{ch,max} \\ 0 \leq P_{dis} \leq P_{dis,max} \end{cases} \tag{9.8}$$

where $S_{OC,min}$ and $S_{OC,max}$ are respectively the minimum allowable value and maximum allowable value of SOC, C_{min} and C_{max} are respectively the minimum allowable value and maximum allowable value of capacity, r_{ch} and r_{dis} are respectively the charging rate and discharging rate of electrochemical energy storage, $r_{ch,max}$ and $r_{dis,max}$ are respectively the maximum allowable charging rate and maximum allowable discharging rate, I_{max} is the maximum allowable charging current and discharging current, and $P_{ch,max}$ and $P_{dis,max}$ are the maximum allowable charging power and maximum allowable discharging power, respectively.

9.2.3.2 Life Model

9.2.3.2.1 Weighted Ampere-Hour Model

In the method given in [12] it is assumed that, within the life cycle of the battery, the ampere-hour throughput capacity is a fixed value. The utilized ampere-hour throughput capacity is used to calculate the battery life, and an effective weighting factor is used to revise the utilized ampere-hour throughput capacity, as discussed in the following.

During a certain time period, the cumulative ampere-hour capacity of the battery can be used to assess the level of life loss. This can be expressed as

$$L_{\mathrm{loss},h} = \frac{A_h}{A_{\mathrm{total}}} \tag{9.9}$$

where $L_{\mathrm{loss},h}$ is the hourly battery age ratio, A_h is the equivalent ampere-hour throughput capacity in a certain time period (here is an hour), and A_{total} is the total ampere-hour throughput capacity within the life cycle of the battery.

A_h can be calculated using

$$A_h = \lambda_{\mathrm{soc}} A_h' \tag{9.10}$$

where A_h' is the un-normalized ampere-hour throughput capacity in an hour and λ_{soc} is the effective weighting factor, defined by

$$\lambda_{\mathrm{soc}} = \begin{cases} a & 0 < S_{\mathrm{oc}} \leq 0.5 \\ bS_{\mathrm{oc}} + c & 0.5 < S_{\mathrm{oc}} \leq 1 \end{cases} \tag{9.11}$$

where a, b, and c can be obtained from experimental data or provided by the manufacturer. The relationship between the effective weighting factor and SOC of a lead–acid battery is shown in Figure 9.3.

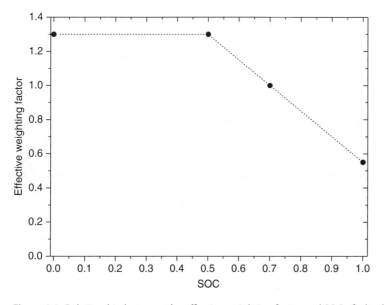

Figure 9.3 Relationship between the effective weighting factor and SOC of a lead–acid battery.

The life span of the energy storage can be expressed by

$$Y = \frac{1}{\sum_{h=1}^{8760} L_{loss}} \tag{9.12}$$

where Y is the life span of the energy storage, with the unit being years.

9.2.3.2.2 Total-Energy-Throughput Model

According to the total-energy-throughput model [13], under standard conditions (at the standard temperature, within the range of the depth of discharge η_{DOD} and below the maximum charging and discharging power P_{max}), the actual cycle times of the energy storage can reach the nominal cycle number n provided by the manufacturer. It is assumed that the total energy released is a fixed value, which can be obtained from the data under the above conditions; namely:

$$ET_{tot} = NC \times nn_{DOD} \tag{9.13}$$

where ET_{tot} is the total energy released from the battery, NC is the effective capacity of the energy storage, n is the number of cycles, and η_{DOD} is the depth of discharge.

The service life of the energy storage is

$$Y = \min\left(\frac{ET_{tot}}{ET_{sim}}, Y_{cal}\right) \tag{9.14}$$

where ET_{sim} is the simulated annual energy-throughput and Y_{cal} is the service life provided by the manufacturer.

9.3 Application of Energy Storage Technology in Distribution Networks with High Photovoltaic Penetration

An energy storage system is one of the effective measures for solving the adverse effects in a distribution network with high PV penetration. In this section, based on the case cited in Section 4.5, how the energy storage system (taking lead–carbon battery as an example) is utilized to improve the PV hosting capacity of the distribution network will be analyzed.

9.3.1 Siting and Sizing Methods for Energy Storage System

9.3.1.1 Siting of Energy Storage System

Owing to the intermittence and fluctuation of PV output power, high PV penetration may lead to voltage issues. As a stability margin index that evaluates the voltage stability and voltage collapse distance of each node in the feeder system, the local L index L_j [14, 15] can be used to determine the weakest-voltage node of the distribution network after the PV installation. The installation of the energy storage system at this node is supposed to improve the voltage stability of the system. The value range of L_j is $[0, 1]$. The closer to 1 that L_j is, the closer the system to the voltage collapse point. The derivation process of the local L index is as follows.

The system node voltage equation can be expressed as

$$I = YV \tag{9.15}$$

where I is the vector of the current injected into the system, Y is the system admittance matrix, and V is the node voltage vector.

Dividing the system nodes into generator nodes and load nodes, Equation 9.15 can be transformed into

$$\begin{vmatrix} I_L \\ I_G \end{vmatrix} = \begin{vmatrix} Y_{LL} & Y_{LG} \\ Y_{GL} & Y_{GG} \end{vmatrix} \begin{vmatrix} V_L \\ V_G \end{vmatrix} \tag{9.16}$$

where I_L and I_G are the vectors of the currents injected into the load nodes and generator nodes respectively, V_L and V_G are the vectors of the voltages at the load nodes and the generator nodes respectively, and Y_{LL}, Y_{LG}, Y_{GL}, and Y_{GG} are the sub-matrixes of the system admittance matrix.

Equation 9.16 can be expressed as

$$\begin{vmatrix} V_L \\ I_G \end{vmatrix} = \begin{vmatrix} Y_{LL}^{-1} & -Y_{LL}^{-1}Y_{LG} \\ Y_{GL}Y_{LL}^{-1} & Y_{GG} - Y_{GL}Y_{LL}^{-1}Y_{LG} \end{vmatrix} \begin{vmatrix} I_L \\ V_G \end{vmatrix} \tag{9.17}$$

The local L index can be expressed as

$$L_j = \left| 1 - \frac{\sum_{i \in \alpha_G} F_{ji} \dot{V}_i}{\dot{V}_j} \right|, \quad j \in \alpha_L \tag{9.18}$$

where L_j is the local L index at load node j, α_G and α_L are the set of generator nodes and the set of load nodes of the distribution network, $\dot{V}_i$ and $\dot{V}_j$ are respectively the voltage at the ith generator node and the voltage at the jth load node, and F_{ji} is the jith element of the $-Y_{LL}^{-1}Y_{LG}$ matrix.

The global L index is the maximum value of all the local L indices in the distribution network and can be expressed as

$$L = \max(L_j) \tag{9.19}$$

9.3.1.2 Sizing of the Energy Storage System

9.3.1.2.1 Mathematical Model

The objective function is to minimize the capacity of the lead–carbon energy storage system when various constraints are met; namely:

$$\min F = \mathrm{Cap}_{\mathrm{stor}} \left(\sum P_{\mathrm{PV}}^N, V_i, P_{\mathrm{char}}, P_{\mathrm{dischar}} \right) \tag{9.20}$$

where $\mathrm{Cap}_{\mathrm{stor}}$ is the capacity of the energy storage device.

Constraints are grouped into equality constraints and inequality constraints. Equality constraints are the power flow balance equations of the distribution network :

$$\begin{cases} P_{Gi} - P_{Li} = V_i \sum_{j=1}^{N} V_j(G_{ij} \cos \delta_{ij} + B_{ij} \sin \delta_{ij}) \\ Q_{Gi} - Q_{Li} = V_i \sum_{j=1}^{N} V_j(G_{ij} \sin \delta_{ij} - B_{ij} \cos \delta_{ij}) \end{cases} \tag{9.21}$$

where P_{Gi} and Q_{Gi} are respectively the active and reactive power output at node i, P_{Li} and Q_{Li} are respectively the active and reactive load at node i, V_i and V_j are the respective voltage amplitudes at nodes i and j, and N is the total number of nodes. G_{ij}, B_{ij} and δ_{ij} are respectively the conductance, susceptance, and voltage phase angle difference between nodes i and j.

Inequality constraints can be divided into the PV hosting capacity of the feeder, PV node voltage constraint, lead–acid battery SOC constraint, and charging and discharging power constraint:

$$\begin{cases} \sum P_{PV}^N \leq P_{PV}^{Ref} \\ V_i^{PV} \leq 1.05 \\ 0.2 \leq S_{oc} \leq 1 \\ 0 \leq P_{char} \leq P_{char}^N \\ 0 \leq P_{dischar} \leq P_{dischar}^N \end{cases} \qquad (9.22)$$

where $\sum P_{PV}^N$ is the sum of the PV output power at each PV node in the feeder; P_{PV}^{Ref} is the planned PV installation capacity of this feeder, V_i^{PV} is the per unit value of the voltage at node i ($i = 1,2, \dots, 10$), S_{oc} is the value of SOC of the lead–acid energy storage device, P_{char} and $P_{dischar}$ are the charging power and discharging power of the energy storage device respectively, and P_{char}^N and $P_{dischar}^N$ are the rated charging power and discharging power of the energy storage device respectively.

9.3.1.2.2 Algorithms

The optimal installation capacity of the energy storage is obtained by adopting the particle swarm optimization (PSO) algorithm. PSO is one of many heuristic algorithms; it shares many similarities with evolutionary computation techniques such as genetic algorithms (GAs). However, in some cases it is easier and faster to implement a PSO than a GA due to inherent difficulties in data representation that evolutionary operators such as crossover and mutation utilize. Since introduced in 1995, PSO has been used in many areas, including nonlinear optimization, control, and artificial intelligence.

PSO is a population-based stochastic optimization technique, inspired by social swarming behavior, such as bird flocking or fish schooling [16]. Each particle in the population represents a candidate solution. All the particles start with randomly initialized positions and then "fly" throughout the search space to find the best possible solution to the problem. During the process, the particles communicate with each other and promulgate the best local solutions/positions that each of them has achieved. Then, based on the global and local information obtained, each particle updates its position toward a desired global optimum. In each iteration, each particle updates its speed and location based on its personal extreme value p_{Best} and global extreme value g_{Best}:

$$\begin{cases} v_{t+1} = \varphi_0 v_t + \varphi_1 r_1(t)(p_{Best} - x_t) + \varphi_2 r_2(t)(g_{Best} - x_t) \\ x_{t+1} = x_t + v_{t+1} \end{cases} \qquad (9.23)$$

where t is the number of iterations, x_t is the space location of the particle in the tth iteration, v_t is the speed of the particle in the tth iteration, φ_0 is the inertia constant, φ_1 and φ_2 are the learning factors, $r_1(t)$ and $r_2(t)$ are the random numbers in $(0, 1)$.

9.3.2 Case Simulation

Based on the distribution network given in Section 4.5.2, the distributed PV integration plan of the feeder is shown in Figure 9.4. As discussed, the maximum allowable PV

Figure 9.4 Plan for integrating PVs into the distribution network.

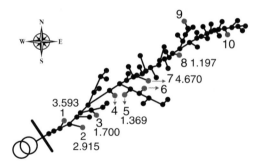

integration capacity is calculated to be 12.355 MW to avoid any overvoltage problems. When the grid-connected PV capacity reaches 15.444 MW (i.e., >12.355 MW), adverse impacts on the distribution network would be derived, such as overvoltage at a certain node. Energy storage is a great choice to address the issue.

Figure 9.5 shows the PV output profile of 8760 hours in one year based on the PV integration plan in Figure 9.4. In Figure 9.5, the PV active power output in the distribution network is higher in mid-year and comparatively lower at the beginning and end of the year, which means the PV output is higher in summer, especially in July.

Considering the economic efficiency of the energy storage system and the maximum PV installation capacity at the node, the rated power of the energy storage system is chosen as 4000 kW, and the other parameters are chosen as follows: the reserve capacity is 20% of the rated capacity; both the charge and discharge efficiency are set to 90%, and the self-discharge rate is set to 1%. The charging of the energy storage happens when the PV output is greater than the load demand in the distribution network; the discharge takes place when the PV output is less than the load demand in the distribution network.

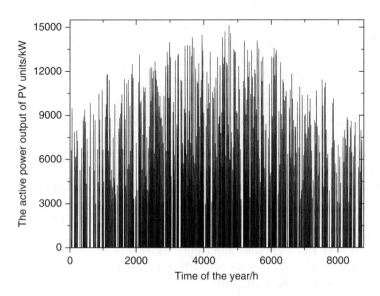

Figure 9.5 The PV active power profile in the distribution network.

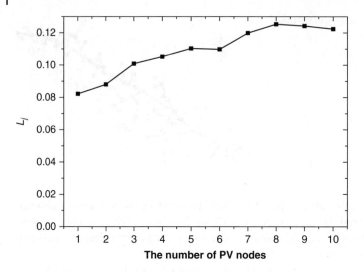

Figure 9.6 Local *L* indices at PV nodes in the distribution network.

According to the PV integration plan in Figure 9.4, the values of the local *L* indices at the PV nodes in the distribution network are obtained according to Equation 9.18 as shown in Figure 9.6.

In Figure 9.6, the value of *L* at the distributed PV integrating node 8 is 0.125, which is the largest, and the voltage at this node is most likely to collapse. According to the siting principle of the energy storage system, the energy storage system is installed at PV node 8. Meanwhile, the PSO is used to obtain the optimum capacity of the energy storage system, which is 19.4 MW h.

Installing a 4 MW/19.4MWh energy storage system at node 8, the method for evaluating the PV hosting capacity of the distribution network given in Section 4.5.1 is used. The comparison of the PV hosting capacity of the distribution network before and after the installation of the energy storage unit is shown in Figure 9.7.

In Figure 9.7, "+" and "•" represent the PV hosting capacity of the distribution network before and after the installation of the energy storage system, respectively. It can be seen from Figure 9.7 that, after the installation of the energy storage system, the PV hosting capacity expands from Zone A to Zone B. Hence, the distribution network meets the requirement for the integration capacity of 15.444 MW, which indicates the integration of the energy storage greatly increases the distributed PV hosting capacity of the distribution network.

In order to further illustrate the voltage stability of the distribution network after the installation of the energy storage system, the local *L* curves in three cases – namely, when PV systems are integrated into the distribution network, when no PV system is integrated into the distribution network, and when the energy storage is integrated into the distribution network with PVs – are shown in Figure 9.8. From Figure 9.8 it can be seen that the global *L* indices in the three cases are 0.018, 0.125, and 0.106 respectively. Comparing the three indices, it can be found that:

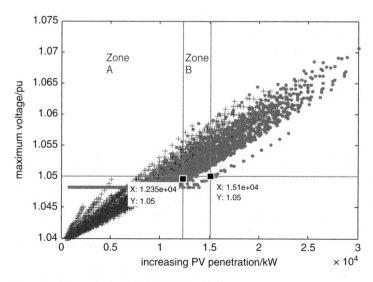

Figure 9.7 Comparison of PV hosting capacity.

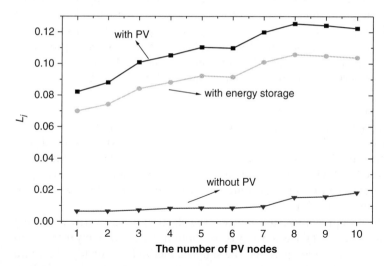

Figure 9.8 Comparison of *L* indices in three cases.

1) Whether the energy storage is installed or not, the integration of PVs can reduce the voltage stability margin of the distribution network.
2) The installation of the energy storage increases the voltage stability of the distribution network by 15.2%, which means it can greatly stabilize the voltage of the distribution network in operation.

This analysis shows the energy storage system is an effective tool to reduce the adverse effects of high PV penetration in distribution network, and greatly improve the PV hosting capacity of the distribution network. However, the current high electrochemical energy storage cost leads to poor economic efficiency after the installation of the energy storage system. As a result, it has been mostly studied in some demonstration projects. With improvement in the technology, battery energy storage would be widely used in the future when the costs are significantly reduced.

9.4 Demand Response

9.4.1 Introduction

DR provides a great opportunity for consumers to take part in the grid operations by reducing or shifting their consumptions during peak periods in response to various kinds of incentives. DR seeks to adjust the demand, not the supply, to achieve the balance between power supply and demand. Influencing the electricity consumption behavior of consumers through a comprehensive DR approach is one of the best possible avenues to achieve energy conservation. DR programs can be divided into two categories [17]. One is called schedulable DR, also known as incentive-based DR. This refers to the DR in which the utility has some capability to control the electricity consumption behavior of users. The other is called nonschedulable DR, also known as price-based DR, which refers to the DR in which the users determine their electricity consumption behavior according to the changes in the price of electricity. In essence, these two kinds of DR are aimed to guide users to change their electricity consumption behavior by providing users with certain economic benefits. Therefore, DR could effective change the load profiles to match the characteristics of PV output and improve the PV hosting capacity of the distributed network, which would simultaneously reduce the cost.

9.4.1.1 Price-Based Demand Response

Price-based DR refers the time-based price mechanism. At present, there are three kinds of time-varying price-based DR programs in the world: time of use pricing, critical peak pricing, and real-time pricing. Users can decide whether to participate in the dynamic pricing program based on their actual electricity consumption situation. Users who participate in the program can freely choose different dynamic prices and adjust the load to reduce the electricity expenditure.

Time of use pricing is a price mechanism where the electricity price varies according to the hour, date, and season of the electricity consumption. There are many methods to divide the electricity consumption periods. The simplest method is to divide 1 year into two or more seasons of electricity price; for example, wet-season and dry-season prices, dry-season and rainy-season prices, heating-season and nonheating-season prices; or more specifically, days can be categorized into working-day price, weekend and holiday price, and so on. For users with a time sharing metering device installation, the 24 h in a day can be further divided into peak price and valley price or peak price, flat price, and valley price.

Critical peak pricing is a relatively new dynamic pricing mechanism in which an extra peak electricity price is set in additional to the general electricity price or the time of use price. The critical peak pricing period would be triggered by system emergency or extremely high electricity purchasing price. The peak load periods of normal time of use pricing usually are certain, but the triggering of the critical peak pricing is uncertain. Instead, when needed, it is temporarily determined through an advance notice within a limit of time. Moreover, critical peak pricing exists on a limited number of days or periods of time in each year some critical-peak-pricing periods are even selected afterwards.

Real-time pricing is different from the aforementioned two strategies because the electricity price is not preset. Instead, it keeps fluctuating and directly reflects the market price. In addition, it is correlated with the day-ahead or real-time market electricity purchasing cost. The real-time price includes the day-ahead real-time pricing and the two-part real-time pricing. The day-ahead real-time price is used to inform the users of the price for the next 24 h a day ahead, so that the user is able to plan and arrange corresponding measures. For example, transfer the load to the valley period, start the on-site standby power generation, or use the contract alternatives available in the market to deal with the peak price period when it is impossible to reduce the demand. Another important form of real-time price is called two-part real-time price, in which a base load curve is first determined based on the user's historic electricity consumption data. For the electricity consumption within the base load curve, the basic fixed price or the peak-valley price will be implemented; for the part outside the base load curve, implementing the real-time price is another good choice. The base load curve is used to provide users with a means to hedge the risk of the fluctuations of the real-time electricity price, and for the remaining part users can obtain the electricity expenditure saving benefits by reducing electricity consumption during the peak load periods.

In the price-based DR programs, the price has the greatest impact on users' electricity consumption behavior. Usually, the price-elasticity matrix is used to quantitatively characterize the impact of the price change on the users' response behavior. The elastic coefficient [18] is often used to reflect the sensitivity of the power consumption demand to price changes. The elasticity of substitution [19] is also used to measure the change ratio of the electricity consumption in the peak price period over that in the valley period.

9.4.1.2 Incentive-Based Demand Response

Incentive-based DR directly uses incentive methods to motivate and guide users to participate in a variety of load reduction programs. Current incentive-based DR programs commonly include direct load control, interruptible load, demand-side bidding, emergency DR, electricity repurchase, and capacity/ancillary service program.

Direct load control is a program where, in case of emergency, utilities adjust or switch off users' electrical appliances through remote control without advance notice or giving users a short notice, at the expense of offering economic incentives to users. The direct load control program is usually used to avoid emergencies at the peak load of the system, and sometimes it is used to avoid high electricity charges and to save electricity purchasing cost. Because in the direct load control program the load reduction time is completely controlled by the program provider without user's intervence, an agreement must be signed with the users in advance to finalize relevant terms.

The *interruptible load* refers to users obtaining financial incentives by agreeing to reduce their loads in the case of a system emergency. If the users fail to implement the load reduction as agreed, they will be punished. However, the interruptible load program is different from the emergency DR program since the interruptible load program is operated by the power supply company, the load service organization, or load reduction service provider. While the emergency DR program is directly operated by the system operating organization, the integrated interruptible load can also get involved in the emergency DR program in the wholesale market.

Demand-side bidding is a new incentive-based DR program that aims to encourage major customers to participate in the bidding. They can show the load amount they are willing to reduce at a certain electricity price or the expected price to reduce a certain amount of load. In this program, when the price in the wholesale market rises, it provides a way to induce the demand side to respond and offers a choice for the users to decide when and how to participate in the real-time and the day-ahead markets. It ensures the users reduce some loads at the request of the market operator and obtain some financial incentives in return. The users submit tenders based on the specific load reduction, duration, and gains, while the market operator selects the submitted tenders according to the market rules. The winning bidder may obtain payment at the highest bid price according to the rules. In some developing demand-side bidding markets the winning bidder may obtain payment according to the previously preset price limit.

In the *emergency DR program*, the system operating organization sets up an incentive payment price so that when a reliability event occurs the users can reduce their loads and obtain corresponding incentive payments. The load reduction is voluntary, and users can ignore the notice and request issued by the system operating organization without being punished. The reliability events are published by the system operating organization in advance according to the safety reliability standards. Generally, users that participate in an emergency DR program will receive notice 24 h before the occurrence of the expected emergency event and receive the confirmation notice when the actual time approaches.

Electricity repurchase is an interruptible and reducible load program operated in the demand-side bidding mode in the electricity retail market. Compared with the emergency DR program in the wholesale market, the two programs operate in a similar way to each other. The difference is that the bidding and contract design are organized by the power sales company in the electricity repurchase program, and the users sign the contract with the power sales company. Nevertheless, in the emergency DR program, the bidding and contract design are organized by the system operating organization, and the consumers sign the contract with the system operating organization. The electricity repurchase program aims to manage the electricity fluctuation risk for the power sales company, and its goal is to share the risk management benefits among the power sales company and users. On the other hand, the emergency DR program aims to manage the system and market stability risk for the system operating organization and its essence is that the system operating organization transfers the risk management benefits to users.

Capacity absorption planning (CAP) can be regarded as the combination of the interruptible load program and the emergency DR program. In the CAP program, users promise that in the case of a system emergency they will implement the predetermined load reduction and obtain financial incentives in return. The CAP program can be regarded as a kind of insurance that no matter whether the reliability events occur or not, the participants can obtain a fixed payment that is similar to the premium income, although in some years they will not be notified to reduce loads.

The *ancillary service market* allows the user to participate in the ancillary service market with their reducible loads as the operating reserve. If the user's competitive tender is accepted, their reducible loads are used as the operating reserve and meanwhile the user obtains a payment of the same market price as that on the supply side; after the user's load reduction is really scheduled, users will once again receive a payment of the electricity price in the spot market. The ancillary service market makes higher requirements for the participating users than all the aforementioned programs. First, there is a high requirement for response speed and, different from the aforementioned programs in which the response time is calculated by hours, it is calculated by minutes. Second, it requires higher minimum capacity. Third, the participating users must have an advanced real-time remote signaling device installed. The ideal participating loads are some pump load, electric arc furnace load, and remote control air conditioning and thermal load, and so on.

In the incentive-based DR program, generally the response quantity, response speed, response duration, response frequency, response time interval, responsibility, response notification time, and other characteristics can be used for modeling of the response characteristics of programs, such as interruptible load [20] and direct load control [21].

Before the concept of DR was proposed, the USA used the direct load control and interruptible load programs as measures to ensure the system reliability. Although the demand-side bidding, emergency DR, and capacity/ancillary service planning programs appeared later, they have advantages in the system monthly capacity planning, day-ahead market economic scheduling, real-time market economic scheduling, and reserve services. Compared with the price-based DR, the incentive-based DR can be more flexible, more conveniently in realizing peak load shifting, and more effective in improving the PV hosting capacity of the distribution network.

9.4.2 Load Characteristics of Demand Response

In order to improve the effectiveness of DR, it is of significant importance to understand the electricity consumption behaviors of different users and load characteristics. The load characteristic curves of different users are shown in Figure 9.9.

It can be seen from Figure 9.9 that the peak period of electricity consumption for industrial users (which is defined as more than 70% of the maximum load) appears

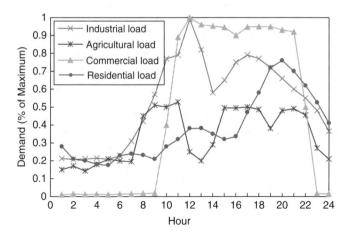

Figure 9.9 Load curves of different users.

between 10:00 and 18:00. The most typical commercial load is commercial load. The peak period of commercial load appears between 10:00 and 20:00. The peak period of agricultural load appears between 10:00 and 14:00. The peak period of residential load appears between 18:00 and 22:00. It is worth noting that the load curves in Figure 9.9 are the weighted averages of all the loads. Taking the industrial load curve as an example, large industrial users work in a three-shift schedule, resulting in a fairly constant load curve. However, some industrial users operate in a regular dayshift work mode; therefore, the peak of the load curve appears during the daytime. For residential users, the load curve is greatly impacted by seasons and holidays. For example, in summer and winter, owing to the use of air conditioners, the load demand in the early morning is higher than that in spring and autumn. During holidays, the load demand in the daytime is significantly higher than that during working days. As a result, Figure 9.9 only roughly reflects the electricity consumption of four different types of load. In order to get more accurate load curves, the behaviors of different users at different times should be analyzed.

Due to the different characteristics of different loads, the suitability levels of different DR programs also vary. Generally speaking, the rights and obligations of users to participate in the programs in most incentive-based DR programs are agreed in the form of a one-to-one contract. Thus, incentive-based DR programs are usually intended for large industrial users. However, as the number of commercial users and residential users is too enormous to adopt one-to-one mode, the relatively flexible price-based DR programs can be used instead, in which users can decide by themselves whether to participate in them based on the changes in the electricity price.

In addition to electricity price and incentives, other factors that affect users' electricity consumption characteristics mainly include industry type, production shift system, the proportion of total electricity expenditure in the total cost, and users' behaviors. For example, most of the users in the cement manufacturing industry are three-shift enterprises and the electricity expenditure accounts for about 15% of the total cost; if the peak price exceeds valley price much, these kinds of users would prefer to move the load/work from the peak or flat period to the valley period at night. For the iron and steel manufacturing industry, the cost of power consumption accounts for about 15% of the total cost, and owing to the large amount of continuous production equipment and high load rate, it requires high power quality and reliability of power supply. Therefore, there is little room for implementing the peak–valley price load response. For commercial and residential users, the air-conditioning and lighting load accounts for a large proportion. In general, these two kinds of users have strong electricity conservation awareness and are willing to change their behavior with a large number of these users there is considerable potential for DR.

Hence, in order to promote the development of distributed PVs, it is essential to also fully utilize the potential of DR. Considering the strong randomness of PV output, it is an indispensable step to incorporate the scientific and rational use of resources on the demand side for improving the PV hosting capacity without curtailment and promoting the overall benefits of PVs via a source-grid-load approach.

9.5 Application of Demand Response in Distribution Networks with High Penetration of Distributed Photovoltaics

9.5.1 Incentive-Based Demand Response Optimization Model

As discussed previously, the incentive-based DR can improve the PV hosting capacity of the distribution network by changing the load curves. The incentive-based DR cycle is set to 24 h, and the shiftable load demand should be met in the response cycle. Each hour is taken as a response period, and the shiftable load is categorized by the load operation duration and power rate.

The users set the information such as whether the shiftable load operates in the next cycle and decide the initial operation time. First of all, the system needs to collect the information from the users, obtain the load and the PV data, and use the incentive-based DR algorithm (see Section 9.5.2) to reset a new operation time for the load to be transferred. After receiving the system's command signal, the shiftable load changes the initial operation time and automatically operates in the time period specified by the system. If no signal is received, the shiftable load operates according to the initial operation time. In some peak periods the shiftable load will be transferred to the periods when there is sufficient PV output so that the load curve can better match the PV generation curve. The coordination between PV output and DR is illustrated in Figure 9.10. Users who participate in the DR program and change the load usage time according to the system's command will obtain financial incentives accordingly. The incentive-based DR can be used to change the load characteristics and effectively reduce the impact of the high PV penetration on the distribution network.

9.5.1.1 Incentive-Based Demand Response Model

See Logenthiran *et al.* [22].

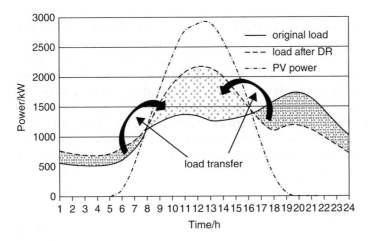

Figure 9.10 Incentive-based DR effect.

9.5.1.1.1 Load Transfer Objective Function

The load transfer objective is to maximize the match between the load curve and the PV generation curve, which can be expressed as

$$
\begin{cases}
\min \sum_{t=1}^{T} |L(t) - P_{\text{new}}(t)| \\
L(t) = L_{\text{befo}}(t) + L_{\text{SLin}}(t) - L_{\text{SLout}}(t)
\end{cases}
\tag{9.24}
$$

where T is the scheduled cycle (set to 24 h in the study) and $P_{\text{new}}(t)$ is the PV output power during the time period t; $L_{\text{befo}}(t)$, $L(t)$, $L_{\text{SLin}}(t)$, and $L_{\text{SLout}}(t)$ respectively represent the load before the response, the load after the response, the load transferred in, and the load transferred out during the time period t.

9.5.1.1.2 Load Transfer Model

The load transfer-in and transfer-out model is as follows:

$$
\begin{cases}
L_{\text{SLin}}(t) = \sum_{k=1}^{N_{\text{SL}}} x_k(t) P_{l,k} + \sum_{h=1}^{h_{\max}-1} \sum_{k=1}^{N_{\text{SLa}}} x_k(t-h) P_{(h+1),k} \\
L_{\text{SLout}}(t) = \sum_{k=1}^{N_{\text{SL}}} y_k(t) P_{l,k} + \sum_{h=1}^{h_{\max}-1} \sum_{k=1}^{N_{\text{SLa}}} y_k(t-h) P_{(h+1),k}
\end{cases}
\tag{9.25}
$$

where N_{SL} is the number of the types of shiftable load, N_{Sla} is the number of the types of shiftable loads whose operation durations are greater than one scheduled period, $h_{\max}$ is the maximum power duration of a shiftable load unit, $x_k(t)$ and $y_k(t)$ are respectively the number of the transferred-in and transferred-out units of the kth type of load with operation starting in time period t, and P_{lk} is the power of the kth type of shiftable load during time period l.

9.5.1.1.3 Customer Satisfaction

Load transfer will affect customer satisfaction, and so a satisfaction index is developed. Customer satisfaction is composed of power purchase satisfaction and power supply satisfaction. Power purchase satisfaction requires the load demand to be met timely. The less the load is transferred, the higher the customer satisfaction is. Power supply satisfaction is based on the promotion of renewable energy development and the PV development policy of "consuming self-generated PV power and integrating the surplus PV power into the power grid." It is desirable to use the PV power locally and reduce the need of power transmission and/or PV power curtailment. If all the PV power is used to supply local users, the power supply satisfaction will be the highest. The customer satisfaction S_{user} is

$$
\begin{cases}
S_{\text{user}} = \frac{1}{2} S_{\text{load}} + \frac{1}{2} S_{\text{PV}} \\
S_{\text{load}} = 1 - \frac{c_{\text{load,shift}}}{c_{\text{load,all}}} \\
S_{\text{PV}} = \frac{c_{\text{PV,one}}}{c_{\text{PVpow,all}}}
\end{cases}
\tag{9.26}
$$

where S_{load} and S_{PV} are the power purchase satisfaction and the PV power supply satisfaction respectively, $c_{\text{load,shift}}$ and $c_{\text{load,all}}$ are the amount of the shiftable load and the total demanded load respectively, and $c_{\text{PVpow,all}}$ and $c_{\text{PV,one}}$ are the total amount of PV generation and the amount of PV electricity directly consumed by the load respectively

9.5.1.2 Constraints

9.5.1.2.1 Transfer Period Constraint
The load can only be transferred in the same cycle; namely:

$$t \in T_n, \quad t' \in T_n \quad \forall x_{t,t'} \tag{9.27}$$

where t is the transfer-in period, t' is the transfer-out period, and T_n is the nth response cycle.

9.5.1.2.2 Transferred Load Capacity Constraint
The capacity of the load that participates in the DR is limited. The actual transferred load in each time period should not exceed the available shiftable load capacity, and the total load in the scheduling period before and after the transfer should remain the same:

$$\begin{cases} x_{SL}(t) \leq X_{SL}(t) \\ L_{SLin}(T) = L_{SLout}(T) \end{cases} \tag{9.28}$$

where, $x_{SL}(t)$ is the actual transferred load in time period t, $X_{SL}(t)$ is the shiftable load capacity in time period t, and $L_{SLin}(T)$ and $L_{SLout}(T)$ are respectively the total load transferred in and transferred out in one cycle.

9.5.2 Incentive-Based Demand Response Algorithm

It can be seen from the previous sections that the load transfer model is a nonlinear, discrete model and cannot be directly solved with the classical programming algorithm. It is a combinatorial optimization problem, and usually with a large number of local extreme points involved. As it is generally a nondifferentiable and highly nonlinear multidimensional NP complete (difficult) problem, the derivative couldn't solve it accurately. The combinatorial optimization solution can be divided into the accurate solution and approximate solution. In the accurate solution, the optimal solution is obtained by traversing all combinations. It is applicable to small-scale problems. However, in the case of large-scale problems, the optimal solution cannot be obtained within a limited time. Therefore, for a combinatorial optimization problem, usually an approximate solution, called the heuristic algorithm, is used.

In this section the PSO algorithm is adopted. The coding of a particle should include information such as the types of load transferred in and transferred out in each period and the number of units. The coding of particles will be very complicated and the calculation time cycle of each iteration is very long, as generally it requires a large number of iterations to find the satisfactory solution. As long as the total transferred load remains the same, it will not affect the value of the objective function. As a result, the range of the feasible solution set S can be narrowed first to determine the transferred-in and transferred-out times, the types of load transferred in each period, and the number of units. Then the PSO algorithm is used to calculate the load transfer results. The solution obtained through the heuristic algorithm may not be exact the same with the optimal solution, but it can be accurate enough and the computation time can be greatly reduced.

The load transfer algorithm implementation process is shown in Figure 9.11 and the specific procedures are as follows:

1) **Input basic data.** Input PV data, load data and the information of the shiftable load set by the user, including the shiftable load capacity and original operation schedule within the response cycle.

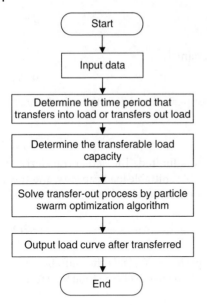

Figure 9.11 Load transfer algorithm implementation flow chart.

2) ***Determine the transfer-in and transfer-out time period of the shiftable load.***
Calculate the PV output, load, and shiftable load in each time period. Compare the PV output and load in each time period, and record the time periods of $P_{new} > P_{Load}$ as the roll-in time periods, and the other periods as the roll-out time periods. If there are all roll-in periods (or just all roll-out periods) in the whole cycle, no response is needed in this cycle and it directly enters the next cycle. The process is shown in Figure 9.12.

3) ***Determine shiftable load capacity.*** Calculate the sum of the difference between the PV output and the load in the roll-in periods and record it as the total roll-in load L_{in}. Record the sum of the difference between the load and the PV output in the

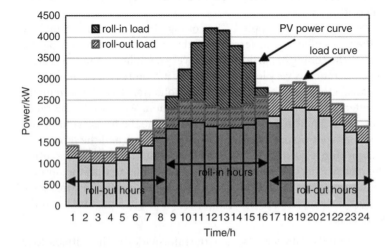

Figure 9.12 Schematic diagram of DR solving process.

roll-out periods as the total roll-out load L_{out}. Take the smaller one of L_{in} and L_{out} as the total shiftable load capacity L_{SL} in this cycle. The shiftable load capacity in each time period is determined according to a certain proportion.

4) **Load transfer-in solving process.** The PSO algorithm is used to obtain the roll-in load in each roll-in period and the specific steps are as follows:

   i) In an initial range, carry out random initialization of the population to generate the initial group G with size n. Each particle contains the information of location and speed. Location information contains the types of roll-in load in each roll-in time period and the number of units. The initial range is determined by the total roll-in load L_{SL} and the shiftable load in each roll-in time period.

   ii) Evaluate the fitness of each particle; store the location and fitness value of each particle in p_{best} of each particle. Store the location and fitness value of the best individual in the p_{best} of all the particles in G_{best}.

   iii) Update the speed and location of each particle according to the particle swarm formula.

   iv) Evaluate the fitness value of each particle and compare it with the fitness value of the best location it has experienced. Update p_{best} if the current fitness value is better.

   v) Compare all the current p_{best} values and G_{best} values and update G_{best}.

   vi) If none of the termination conditions is met, go to step 3; otherwise output the location and end the process.

5) **Update the load curve after the load transfer.**

9.5.3 Case Study

In this section a 10 kV feeder in a certain region will be taken as an example. The feeder has an installed PV capacity of 5500 kW. The total annual electricity consumption is 8640 MW h, with an average hourly load power of 986.30 kW. This region has a relatively low load demand but high PV penetration. The annual load curve and the annual solar irradiance curve in this region are shown in Figure 9.13. According to the Figure 9.13a, the daily average solar irradiance is 4.13 kW h/(m² day) and the annual PV generation is about 6636 MW h. It is assumed that the users participating in the incentive-based DR program described earlier are offered RMB 0.4 for each kilowatt-hour as compensation.

Figure 9.14 shows the load curves under typical weather conditions when the incentive-base DR is introduced and the percentage of the shiftable load capacity is set to be 10% of the total load. Figure 9.14a shows the load curves in typical overcast and rainy weather. Under this circumstance, the solar irradiance is weak. Since the daily PV output is always smaller than the load demand, load shift is not needed. In Figure 9.14b the load curves represent a typical day when the solar irradiance is in medium level and in some periods the PV output is greater than the load. After the load shift via the DR, the load curve gets matched with the PV output curve. Figure 9.14c shows the load curves for typical sunny weather. In the day, the solar irradiance is so strong that even when the load from other time periods has been transferred to noon the PV output still exceeds the load. In Figure 9.14, in the case of load shifting, the overall trend shows the load is transferred from other time periods (especially at peak load times) to the period around noon when the PV reaches its peak. This effectively improves the load

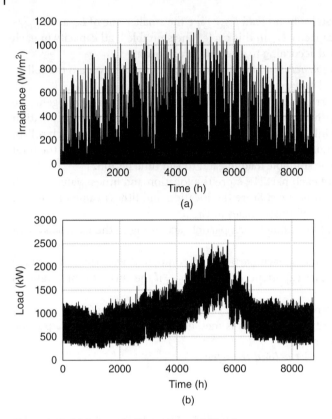

Figure 9.13 (a) Solar radiation curve and (b) load curve.

characteristics, fully responds to the characteristics of PV output, and improves the PV hosting capacity of the distribution network.

Figure 9.15 shows the variation of the PV penetration and customer satisfaction when the proportion of the shiftable load in the total load increases. It can be seen from Figure 9.15 that the PV penetration increases with the increase of the shiftable load capacity. In other words, load shifting can effectively improve the PV hosting capacity, and the electricity purchased from the grid will also be reduced. Owing to load shifting, the power purchase satisfaction declines while the power supply satisfaction increases. As a result, the overall customer satisfaction remains almost unchanged.

9.6 Cluster Partition Control

The aforementioned energy storage technologies and DR methods can be regarded as a "passive change of the distribution system" to accommodate more PVs instead of giving full play to the regulation ability of the PVs to actively support the distribution operation. The target of the distributed PVs is to make PVs have the same controllable and schedulable characteristics as conventional thermal units and realize their schedulability in steady state and dynamic support for the distribution system during transients. This

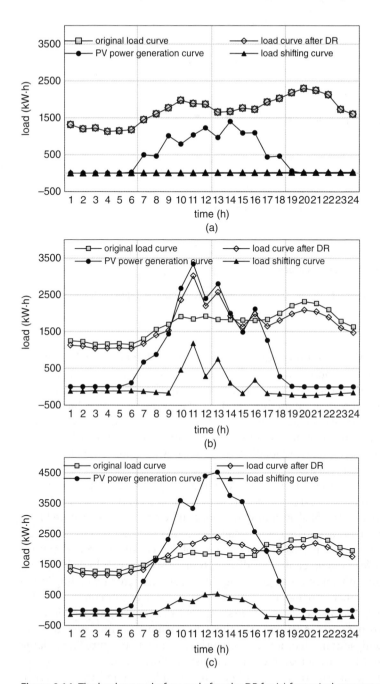

Figure 9.14 The load curves before and after the DR for (a) for typical overcast and rainy weather, (b) typical cloudy weather, and (c) typical sunny weather.

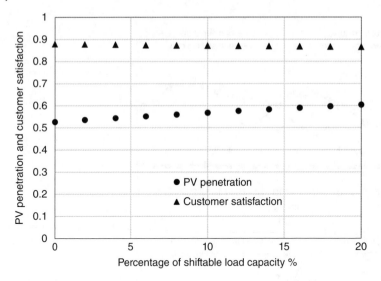

Figure 9.15 PV penetration and customer satisfaction with different percentages of shiftable load capacity.

will ultimately change the operation of distributed PVs from "fit and forget" to "fit and rely upon." As a result, it is important to develop efficient, fast, and fair regional PV regulation strategies, ensure the safe and economic operation of the distribution network, and to study the optimal regulation methods and control strategies of distributed PVs.

Now PV inverters becomes intelligent and have more functions, the PV systems have more flexible active and reactive power control (RPC) functions. The distributed PVs have changed from a passive "negative load" to an "active power supply" and play a more important role in voltage regulation.

A distribution network with high PV penetration is a complex system. It is difficult to control all the PV systems and carry out real-time control in a centralized way. Therefore, it is desirable to have regional autonomous cooperative control of distributed PVs, and the overall system voltage management can be carried out in a hierarchical way. In this chapter, a community partition-based zonal voltage coordination and control scheme is proposed for voltage control of distribution networks with high PV penetration. Grid partition control refers to grouping the nodes with strong coupling in the distribution system together into the same sub-network (cluster, or sub-community) for voltage management based on the node voltage sensitivity and community partition theory. Within each sub-network, independent voltage management is implemented and there is a weak coupling among all other sub-networks. As shown in Figure 9.16, a large number of distributed PVs are integrated into the distribution system containing several feeders so that the distribution system becomes complicated and the difference among node voltages can be significant. As a result, it is difficult to realize optimal control of all the node voltages by only relying on the traditional voltage regulation method.

At this point, the proposed method utilizes the voltage regulation ability of each distributed PV system, and meanwhile it need to have an optimal partition algorithm to

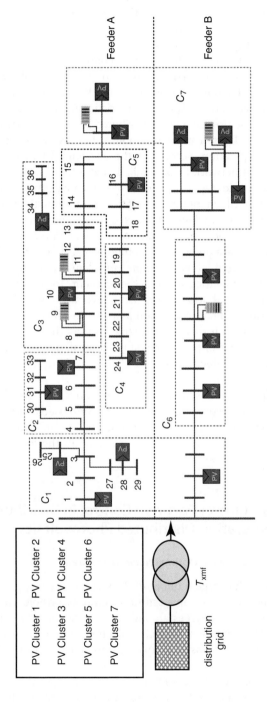

Figure 9.16 Basic framework of distributed PV cluster control in distribution network.

virtually divide a large distribution network into an optimal set of sub-networks. For example, the nodes in Feeder A and Feeder B are divided into seven PV clusters and carry out active and reactive power optimal settings for each PV inverter in each cluster, which can realize real-time control of the local voltage.

There are several advantages for zonal voltage control of distributed PVs via network partition. First, the distributed PV cluster partition control can integrate geographically dispersed PVs related to the same characteristics to effectively stabilize the randomness, and intermittence of PV output. This makes the PV system similar to a conventional power supply in external regulation characteristics and have the ability to flexibly respond to power grid regulation demand. Distributed PV cluster partition control not only can maximally utilize PV self-regulation ability, but also improve the safety and reliability of the system operation. Second, as the distributed PVs are geographically dispersed, the distribution network to be studied is at a large scale. Therefore, if the distribution network is considered as a whole, the nonconvexity and nondeterministic polynomial-hard (NP-hard) characteristics of this complex mathematical model will make it difficult to solve the optimal model in a reasonable time. However, a reasonable partitioning algorithm can divide the distribution network into several sub-networks for zonal control based on parameters such as topological structure and node voltage sensitivity of the power injected. Thus, a complex problem is decomposed into several simple subproblems to facilitate the solving of the problem. Third, the essence of partition control is to divide the distribution network into several independent sub-networks. Consequently, in the case of voltage regulation in a certain sub-network, the impact of the change of the parameters in this sub-network on the other sub-networks can be neglected. And last, in the case of partition voltage control, parallel computing can be used for all the sub-networks, which can effectively reduce the response time in voltage control.

In general, network partition-based zonal voltage control of a distribution network with PVs includes two processes: community-partition-based cluster optimization and PV active and reactive optimization control in each sub-network. The details of network-partition-based zonal voltage control are given in the following application example.

9.7 Application of Cluster Partition Control in Distributed Grid with High-Penetration Distributed Photovoltaics

9.7.1 Community-Detection-Based Optimal Network Partition

A distribution network with a large number of PV systems is a complex network. It is difficult to control all the PV systems in a centralized way as a whole when the network is large. Based on the coupling degrees among different nodes, it is more reasonable to group strongly coupled nodes as a cluster, or a sub-community, for voltage control. The nodes in a group are more closely coupled compared with the rest nodes in the network.

Girvan and Newman proposed a community detection algorithm for partitioning complex networks based on a modularity index ρ to measure the quality of a partition [23]. As opposed to other clustering methods, the community detection algorithm based on the modularity index can obtain the optimal number of communities without predetermining the number of clusters in a data set. The modularity index is defined as follows [24]:

$$\rho = \frac{1}{2m} \sum_i \sum_j \left(A_{ij} - \frac{k_i k_j}{2m} \right) \delta(i,j) \tag{9.29}$$

where A_{ij} is the weighted adjacency matrix of the network, which represents the weighted value of the edge between nodes i and j. $A_{ij} = 0$ if no link exists between the two nodes, otherwise $A_{ij} = 1$. $k_i = \sum_j A_{ij}$ represents the sum of all the weighted values of the links connected with node i; $m = \frac{1}{2} \sum_i \sum_j A_{ij}$ is the total edge weight; and $\delta(i,j) = 1$ if nodes i and j are in the same sub-community, and $\delta(i,j) = 0$ otherwise.

In this example, the edge weight A_{ij} of a distribution network is determined by the reactive power–voltage (Q–V) or real power–voltage (P–V) sensitivity matrix. The relationships between the voltages (i.e., incremental change in voltage magnitude ΔV and incremental change in phase angle $\Delta \delta$) and the changes in the power injections (i.e., active power ΔP and reactive power ΔQ) can be represented by the following sensitivity matrix:

$$\begin{bmatrix} \Delta \delta \\ \Delta V \end{bmatrix} = \begin{bmatrix} S_{\delta P} & S_{\delta Q} \\ S_{VP} & S_{VQ} \end{bmatrix} \begin{bmatrix} \Delta P \\ \Delta Q \end{bmatrix} \tag{9.30}$$

where $S_{\delta P}$ and $S_{\delta Q}$ represent the voltage angle sensitivity with respect to active and reactive power respectively, S_{VP} and S_{VQ} represent the voltage magnitude sensitivity with respect to active and reactive power respectively. In order to describe the coupling degree of two nodes, an average edge weight is used to represent the V–Q weight A_{ij}^{VQ}; that is:

$$A_{ij}^{VQ} = \frac{S_{VQ}^{ij} + S_{VQ}^{ji}}{2} \tag{9.31}$$

where S_{VQ}^{ij} and S_{VQ}^{ji} represent the (i,j) and (j,i) elements of S_{VQ}. In this way, the weight matrix is symmetric.

In order to implement community detection, an improved modularity index that considers local reactive power balance is introduced to partition a distribution network into several clusters/communities. A higher modularity index ρ represents a more reasonable partition of the network than the partition with a lower modularity index. The modularity index ρ, the average sensitivity value S_{C_k}, and the reactive power balance degree ψ_{C_k} of each sub-community C_k are selected in this study as the indices to quantify the coupling degree of nodes and how good a sub-community partition is. The average sensitivity value S_{C_k} and the reactive power balance degree ψ_{C_k} are defined as

$$S_{C_k} = \text{avg} \left(\sum_{i,j \in C_k} \left(A_{ij}^{VQ} \right) \right) \tag{9.32}$$

where the function avg() obtains the mean value of all the weighted edges of sub-community C_k. A larger value of S_{C_k} means that the voltage profile in sub-community C_k is more sensitive to the reactive power injection in the region:

$$\psi_{C_k} = \begin{cases} 1, & Q_{\text{supplied}} \geq Q_{\text{needed}}, \text{ or } Q_{\text{needed}} = 0 \\ \left| \frac{Q_{\text{supplied}}}{Q_{\text{needed}}} \right|, & \text{otherwise} \end{cases} \tag{9.33}$$

where $Q_{supplied}$ is the total reactive power that can be supported by the PVs in C_k. When the reactive power support capability is greater than needed or there is no need for reactive power support in the sub-community, $\psi_{C_k} = 1$. For the other situations, $\psi_{C_k} < 1$. The minimum reactive power needed in C_k is given by

$$Q_{needed} = \sum_{i \in C_k} \frac{\Delta V_i}{S_{VQ}^{ii}} \tag{9.34}$$

where ΔV_i is the incremental voltage change of node i and S_{VQ}^{ii} denotes the reactive-voltage sensitivity of the ith PV unit on the ith node in cluster C_k. It is obvious that a local node has the highest sensitivity to itself. As a result, PV systems, if any, installed at a local node are supposed to absorb the minimum amount of reactive power (i.e., provide the minimum amount of reactive power compensation) to regulate the voltage.

If the network is divided into N clusters, then the improved modularity index ρ_{im} can be expressed as

$$\rho_{im} = \rho - \frac{1}{N} \sum_{k=1}^{N} (\psi_{C_k} + S_{C_k}) \tag{9.35}$$

The average sensitivity value S_{C_k} and the reactive power balance degree ψ_{C_k} are integrated in the improved modularity index. The improved index takes both the network topology and the reactive power compensation capability into consideration.

To obtain the improved modularity index ρ_{im}, the greedy network partition optimization algorithm is used. It is assumed that the method starts with a weighted network of n nodes, and the average adjacency matrix is calculated using Equation 9.31. The network partition optimization algorithm is summarized as follows:

Step 1: Initialize the community partition with one node per community and calculate the initial improved modularity ρ_{im}^0 of each community using Equations 9.31–9.35.

Step 2: For each node i, another node j is randomly selected from the remaining nodes to form a two-node community com(i, j). The new community's modularity ρ_{im}' is calculated. Measure modularity variation $\Delta\rho = \rho_{im}' - \rho_{im}^0$ for each candidate partition where a pair of nodes is merged. If $\Delta\rho < 0$, another node will be selected to form a new sub-community and the modularity variation will be calculated . The nodes i and j will be grouped into the same sub-community until the modularity variation reaches the highest value and the total value of modularity $\rho_{im}' = \rho_{im}^0 + \Delta\rho$ will be updated. This greedy community merge process will be carried out iteratively for all the nodes until no further improvement of the modularity can be achieved; that is, $\Delta\rho$ reaches its maximum positive value.

Step 3: Treat the newly formed sub-communities (or clusters) as individual nodes and conduct the community merge process by repeating step 2. The network is updated with the newly formed sub-communities.

Step 4: The process is stopped when no node can be merged into any community anymore and the *modularity* does not increase. The optimal partition and the community sets and the associated *modularity* indices are obtained.

Although the partition of a distribution network is mainly determined by the topology of the network, the time-varying load demand and PV output will also have impacts on the partition result. The proposed partition method is scalable and adaptive to reflect the change in adding or removing any PV nodes and can be updated by time-varying load demands and PV output, resulting in a dynamic partition scheme. However, in order to prevent the clustering scheme from being updated too frequently under varying conditions, a typical partition scheme based on the most critical scenario operation condition can be predetermined and updated over a certain time period (e.g., a month, a quarter of year, and a year) as needed. The maximum real-time PV penetration situation is selected as the most critical scenario. The real-time PV penetration $R_{PV}(t)$ is defined as

$$R_{PV}(t) \ (\%) = \frac{P_{PV}(t)}{P_{load}(t)} \times 100 \tag{9.36}$$

where $P_{PV}(t)$ and $P_{load}(t)$ denote the PV power and the total load demand at time t ($t = 1, 2, \ldots, 8760$) respectively. The network partition at the maximum PV penetration can be regarded as the most representative scenario because it withstands the most severe overvoltage problem. Some other possible partition schemes based on typical scenarios, such as under different loading conditions (peak, shoulder, and normal), different seasons and different days (weekdays, weekends/holidays), and so on, can also be used to predetermine the partition.

9.7.2 Sub-community Reactive/Active Power-Voltage Control Scheme

Assume that a distribution network has been partitioned into N optimal clusters/sub-communities by the network partition optimization algorithm described earlier. The sub-communities are denoted as $\{C_1, C_2, \ldots, C_k, \ldots, C_N\}$, and there is no overlap among the N regions. Owing to the weak coupling among the different sub-communities, the voltage of each sub-community is controlled independently. The active and reactive power–voltage control scheme is demonstrated in detail in the following by taking sub-community C_k as an example.

Phase 1. The critical load nodes set $V_{C_k}^{cr}$, the normal load nodes set $V_{C_k}^{norm}$, and the PV control nodes set PV_{C_k} are first determined by carrying out power flow studies for the network. The nodes whose voltages are out of the normal range are recorded as the critical load nodes set while the set of nodes whose voltages are within the normal range is denoted as the normal loads set. The PV control nodes set contains the nodes with controllable PV units. These three node sets are illustrated in Figure 9.17. It is noted that the critical and normal sets in Figure 9.17 are exclusive from each other, but the PV control nodes can overlap the two nodes sets. The voltage sensitivity matrices S_{VQ} and S_{VP} are calculated and the current reactive/active power outputs of each PV inverter are also recorded through the power flow simulation.

Phase 2. When the network partition is done, the sub-community with the highest voltage magnitude representative node will be selected as the first sub-community to conduct the voltage control process. In order to achieve the optimal voltage control in the sub-community, an intelligent method based on PSO is employed to determine the optimal absorbed reactive power and curtailed active power for each PV control node. Here, the PV control nodes indicate that the PV systems are controllable and have sufficient reactive/active power regulation capability under current operation situation. Two objective functions and the corresponding constraints will be considered and are

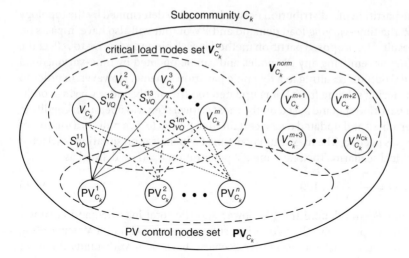

Figure 9.17 Network partition and process of voltage control.

given as follows:

$$\text{obj1} = \min\left(\sum_{d=1}^{n} \Delta Q_d\right) \tag{9.37}$$

$$\text{obj2} = \min\left(\sum_{d=1}^{n} \Delta Q_d + \sum_{d=1}^{n} \Delta P^d_{\text{curt}}\right) \tag{9.38}$$

$$\text{s.t.}\begin{cases} P_{Gi} - P_{Di} = V_i \sum_{j=1}^{N_{C_k}} V_j(G_{ij}\cos\theta_{ij} + B_{ij}\sin\theta_{ij}) \\[2mm] Q_{Gi} - Q_{Di} = V_i \sum_{j=1}^{N_{C_k}} V_j(G_{ij}\sin\theta_{ij} - B_{ij}\cos\theta_{ij}) \\[2mm] 0.95 \le V^i_{C_k} \le 1.05 \text{ p.u.} \\[2mm] 0 \le \Delta P^d_{\text{curt}} \le P^d_{\text{out}} \\[2mm] 0 \le \Delta Q_d \le Q^d_{\text{absorb}} \quad i = 1, 2, \dots, N_{C_k}, \;\; d = 1, 2, \dots, n \\[2mm] -a\,\cos(0.9) \le \varphi_{\text{PV}_d} \le a\,\cos(0.9) \end{cases} \tag{9.39}$$

where P_{Gi} and Q_{Gi} are the active and reactive power generated by conventional generators and PV units respectively, P_{Di} and Q_{Di} are the load active and reactive power demand respectively, V_i and V_j are the voltage magnitudes of the node i and node j respectively, B_{ij} and G_{ij} are respectively the susceptance and conductance of the branch consisting of nodes i and j, θ_{ij} is the phase difference of node i and node j, ΔP^d_{curt} and ΔQ_d are respectively the curtailed active power and absorbed reactive power of the dth PV system, $V^i_{C_k}$ is the voltage magnitude of node i in C_k, P^d_{out} and Q^d_{absorb} are respectively the maximum active power and absorbable reactive power of the dth PV system, and φ_{PV_d} is the power factor angle of dth inverter.

First, the optimal reactive power is determined based on the objective function in Equation 9.37 using the RPC strategy. If the voltages are in the acceptable range after the optimization of reactive power compensation, then the PV inverters have sufficient reactive power–voltage control capability, and thus no real power curtailment is needed. Otherwise, an active power control is needed, based on another objective function (Equation 9.38), to meet the voltage regulation requirement.

Phase 3. The node voltage inequality constraint in Equation 9.39 can be verified by carrying out full power flow calculations with the updated reactive and active power change vectors $\Delta Q_{\mathrm{PV}} = [\Delta Q_d]_{1 \times n}$ and $\Delta P_{\mathrm{PV}} = [\Delta P^d_{\mathrm{cut}}]_{1 \times n}$. To reduce the computation burden of frequent power flow calculations, a voltage update process based on a voltage sensitivity matrix is used. This is a reasonable tradeoff between speed and accuracy, since the voltage sensitivity matrix of a distribution network is not sensitive to operation conditions. Hence, the voltage sensitivity matrices S_{VQ} and S_{VP} obtained in phase 1 are used to update the voltages, and this is done as follows:

$$V^i_{C_k} = V^0_i + \sum_{j=1}^{n} S^{ij}_{VP} \Delta P^d_{\mathrm{curt}} + \sum_{j=1}^{n} S^{ij}_{VQ} \Delta Q_j \tag{9.40}$$

where V^0_i is the voltage of node i without PV regulation and S^{ij}_{VQ} and S^{ij}_{VP} are the (i,j)th element of matrices S_{VQ} and S_{VP} respectively. The updated voltages are denoted as $\{V^{1'}_{C_k}, V^{2'}_{C_k}, \ldots, V^{N_{C_k}'}_{C_k}\}$, and the corresponding updated sets are denoted as $V^{\mathrm{cr}'}_{C_k}$, $V^{\mathrm{norm}'}_{C_k}$, and $\mathbf{PV}'_{C_k}$.

Phase 4. In order to address the interaction between different sub-communities in a distribution network, the proposed method needs to do an overall network power flow calculation after the optimal voltage control for sub-community C_k has finished. The new voltages of nodes for other sub-communities will be obtained and phases 2 and 3 run repeatedly for all the partitioned communities until the voltages of all the critical load nodes are regulated to within the normal range or all the PVs have no more available capability to provide voltage support. The proposed sub-community zonal voltage control scheme is summarized in Figure 9.18.

9.7.3 Case Study

A practical three-phase balanced 10 kV radial feeder located in Zhejiang Province, China, is utilized to verify the proposed approach. As shown in Figure 9.19, there are 36 load nodes with a total nominal load of 26.464 MW + j12.023 MVar along the feeder. Each PV unit in the system is integrated into the feeder through a step-up transformer.

For the practical system, two 500 kW PV systems have been installed at node 1, three 500 kW PV systems at node 25, and four 500 kW PV systems at node 31. There has been no overvoltage issue reported during its past operations under this situation. According to the future PV installation plan, a total of 18 315 kW PV capacity will be added to this feeder. The future PV installation capacity of each new PV unit together with the existing ones is shown in Figure 9.20. This feeder system has both overhead lines and underground cables.

In order to implement the method, a corresponding simulation model is established using the OpenDSS program. The base voltage and the base power are set as 10 kV and 100 MVA respectively. Bus 0 is the reference node with a voltage of 1.04 p.u. The PV

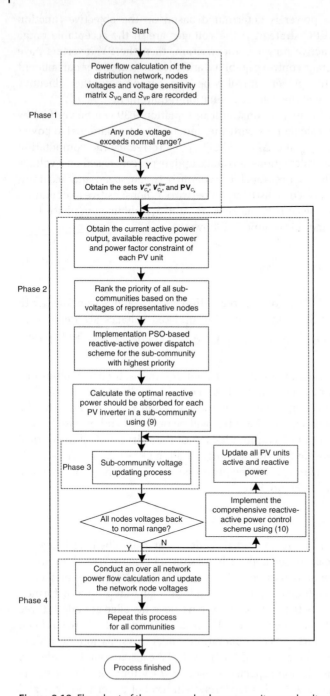

Figure 9.18 Flowchart of the proposed sub-community zonal voltage control.

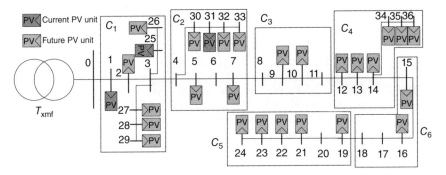

Figure 9.19 Topology of the real feeder with future PV installation.

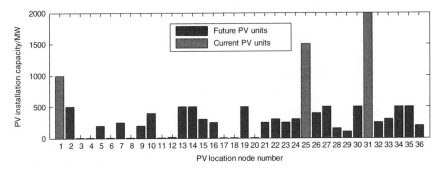

Figure 9.20 Current and future PV installations in the feeder.

model in Ref. [25] is used in the simulation studies. The normalized annual solar irradiance (with respect to the reference irradiance of 1000 W/m^2) and the load profile of year 2012 are given in Figure 9.21. The ambient temperature is set as 25 °C in the simulation studies. The efficiency curve for the inverter is shown in Figure 9.22. Each inverter has an operational power factor range of 0.95 lagging to 0.95 leading. Both the cut-in power and cut-out power are set as 10% of the inverter's nominal capacity. In order to guarantee sufficient reactive power support, the inverter capacity is usually set a little larger than the DC capacity of the PV panels [26]. In this case, it is set to 105% of PV installation capacity. In this study, the 1140th hour in the year is chosen as the representative scenario because the load has 0.1126 times the total load demand and 0.8295 times the PV production (i.e., the highest PV penetration scenario). All the PV units operate at unity power factor at the beginning, and the solar irradiance has an intensity of 900 W/m^2 at this moment. In the PSO-based reactive/active dispatching process, the problem is of dimensions in which both the absorbed reactive power vector and the curtailed active power vector have a dimension of N_{C_k}. The parameters of the PSO algorithm are set as follows: population size Nsize = 20, the cognitive learning factors c1 = 2 and social learning factor c2 = 2, constant inertial weight w = 0.8, and the maximum number of iterations Nmax = 1000. It is noted that the nonlinear inequality constraints for power factor of the inverter and voltage permissible range should be tested in the initialization and the position update process of each particle in order to properly solve this constrained nonlinear optimization problem.

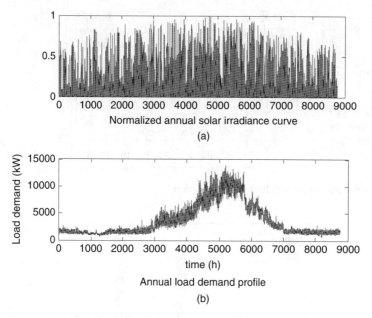

Normalized annual solar irradiance curve

(a)

Annual load demand profile

(b)

Figure 9.21 Annual solar irradiance and load profiles (i.e., in 8760 h).

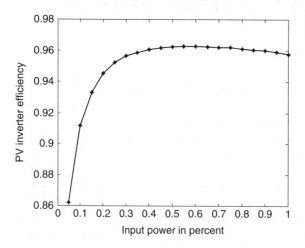

Figure 9.22 PV inverter efficiency versus per unit input power curve.

If all the planned 18 315 kW PV units are integrated into the feeder, the annual time-varying simulation result in Figure 9.23 shows that overvoltage will occur during summer or holiday days due to the large excess PV output in those periods. In Figure 9.23, each color curve represents the annual voltage profile of a node load. It can be seen from Figure 9.23 that the voltage variations in the 37-node system are severe, with a maximum value of 1.067 p.u. and a minimum value of 0.952 p.u. It will become very challenging to regulate the voltage for a good voltage quality by just changing the tap position of the transformer in the substation.

The proposed method is used to address the overvoltage issue. Figure 9.24 shows the optimal partition for the practical network. The modularity function reaches its

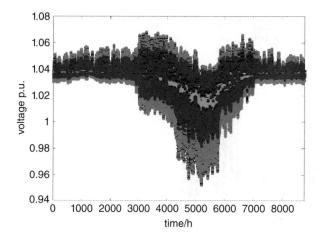

Figure 9.23 Annual voltage profiles for all nodes.

Figure 9.24 Modularity ρ function versus the number of clusters.

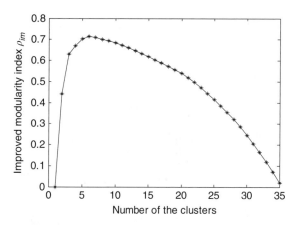

maximum value $\rho_{max} = 0.7144$ when the feeder is partitioned into six clusters. The optimal partition process based on the maximum PV penetration partition method has been conducted and the final partition result is shown in Figure 9.19 where the clusters are denoted as $\{C_1, C_2, C_3, C_4, C_5, C_6\}$. The voltages of nodes in sub-communities C_1 and C_2 are all in the normal range, while sub-communities C_3, C_4, C_5, and C_6 have overvoltage issues. The priorities are ranked as C_6, C_5, C_4, and C_3, and their representative nodes are nodes 24, 18, 36, and 11 respectively.

For the sub-community set of $\{C_6, C_5, C_4, C_3\}$, the proposed zonal voltage control scheme is first used to regulate the node voltages for sub-community C_6, then the voltages of nodes in sub-communities C_5, C_4, and C_3 are recalculated for further optimal voltage control. When all overvoltage issues are eliminated, the curtailed active power and the compensated reactive power are as given in Figure 9.25. It can be seen that not all the PVs are used for voltage support. Some PVs still operate at unity power factor without reactive–active power control.

The node voltage profiles of three cases – namely, the base case, the future planned system without voltage control, and the future planned system using the proposed voltage

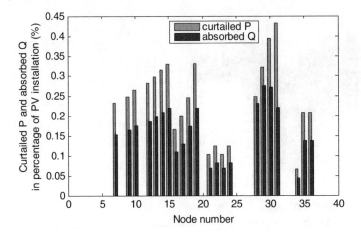

Figure 9.25 The curtailed active power and the absorbed reactive power of the PV units using the proposed scheme.

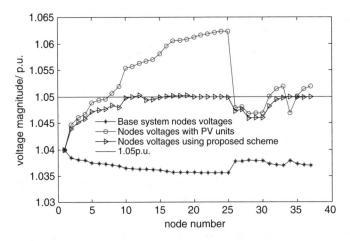

Figure 9.26 Voltage profiles under different scenarios.

control scheme – are shown in Figure 9.26. As shown in the figure, the proposed method can regulate the voltages within the acceptable range by absorbing 1302.48 kVar reactive power and curtailing 2161.35 kW active power in total.

The sub-community C_4 is taken as an example to illustrate the process of the proposed method. Table 9.4 shows the voltage control result of sub-community C_4. The goal of voltage regulation cannot be achieved by using only RPC due to the limited reactive support capability of each PV inverter. Hence, the proposed reactive–active power combined control scheme is needed to bring the voltage back to the normal range.

In order to illustrate the feasibility of the proposed method, the simulation results of a non-partition management scheme are given as a comparison.

In this control strategy, all the PV systems are treated as a whole cluster and an optimal reactive–active power management is determined by the PSO algorithm. Using this scheme, the absorbed reactive power and the curtailed active power as a percentage for

Table 9.4 Voltage control of sub-community C_4.

Node number	Original system without voltage regulation			Voltage after the implementation of RPC scheme using (9.37)		Voltage after the implementation of proposed reactive–active power control scheme (9.38)		
	PV P (kW)	PV Q (kVar)	Voltage (p.u.)	Q absorption using (9.37) (kVar)	Voltage (p.u.)	P curtailment using (9.38) (kW)	Q absorption using (9.38) (kVar)	Voltage (p.u.)
12	15.2300	0.0000	1.0567	9.37422	1.0526	4.5662	3.4720	1.0495
13	432.5817	0.0158	1.0570	127.3185	1.0537	126.4700	163.6100	1.0497
14	432.6530	0.0177	1.0581	136.1720	1.0544	131.2000	151.3207	1.0499
34	693.4481	0.0283	1.0501	87.3536	1.0500	196.5989	257.3512	1.0438
35	693.7862	0.0297	1.0515	102.4112	1.0500	195.5510	256.9605	1.0437
36	862.8365	0.0315	1.0520	192.6520	1.0506	248.0976	328.5714	1.0418

each PV system is given in Figure 9.27. The total active power control is 1245.04 kW and the total RPC is 2003.41 kVar. It can be seen that the non-partition method results in an equal ratio of active curtailment over reactive power absorption. Whereas, the non-partition management scheme requires a whole optimization process for all the PV inverters once the operation condition varies. The set points of reactive and active power for all inverters are the variants to be optimized without considering the adjustment sensitivity of each inverter. Hence, it has a poor dynamic response to fast transients, such as a transient weather conditions introduced by passing cloud shadows. Moreover, it will not be cost-effective in practice and the communication network will become considerably complex when there are a large number of nodes. The most significant

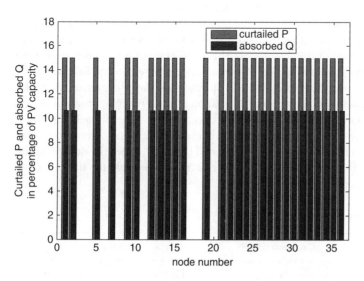

Figure 9.27 Curtailed active and absorbed reactive power as a percentage under the non-partition management scheme.

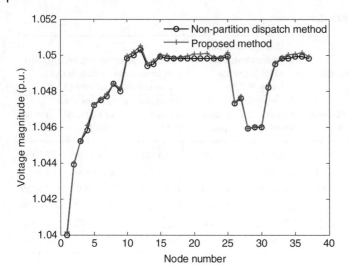

Figure 9.28 Voltage magnitude profiles under the two different control schemes.

advantage of the community-based zonal voltage control is that it only needs to regulate local inverters to control voltage for the communities; and the inverters in other communities can still maintain their operation strategies. In this way, a complex problem is divided into several relatively simple sub-problems that can be solved easily and fast. Furthermore, if only limited communication links are available between PVs, the proposed method based on sub-communities can still achieve the voltage regulation goal.

Figure 9.28 illustrates the regulated steady-state voltage profiles under the two schemes; namely, the proposed zonal voltage control scheme and the non-partition management scheme. Obviously, the non-partition management scheme would have the best steady-state voltage profile after regulation, and it can be regarded as the steady-state benchmark for the effectiveness evaluation of other voltage control methods. It can be concluded that the proposed scheme can regulate the voltage within the specified range, and it has nearly the same control effect compared with the non-partition management scheme. Moreover, as discussed in the following, the proposed method has a much better dynamic response than the non-partition scheme.

Table 9.5 shows the detailed active power curtailments and reactive power supports for the two management schemes. The average computation time of each scheme is also listed. Owing to the fact that the proposed scheme partitions the complex network into smaller networks, a much shorter time is needed than that under the non-partition control scheme. Note that the computation time listed in the table for the proposed method is the longest action time among all the clusters.

Table 9.5 Comparison of the two schemes.

Schemes	Q support (kVar)	P curtailment (kW)	Time (s)
Proposed scheme	2161.35	1302.48	1.48
Non-partition scheme	2003.41	1245.04	9.67

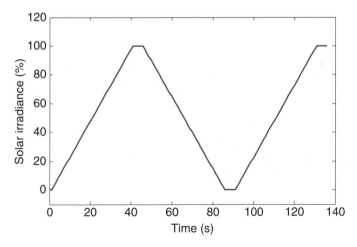

Figure 9.29 Varying solar irradiance profile used in the dynamic simulations.

The effectiveness of the proposed reactive/active power management method in terms of having fast response time is verified through dynamic simulations. In order to illustrate the dynamic response of the proposed method, the system in Figure 9.19 is investigated under rapid solar irradiance changes due to cloud dynamics.

The varying solar irradiance profile shown in Figure 9.29 is used for the dynamic simulation studies. In this study, 100% solar irradiance means 1000 W/m². As shown in the figure, the solar irradiance starts to rise from 1% at $t = 1$ s and reaches 100% at $t = 41$ s; then it stays at the peak for 5 s and then takes 40 s to decline back to 1%; 5 s later it starts rising to 100% again. By using a very high ramp rate, (i.e., from 1% to 100% in 40 s), the performance of the proposed method is tested under a relatively

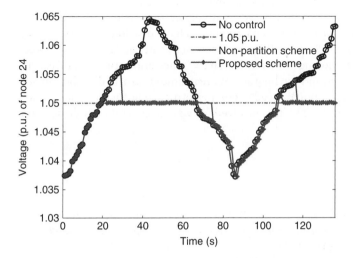

Figure 9.30 Voltage profile of node 24 under three conditions: no control, proposed method, and non-partition dispatch scheme.

tough condition. Strictly speaking, the communication delay should also be considered, whereas it is ignored for simplicity.

In order to illustrate the voltage control capability of the proposed method, node 24, which can have severe overvoltage issues when the PV systems are operated at their normal capacities, is selected for the simulation studies. The voltage profiles of node 24 under three conditions of no control, control using the proposed method, and the non-partition scheme are shown in Figure 9.30. It can be seen from the figure that, because the proposed method has a much shorter computation time, it has a quicker response in voltage control. Consequently, the proposed method achieves a better voltage regulation than the non-partition control method. It is noted that the action time of the inverter in the two methods is assumed zero in the simulation studies.

9.8 Summary

With the integration of a large number of PVs into the distribution network, the high PV penetration impacts a lot to the safe, economic, and reliable operation of the distribution network. In this chapter, some enabling technologies have been introduced to effectively reduce the adverse effects of the integration of the high-penetration PVs into the distribution network and improve the PV hosting capacity of the distribution network. The following is a summary for the chapter.

1) Energy storage can effectively improve the PV hosting capacity of the distribution network and take full advantage of the economic and environmental benefits of the PVs. However, owing to the high investment cost of the energy storage system, it is yet to be improved and is still in the demonstration application stage for most cases.

2) As one of the most economically feasible existing technical means, DR can be used to change the load curve, to match the PV output characteristics, and to realize the full potential of PVs via the interaction of source-network-load.

3) A community detection algorithm based on a voltage sensitivity matrix has been discussed for zonal voltage control by partitioning a large distribution network into smaller sub-communities/clusters. There has been an urgent need for real-time solutions to manage distribution networks with high PV penetration. The method exactly targets to address the need and can help achieve real-time control of distribution networks. The method has a similar steady-state performance but with a much faster response and a lot fewer PV inverters involved for voltage control than those of the non-partition method.

References

1 Evans A., Strezov V., and Evans T.J. (2012) Assessment of utility energy storage options for increased renewable energy penetration. *Renewable and Sustainable Energy Reviews*, **16**(6), 4141–4147.

2 Ibrahim H., Ilinca A., and Perron J. (2008) Energy storage systems – characteristics and comparisons. *Renewable and Sustainable Energy Reviews*, **12**(5), 1221–1250.

3 Zhao H., Wu Q., Hu S., *et al.* (2015) Review of energy storage system for wind power integration support. *Applied Energy*, **137**, 545–553.

4 Chen H., Cong T.N., Yang W., *et al.* (2009). Progress in electrical energy storage system: a critical review. *Progress in Natural Science*, **19**(3), 291–312.

5 Luo X., Wang J., Dooner M., and Clarke J. (2015) Overview of current development in electrical energy storage technologies and the application potential in power system operation. *Applied Energy*, **137**, 511–536.

6 Ferreira H.L., Garde R., Fulli G., *et al.* (2013) Characterisation of electrical energy storage technologies. *Energy*, **53**(5), 288–298.

7 Hadjipaschalis I., Poullikkas A., and Efthimiou V. (2009) Overview of current and future energy storage technologies for electric power applications. *Renewable and Sustainable Energy Reviews*, **13**(6–7), 1513–1522.

8 Fossati J.P., Galarza A., Martín-Villate, A., and Fontán L. (2015) A method for optimal sizing energy storage systems for microgrids. *Renewable Energy*, **77**, 539–549.

9 Bortolini M., Gamberi M., and Graziani A. (2014) Technical and economic design of photovoltaic and battery energy storage system. *Energy Conversion and Management*, **86**(10), 81–92.

10 Chen S.X., Gooi H.B., and Wang M.Q. (2012) Sizing of energy storage for microgrids. *IEEE Transactions on Smart Grid*, **3**(1), 142–151.

11 Zheng M., Meinrenken C.J. and Lackner K.S. (2014) Agent-based model for electricity consumption and storage to evaluate economic viability of tariff arbitrage for residential sector demand response. *Applied Energy*, **126**, 297–306.

12 Zhao B., Zhang X., Chen J., *et al.* (2013) Operation optimization of standalone microgrids considering lifetime characteristics of battery energy storage system. *IEEE Transactions on Sustainable Energy*, **4**(4), 934–943.

13 Zheng M., Meinrenken C.J. and Lackner K.S. (2015) Smart households: dispatch strategies and economic analysis of distributed energy storage for residential peak shaving. *Applied Energy*, **147**, 246–257.

14 Chen G., Liu L., Song, P., *et al.* (2014) Chaotic improved PSO-based multi-objective optimization for minimization of power losses and *L* index in power systems. *Energy Conversion and Management*, **86**(10), 548–560.

15 Kessel P. and Glavitsch H. (1986) Estimating the voltage stability of a power system. *IEEE Transactions on Power Delivery*, **PER-6**(3), 346–354.

16 Devi S. and Geethanjali M. (2014) Optimal location and sizing determination of distributed generation and DSTATCOM using particle swarm optimization algorithm. *International Journal of Electrical Power and Energy Systems*, **62**, 562–570.

17 US DoE. (2006) Benefits of Demand Response in Electricity Markets and Recommendations for Achieving Them. Technical Report, US Department of Energy, Washington, DC.

18 Kirschen D. S., Strbac G. and Cumperayot P., *et al.* (2000) Factoring the elasticity of demand in electricity prices. *IEEE Transactions on Power Systems*, **15**(2), 612–617.

19 Hopper N., Goldman C., and Neenan B. (2006) Demand response from day-ahead hourly pricing for large customers. *The Electricity Journal*, **19**(3), 52–63.

20 Aalami H.A., Moghaddam M.P., and Yousefi G.R. (2010) Demand response modeling considering interruptible/curtailable loads and capacity market programs. *Applied Energy*, **87**(1), 243–250.

21 Molina A., Gabaldon A., and Fuentes J.A. (2003) Implementation and assessment of physically based electrical load models: application to direct load control residential programmes. *IEE Proceedings – Generation Transmission and Distribution*, **150**(1), 61–66.

22 Logenthiran T., Srinivasan D., and Shun T.Z. (2012) Demand side management in smart grid using heuristic optimization. *IEEE Transactions on Smart Grid*, **3**(3), 1244–1252.

23 Girvan M. and Newman M.E.J. (2002) Community structure in social and biological networks. *Proceedings of the National Academy of Sciences of the United States of America*, **99**(12), 7821–7826.

24 Newman M.E.J. (2004) Fast algorithm for detecting community structure in networks. *Physical Review E*, **69**(6), 066133.

25 Smith J.W., Dugan R., and Sunderman W. (2011) Distribution modeling and analysis of high penetration PV. In *2011 IEEE Power and Energy Society General Meeting*, Detroit, MI; pp. 1–7.

26 Fuller J.C., McHann S.E., and Sunderman W. (2014) Using open source modeling tools to enhance engineering analysis. In *2014 58th IEEE Rural Electric Power Conference (REPC)*, Fort Worth, TX, pp. 41–45.

10

Design and Implementation of Standalone Multisource Microgrids with High-Penetration Photovoltaic Generation

10.1 Introduction

In the previous chapters the focus has been given to the operation and control of grid-connected distributed PV generation, particularly for distribution networks with high penetration of distributed PV generation. A variety of benefits can be achieved from grid-connected PVs, such as enhanced reliability and resilience by generating electricity from distributed generation sources (DGSs) (including energy storage if installed), increased efficiency by reducing losses, deferred system upgrades by generating locally, and reduced emissions. Nevertheless, standalone PV generation is another important aspect of solar energy utilization. PV generation and other environmentally friendly energy sources (such as wind and fuel cells) can form standalone clean energy microgrids (MGs), which have been considered as an attractive and important way to achieve a self-sustainable electricity supply and are particularly promising to provide electricity for remote areas by eliminating the need of construction of new power lines and by better utilizing the local renewable energy resources [1–4].

MGs have been extensively studied for managing local generation sources and load demands by forming a standalone system and/or interfacing distributed energy resources (DERs) to the distribution networks [5–8]. Just like the concept of a smart grid, different people may have different definitions of what an MG is. The US Department of Energy defines an MG as "a group of interconnected loads and DERs with clearly defined electrical boundaries that acts as a single controllable entity with respect to the grid and can connect and disconnect from the grid to enable it to operate in both grid-connected or island modes" [9]. In short, an MG is a cluster of generation sources and loads that can be controlled and managed locally to form a subset of large (backbone) power systems.

Similar to grid-connected distributed PV generation, a variety of benefits can be achieved from grid-connected MGs as well. However, the total number of grid-connected MGs that have been really implemented and stay in operation is still limited. A possible reason for this is that some utility companies and MG developers are reluctant in pushing forward for renewable energy resources not only because of the relatively higher cost of renewable DGSs, but also the operational challenges introduced by these grid-connected DGSs. However, this concern becomes less critical for standalone MGs, which makes things more attractive in supplying electricity for remote areas.

As already mentioned, standalone multisource MGs play an important role in solving power supply problems in remote areas, where sometimes it is extremely difficult or

Grid-Integrated and Standalone Photovoltaic Distributed Generation Systems: Analysis, Design, and Control,
First Edition. Bo Zhao, Caisheng Wang and Xuesong Zhang.
© 2018 China Electric Power Press. Published 2018 by John Wiley & Sons Singapore Pte. Ltd.

prohibitively expensive to be connected to the main power grid. Diesel generators had been mainly used to provide electricity for those remote areas. However, given high fuel transportation cost, large fuel price volatility, and significant airborne pollutant emissions, diesel generator operation in these areas remains problematic. Alternatively, geographically isolated areas may have great potential of renewable DGSs to substitute diesel generation, particularly when regulatory or legislative policy supports these technologies [10–13]. Therefore, it is no surprise that standalone MGs with renewable DERs are presently considered viable and, in some cases, one of the best solutions to provide electricity supply for remote areas [14–17].

For standalone MGs, particularly MGs with the different DGSs, tools are needed for long-term planning and design, day-ahead and hourly scheduling, real-time management and control of a quite heterogeneous system, and interconnecting a variety of sources, either programmable or stochastic (e.g., intermittent renewable sources). Based on the voltage and current type of the main bus in an MG, MGs can be grouped into AC, DC, and hybrid AC/DC MGs. Planning and design for both AC and DC, or hybrid AC/DC MGs, has been of recent interest in the literature. Extensive studies have been done on the design, planning, and control of AC MGs [18–20]. A good summary on AC and DC MG planning can be found in Ref. [21], where an MG planning model is presented for determining the optimal size and the generation mix of DERs, as well as the MG type (i.e., AC or DC).

In this chapter we will review the configurations of AC, DC, and AC/DC hybrid MGs, system unit sizing of MG components, the control framework, and the implementation of MGs. The optimal design, planning, and control of MGs are given in the chapter through the discussion of the implementation examples of two real standalone MGs. The implementation and operation experiences and lessons learned from the two MGs are also summarized in the chapter.

10.2 System Configurations of Microgrids with Multiple Renewable Sources

Owing to the intermittent nature of many renewable energy sources (RESs – e.g., PV, wind, and ocean wave), the hybrid combination of two or more of their relevant power generation technologies, along with storage and/or alternative energy (AE) power generation, can improve system performance. For example, wind and solar energy resources in a given area are somewhat complementary on a daily and/or seasonal basis. In general, hybrid systems convert all the resources into one form (typically electrical) and/or store the energy into some form (chemical, compressed air, thermal, mechanical flywheel, etc.), and the aggregated output is used to supply a variety of loads. Hybridization and diversification of multiple sources in an MG could result in increased reliability; however, proper technology selection and generation unit sizing are essential in the design of such systems for improved operational performance, and dispatch and operation control [22–26].

10.2.1 Integration Schemes

Different DERs have different operating characteristics; MGs have been regarded as a great platform to integrate various energy sources. A robust MG should also

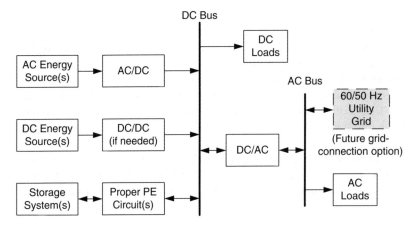

Figure 10.1 Schematic diagram of a DC MG.

have "plug-and-play" operation capability. Adapted from the concept widely used in computer science and technology, plug-and-play operation here means a device (a DG, an energy storage system, or a controllable load) capable of being added into an existing system (MG) without requiring system reconfiguration to perform its designed function, namely, generating power, providing energy storage capacity, or carrying out load control. A suitable system configuration and a proper interfacing circuit (also called a power electronic building block) may be necessary to achieve the plug-and-play function of a DG system [27, 28].

There are many ways to integrate different DGSs to form an MG. The methods can be generally classified into three categories: DC MG (DC-coupled), AC MG (AC-coupled), and DC/AC hybrid MG (hybrid-coupled). The AC MGs can further be classified into power frequency AC (PFAC)-coupled MGs and high-frequency AC (HFAC)-coupled MGs [29, 30]. These methods are briefly reviewed below.

1) **DC MGs.** In a DC MG, shown in Figure 10.1, the different AE sources are connected to a DC bus through appropriate power electronic interfacing circuits. The DC sources may be connected to the DC bus directly if appropriate. If there are any DC loads, they can also be connected to the DC bus directly, or through DC/DC converters, to achieve appropriate DC voltage for the DC loads. The system can supply power to the AC loads (50 or 60 Hz), or be interfaced to a utility grid through an inverter, which can be designed and controlled to allow bidirectional power flow. The DC coupling scheme is simple, and no synchronization is needed to integrate the different energy sources, but it also has its own drawbacks. For instance, if the system inverter is out of service, then the whole system will not be able to supply AC power. To avoid this situation, it is possible to connect several inverters with lower power rating in parallel, in which case synchronization of the output voltage of the different inverters, or synchronization with the grid, if the system is grid-connected, is needed. A proper power sharing control scheme is also required to achieve a desired load distribution among the different inverters [31].

2) **AC MGs.** AC coupling can be divided into two sub-categories: PFAC-coupled and HFAC-coupled MGs. The schematic of a PFAC-coupled systems is shown

in Figure 10.2a, where the different energy sources are integrated through their own power electronic interfacing circuits to a power frequency AC bus. Coupling inductors may also be needed between the power electronic circuits and the AC bus to achieve desired power flow management.

The schematic of an HFAC-coupled MG is shown in Figure 10.2b. In this scheme, the different energy sources are coupled to an HFAC bus, where HFAC loads are connected to. This configuration has been used mostly in applications with HFAC (e.g., 400 Hz) loads, such as in airplanes, vessels, submarines, and in space station applications [32, 33].

In both PFAC and HFAC MGs, DC power can be obtained through AC/DC rectification. The HFAC configuration can also include a PFAC bus and utility grid (through an AC/AC or a DC/AC converter), to which regular AC loads can be connected.

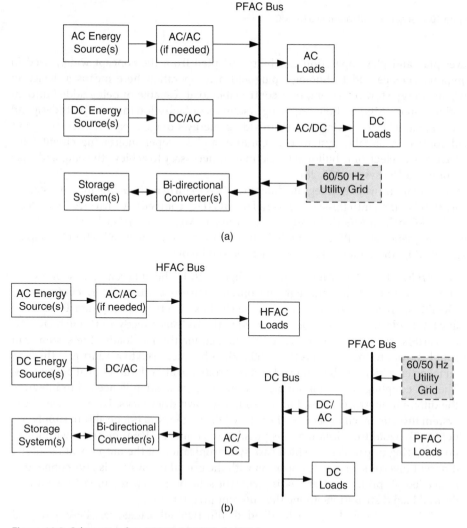

Figure 10.2 Schematic of an AC MG: (a) PFAC; (b) HFAC

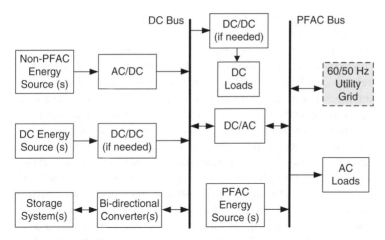

Figure 10.3 Schematic diagram of a hybrid MG.

3) **Hybrid MGs.** Instead of connecting all the sources in an MG to just a single DC or AC bus, as discussed previously, the different sources can be connected to the DC or AC bus in the MG. Figure 10.3 shows a hybrid-coupled system, where the DG resources are connected to the DC bus and/or AC bus. In this configuration, some energy sources can be integrated directly without extra interfacing circuits. As a result, the system can have higher energy efficiency and reduced cost. On the other hand, control and energy management might be more complicated than for the AC or DC MGs.

Different coupling schemes of MGs find their own appropriate applications. If major generation sources in an MG generate DC power, and there are also substantial amounts of DC loads, then a DC MG may be a good choice. On the other hand, if the main power sources generate AC, then an AC MG is a good option. Some interesting discussions on selecting DC MG versus AC MG are given in Ref. [9]. If the major power sources of an MG generate a mixture of AC and DC power, then a hybrid-coupled integration scheme may be considered. It is worth mentioning that the power electronic interfacing circuits in Figures 10.2 and 10.3 can be made as modular building blocks, which will give the systems more flexibility and scalability.

10.2.2 Unit Sizing and Technology Selection

Component sizing of an MG with multiple DGSs is important and has been studied extensively; for example, see Refs [22, 34]. Selection of the most suitable generation technologies (i.e., suitable mix of renewable energy/AE/conventional sources) for a particular application is also equally important. Available application software can be used to properly select generation technologies and their sizes for specific applications. For example, with the aid of HOMER software, developed at the National Renewable Energy Laboratory (http://homerenergy.com/), an MG with hybrid DGSs can be designed; and with the aid of the Distributed Energy Resource-Customer Adaption Model (DER-CAM) software, developed at Lawrence Berkeley National Laboratory (https://building-microgrid.lbl.gov/projects/der-cam), the cost and emission of operating DGSs and combined heat and power systems in a MG can be minimized.

The Microgrid Design Toolkit (http://energy.sandia.gov/download-sandias-microgrid-design-toolkit-mdt/), developed by Sandia National Laboratories, is a decision support software tool for MG designers in the early stages of the design process.

Unit sizing and technology selection can sometimes be as straightforward as meeting certain simple requirements, such as using the available generation technology and not exceeding the equipment power rating, or it can be as complex as satisfying several constraints and achieving several objectives to maximum extent at the same time. Normally, based on available statistical information about generation, load, financial parameters (e.g., interest rate), geographic factors, desired system reliability, cost requirements, emission data and requirements, and other case-specific information, generation technologies and their sizes can be optimized to satisfy specific objective functions, such as minimizing environmental impact, installation and operating costs, payback periods on investment, and/or maximizing reliability. Power system optimization methods such as linear programming [35], interior-point-method [36], mixed-integer nonlinear programming method [37], and heuristic methods such as genetic algorithms (GAs) and PSO can be used for component sizing and energy management of hybrid renewable energy/AE systems [38–43]. These techniques are especially attractive when multiple objectives are to be met, some of which may be conflicting (e.g., minimizing cost, maximizing system availability and efficiency, and minimizing carbon emission). It should be noted that the operating strategy can have a significant influence on the actual performance of systems with different sizes. Hence, the unit sizing and system operational strategy need to be optimized simultaneously. Joint optimization of operation and component sizing for a standalone MG will be discussed in the implementation examples given later in this chapter.

10.3 Controls and Energy Management

Proper control of an MG with different DGSs and loads is critical to achieving the high system reliability and operation efficiency [30, 44]. Typically, a control (or energy management) system needs to determine and dispatch active and reactive output power of each energy source while keeping its output voltage and frequency at the desired level. Generally, the control structure of such systems can be classified into three categories: centralized, distributed, and hybrid control paradigms. In all three cases, each energy source is assumed to have its own local controller which can determine optimal operation of the corresponding unit based on the information available. If multiple (maybe conflicting) objectives need to be met, and all energy sources cannot operate optimally, a compromised (global optimal) operating decision may be achieved. A brief description of each control paradigm follows.

10.3.1 Centralized Control Paradigm

In a centralized control paradigm, the measurement signals of all energy units in an MG are sent to an MG centralized controller (MGCC) [45], as shown in Figure 10.4. Similar to the control scheme of a bulk power system via supervisory control and data acquisition, the centralized controller acts as an energy supervisor and makes decisions on control actions based on all measured signals and a set of predetermined constraints and objectives. The control signals are then sent to the corresponding energy sources

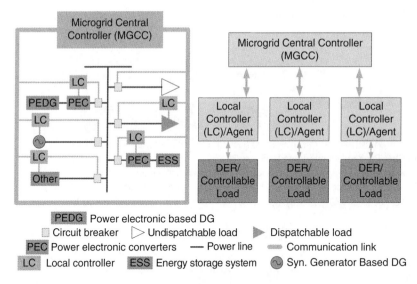

Figure 10.4 Illustration of a centralized control paradigm.

to output proper power. The advantage of this control structure is that the energy management system (EMS) can achieve global optimization based on all available information. However, the centralized control is against the "distributed" nature of distributed generation. Although the centralized control strategy is the most straightforward way to achieve the management and optimization of the overall system, it certainly has several drawbacks:

1) A single-point failure of the central controller may cause a crash of the whole system.
2) The amount of data and communication traffic may quickly exceed the level that can be handled.
3) High investment in communication and data processing.
4) Onerous testing is required for even just a few modifications on the control algorithm.
5) The shutdown of the whole system is required for maintenance.

10.3.2 Distributed Control Paradigm

To follow the distributed nature of distributed generation, an opposite way to centralized control is to develop a decentralized control system. As shown in Figure 10.5, the devices in decentralized methods can be autonomous. In a fully distributed control scheme, the measurement signals of the energy sources of the hybrid system are sent to their corresponding local controller, as shown in Figure 10.6 [30]. The controllers communicate with one another to make compromised (Pareto) operating decisions and achieve global optimization. An advantage of this scheme is the ease of "plug-and-play operation" [30]. With this control structure, the computation burden of each controller is greatly reduced, and there are no single-point failure problems; its disadvantage is still the potential complexity of its communication system.

A promising approach for distributed control problems is a multiagent system (MAS) [44, 46, 47], which is a system containing two or more coordinated agents. MASs have

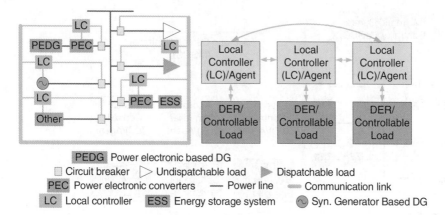

Figure 10.5 Illustration of a distributed control paradigm.

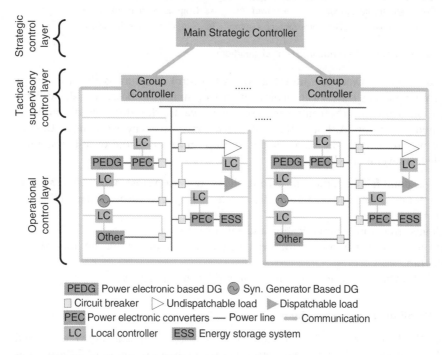

Figure 10.6 Illustration of a hybrid hierarchical control paradigm.

been used, for example, for power system integration, restoration, reconfiguration, and power management of MGs [44, 48–51]. An MAS may be distributed as a coupled network of intelligent hardware and software agents that work together to achieve a global objective. However, as the agents consider their own benefits more than the global optimization, the solution provided by an MAS may be suboptimal.

10.3.3 Hybrid Hierarchical Control Paradigm

Hybrid hierarchical control can be a more practical paradigm, which combines centralized and distributed control schemes and performs control at different levels/layers, as shown in Figure 10.6 [30, 52]. The distributed energy sources and loads are grouped into different groups within an MG; centralized control is used within each group, while distributed control is applied among groups. With such a hybrid energy management scheme, local optimization is achieved via centralized control within each group, while global coordination among the different groups is achieved through distributed control. This way, the computational burden of each controller is reduced, and single-point failure problems are mitigated.

In the top strategy layer of a hybrid hierarchical control scheme, the controller is in charge of the overall operation of the system (e.g., system "startup" or "shutdown"). Based on the strategic orders, controllers in the lower tactical control layer perform different predefined functions to generate optimal settings for the local controllers in the lowest operational control layer. A general hierarchical control structure of an MG was proposed by Guerrero *et al.* [53]. In the proposed scheme, the four layers from the top down are tertiary control layer, the secondary control layer, the primary control layer, and the inner control layer.

10.4 Implementation of Standalone Microgrids

Implementation is a critical step to finally achieve the benefits that MGs can provide. The area of MG implementation is very dynamic, and various demonstration projects on such systems have been (and are currently being) conducted around the world for different purposes, such as electrification of remote villages, or for clean and secure electricity production purposes in urban centers [54–58]. The Office of Electricity Delivery and Energy Reliability (OE), in the US Department of Energy, has a comprehensive portfolio of activities of the development and implementation of MGs in the USA (http://energy.gov/oe/services/technology-development/smart-grid/role-microgrids-helping-advance-nation-s-energy-syst-0). There have also been fast-growing activities in Europe, China, Japan, and other countries on MG development and implementation. The following are some sample applications of standalone MGs with multiple DGSs. For further information about these projects and their actual configuration, the reader is referred to the reference(s) given for each system. The details of the development and implementation of two of the projects that we have been in charge of or involved with are given in the rest of this subsection.

- **Starkenburger Lodge, Austria:** a standalone MG with a hybrid PV-storage system and combined heat and power based engine supplies the power demand of the remote Alpine Lodge in the Austrian Alps [55].
- **The Kythnos Island MG project, Greece:** a single-phase PV-storage–diesel-engine MG installed on the island to provide uninterruptible power to the island residents [15].
- **The Hachinohe MG project, Japan:** the purpose of this renewable-energy-based MG, built in the urban city of Hachinohe, is to investigate its grid integration and stabilization, as well as control and reliable operation in island mode [56].
- **Chicago, IL, USA:** the MG at Illinois Institute of Technology (IIT) has various sustainable energy sources including roof-top solar panels, wind generation units, flow batteries and charging stations for electric vehicles, and smart building automation technology for energy efficiency and demand response [57]. Although the MG is developed as a grid-connected system, it can run off-grid as an islanded system.

- **Dongfushan Island, Zhejiang, China:** In addition to various DGSs (wind, PV, and diesel generator) and energy storage, a seawater desalination system (SWDS) is used as a controllable load in this standalone system [3].
- **Plateau MG, China:** this high-altitude (>4000 m) MG consists of PV panels, small hydro turbine generators, energy storage units, and diesel generators to supply reliable and clean electricity to local residents [58].

10.4.1 Dongfushan Microgrid: Joint Optimization of Operation and Component Sizing

In this subsection we will discuss the implementation of the standalone MG on Dongfushan Island via an approach of joint optimization of operation and component-sizing [3]. The actual system configuration, joint optimal design, operating strategy are discussed in detail, as well as the operational experience regarding the unique MG issues observed and lessons learned that may be useful for future MG development and implementation.

10.4.1.1 System Configuration

Dongfushan Island, distant from the mainland, is the farthest eastern inhabited island in China. The inhabitants of the island used to rely solely on diesel engine (DE) generators for electrical power. In order to improve the quality-of-life for the residents living on the island, the Dongfushan MG Project was started in September, 2010, and completed in April, 2011. As part of improving power delivery and reducing reliance upon diesel generation, the Dongfushan MG Project included renewable energy resources in its design.

A system overview of the MG operating on Dongfushan Island, including the SWDS, is shown in Figure 10.7. It is noted that not only does the SWDS relieve water-supply challenges faced on the island, but it also serves as a controllable load allowing increased

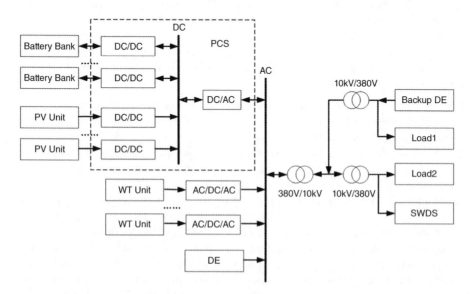

Figure 10.7 Configuration diagram of Dongfushan MG.

Figure 10.8 Photographs of the devices in the Dongfushan MG.

RES utilization. Upon the installation of the MG, the existing DEs became a backup power source on the customer side of the electrical distribution, while a new DE generator was installed as the primary base-load AC power source. As shown in Figure 10.7, an AC/DC hybrid system configuration incorporates the main DE generator and wind turbine (WT) generators on the primary AC bus, while the PV arrays and lead–acid battery storage (BS) banks are connected to the DC bus through appropriate power converting systems (PCS; i.e., power electronic converters). Photographs of the actual system and devices are shown in Figure 10.8.

10.4.1.2   Operating Strategy

The Dongfushan MG operates with a real-time control strategy for the management of the different sources in the system. As shown in Figure 10.9, the system may be operated in different modes, which are mainly determined by the SOC of the BS. In each mode, a master unit is designated that becomes the primary power source under that operating mode. The master unit then takes care of the unbalanced power of the system, namely the net power P_{net}, which fluctuates due to the intermittent output power of the RES units p'_{re} and the load demand p_l. Given that it is difficult to regulate RES power availability, especially the WT units, it is desirable to maintain RES units connected to the system for the longest period possible, thereby minimizing switching operations. This is accomplished through the use of the three operating modes and the employment of changing master units, leading to improved operational efficiency and lengthening service lifetimes.

As shown in Figure 10.9, there are three main operating modes of the Dongfushan MG. In addition, two system conditions are invoked for start-up and standby events. The details of these modes and conditions are:

- **System start-up condition:** the system starts and the operating mode is selected based on the system state.
- **System standby condition:** no main power is available; the system is in standby, waiting for further instructions.

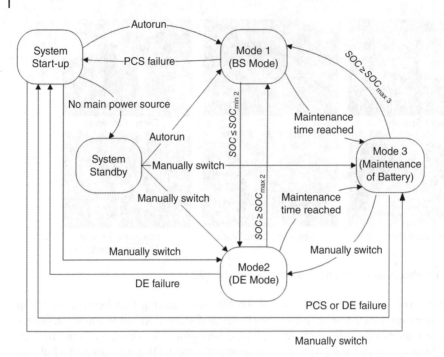

Figure 10.9 System operation modes.

- **Mode 1 "BS as the main power source (i.e., the master unit)":** the BS and the RES units meet the load demand. The DE is OFF.
- **Mode 2 "DE as the main power source":** the BS is being charged and the load demand is supplied by the RES units and the DE. In this mode, the priority is to maximize the utilization of the RES units.
- **Mode 3 "maintenance of BS":** the BS must be operated in accordance with the manufacturer's suggested maintenance procedure. This ensures the reliability and longevity of the BS. At the same time, the RES units and the DE work together to supply power to the load.

In Figure 10.9, SOC_{max2} and SOC_{min2} are the operational upper and lower SOC limits of the BS, while SOC_{max3} and SOC_{min3} are the technical upper and lower SOC limits. Another set of limits, SOC_{max1} and SOC_{min1}, are used to determine the desired (charging or discharging) states of the BS (see Figure 10.10 and the detailed discussion later in this section). These limits stay with the following order: $SOC_{min3} < SOC_{min2} < SOC_{min1} < SOC_{max1} < SOC_{max2} < SOC_{max3}$.

The MG operates in Mode 1 for most of the time. This strategy ensures that energy is derived from the RESs to the maximum extent. In Mode 1, the system first checks the BS SOC and determines the operational state of the BS, namely the "charging desired" or "discharging desired" state. As shown in Figure 10.10, when $SOC \leq SOC_{min1}$, the BS enters its "charging desired" state where the BS is desired to be charged. When sufficient power is available from the RES units, the BS is charged at a power level that is suitable for its healthy operation, until the SOC reaches SOC_{max1}. When $SOC \geq SOC_{max1}$, the BS enters its "discharging desired" state. As shown in Figure 10.10,

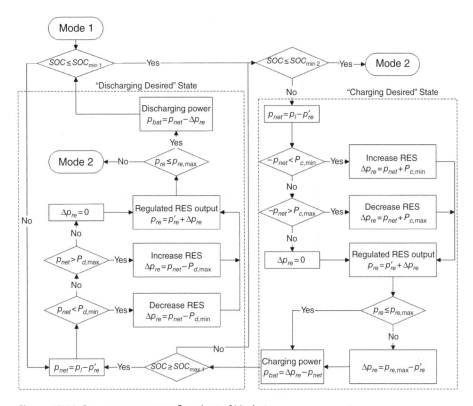

Figure 10.10 Power management flowchart of Mode 1.

the output power of the RES units may be regulated (increased or decreased) to meet the battery charging and discharging requirements when necessary. For example, when the BS is in the "charging desired" state, if net excess power P_{net} is greater than the maximum allowable charging power $P_{c,max}$ of the BS, the RES system will be regulated to reduce its output power. The power outputs from PV are regulated first when the RES needs to be reduced. Further, the WT units are controlled selectively, if needed. It is noted that the BS can be discharged even if it is in the state of "charging desired." When the load is heavy and/or RES output is not sufficient, the BS will be discharged until the SOC reaches SOC_{min2}. When $SOC \leq SOC_{min2}$, the system enters Mode 2.

In Mode 2, to avoid overdischarging the BS, the DE starts up to charge the BS at a constant power. The DE is controlled to utilize the maximum RES power available, which is similar to the goal of Mode 1. If excess RES power is available, the output of the RES is regulated so that the net load is met within the practical power limits of DE, as shown in Figure 10.11. The system will revert back to Mode 1 when the BS SOC reaches its operational SOC upper limit SOC_{max2}; that is, $SOC \geq SOC_{max2}$. The operational strategy for Mode 2 is as a backup of Mode 1. Therefore, the operational SOC limit SOC_{max2} is greater than the desired SOC upper limit SOC_{max1} so that RES is dispatched preferably to the DE during the charging of the BS.

Mode 3 is designed to achieve the proper maintenance of the BS in order to extend its operational life. Mode 3 occurs periodically, a fully discharging and charging cycle,

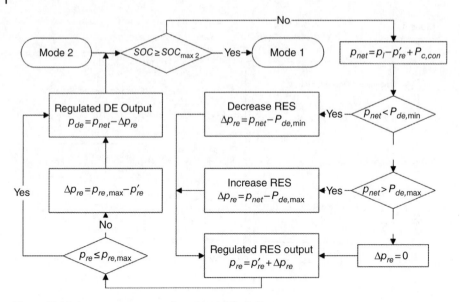

Figure 10.11 Power management flowchart of Mode 2.

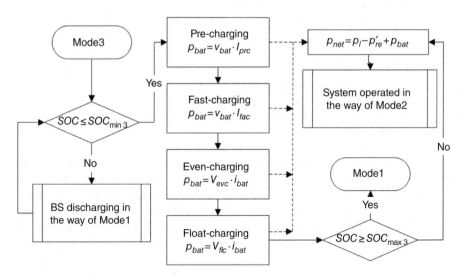

Figure 10.12 Power management flowchart of Mode 3.

as shown in Figure 10.12. When this maintenance mode is required, the BS is first discharged to its technical SOC lower limit SOC_{min3}, similar to that of Mode 1. The BS is then charged to the technical SOC upper limit SOC_{max3} through a consecutive procedure of pre-charging, fast charging, even charging, and float charging. This recharge cycle is specially tuned for the internal physical and chemical characteristics of the lead–acid battery.

In summary, Mode 1 is the main operating mode of the system. In this mode, the total load demand of the island can be supplied by the RES, while the BS is only used to transfer energy. Operational experience has shown that the MG may remain in Mode 1 continuously for a week when the RES is abundant. Mode 2 is a safe mode when poor RES supply is experienced. Mode 3 is specially designed to manage the maintenance requirements of the BS for the purpose of extending its operational life.

10.4.1.3 Optimization Model

10.4.1.3.1 System Sizing

The MG system design must ensure that peak load demand can be met with high reliability, while including a necessary reserve capacity. A safety factor SF is introduced here incorporating the reserve capacity:

$$P_{pl}(1 + SF) = P_{ss} + P_{wt} + P_{pv} + P_{de}$$
$$= (\xi_{ss} + \xi_{wt} + \xi_{pv} + \xi_{de})P_{pl} \tag{10.1}$$

or simplified as

$$SF = \xi_{ss} + \xi_{wt} + \xi_{pv} + \xi_{de} - 1 \tag{10.2}$$

where P_{pl} (kW) is the peak load and P_{ss}, P_{wt}, P_{pv}, and P_{de} are the nominal power of the BS, WT, PV, and DE respectively in the system.

In Mode 1, the load demand is supplied by the BS, the WT, and the PV units. Therefore, Equation 10.2 is modified to

$$SF_1 = \xi_{ss} + \xi_{wt} + \xi_{pv} - 1 \tag{10.3}$$

For Mode 2, the load demand is supplied by the DE, the WT, and the PV units. This modifies Equation 10.2 to

$$SF_2 = \xi_{de} + \xi_{wt} + \xi_{pv} - 1 \tag{10.4}$$

In Mode 3, the full-discharging procedure refers to that of Mode 1 and the full-charging procedure refers to that of Mode 2. The BS should have a sufficient regulation capability, which can be expressed as the duration of its nominal power:

$$d_0 P_{ss} = \eta_{ss} E_{ss} DoD_L \tag{10.5}$$

where d_0 (h) is the duration of nominal power, η_{ss} is the conversion efficiency, E_{ss} (kW h) is the storage capacity of the BS, and DoD_L (%) is the depth of discharge.

10.4.1.3.2 Optimization Goal

The potential benefits of the MG fall into three major categories: cost reduction, fuel saving, and improved environmental emissions. The optimization goal is to maximize these benefits.

Life-Cycle Cost   Recent experience has shown that it is critical to carry out a complete life-cycle cost analysis prior to implementing renewable energy systems. A good example of this type of life-cycle cost analysis is given by Kaldellis *et al.* [59], comparing various energy storage systems for island autonomous power supply systems. The system life-cycle cost normally includes initial cost, maintenance and operation (M&O) cost, replacement cost, fuel cost, and residual value.

The initial investment costs of the DE, the WT, and PV units in the system are mainly determined by their nominal power ratings. The cost of the BS is determined both by its nominal power and storage capacity. The economic subsidies available for RES and BS units are similarly considered to reflect the influence of regulatory or legislative policy:

$$
\begin{aligned}
\text{IC} &= \text{IC}_{ss}(1 - \gamma_{ss}) + \text{IC}_{wt}(1 - \gamma_{wt}) + \text{IC}_{pv}(1 - \gamma_{pv}) + \text{IC}_{de} \\
&= (c_{se}E_{ss} + c_{sp}P_{ss})(1 - \gamma_{ss}) + c_{wt}P_{wt}(1 - \gamma_{wt}) + c_{pv}P_{pv}(1 - \gamma_{pv}) + c_{de}P_{de} \\
&= \left[\left(c_{se}\frac{d_0}{\eta_{ss}\text{DoD}_L} + c_{sp} \right)(1 - \gamma_{ss})\xi_{ss} + c_{wt}(1 - \gamma_{wt})\xi_{wt} \right. \\
&\quad \left. + c_{pv}(1 - \gamma_{pv})\xi_{pv} + c_{de}\xi_{de} \right] P_{pl}
\end{aligned}
\tag{10.6}
$$

where IC is the system initial cost; IC_{ss}, IC_{wt}, IC_{pv}, and IC_{de} are the initial cost of the BS, WT, PV, and DE respectively; and γ_{ss}, γ_{wt}, and γ_{pv} are the subsidy percentages of the BS, WT, and PV units respectively.

The M&O cost is split into the fixed and variable costs. For the BS, the WT, and the PV units, maintenance is scheduled based on certain fixed periods. In general, the annual M&O cost can be expressed as a fraction of the corresponding initial capital costs. The maintenance of the DE is scheduled based on how many hours the unit has been running. It follows, therefore, that the M&O cost of the DE is related to its energy generation. Considering the economic depreciation rate, an equivalent annual increase of the M&O cost is assumed as part of the calculation:

$$
\begin{aligned}
\text{MC} &= \left(\text{IC}_{ss}m_{ss} + \text{IC}_{wt}m_{wt} + \text{IC}_{pv}m_{pv} \right) \sum_{j=1}^{n} \left(\frac{1 + i_{mc}}{1 + i} \right)^{j} \\
&\quad + \sum_{j=1}^{n} (\text{IC}_{de}m_{de} + c_{om}E_{de,j}) \left(\frac{1 + i_{mc}}{1 + i} \right)^{j}
\end{aligned}
\tag{10.7}
$$

where MC is the system M&O cost, m_{ss}, m_{wt}, m_{pv}, and m_{de} (%) are the M&O cost coefficients of BS, WT, PV, and DE, respectively; n (years) is the expected lifetime of system, c_{om} is the M&O cost of DE per unit energy, i_{mc} is the annual increase coefficient of the M&O cost, and $E_{de,j}$ is the total annual energy generation of DE.

The replacement costs mainly depend on the replacement of important components within the units that have a shorter lifetime than the system as a whole. Replacement costs are expressed as a fraction of the initial capital costs of each unit, considering the time value of capital costs and the corresponding technological improvements over time:

$$
\text{VC} = \text{IC}_{ss} \sum_{k=1}^{k_0} r_k \left\{ \sum_{l=0}^{l_k} \left[\frac{(1 + i_{k1})(1 - i_{k2})}{(1 + i)} \right]^{ln_k} \right\}
\tag{10.8}
$$

where VC is the system replacement cost, r_k is the cost percentage of part k, k_0 is the total number of major parts, l_k is the replacement occurrences of part k, n_k is the expected lifetime of part k, i (%) is the discount rate, i_{k1} is the annual price change of part k, and i_{k2} is the technological improvement of part k.

The fuel consumption of the DE in a year can be represented as

$$
\text{FC} = \sum_{j=1}^{n} \left[c_f \sum_{h=1}^{8760} \eta_{de}(p_{de,j,h}) \right] \left(\frac{1 + i_{fc}}{1 + i} \right)^{j}
\tag{10.9}
$$

where c_f is the price of fuel and i_{fc} is the annual price change of fuel, $p_{de,j,h}$ is the hourly energy generation of DE, and $\eta_{de}(p_{de,j,h})$ is a function of fuel consumption with a quadratic term.

Upon reaching the end of the designed life of the system, the residual value of the system can be expressed as

$$RV = \frac{RV_{ss} + RV_{wt} + RV_{pv} + RV_{de}}{(1+i)^n} \qquad (10.10)$$

where RV is the system residual value and RV_{ss}, RV_{wt}, RV_{pv}, and RV_{de} are the residual values of the BS, WT, PV, and DE, respectively.

The total life-cycle cost of the system may be formulated, therefore, as

$$C = IC + MC + FC + VC - RV \qquad (10.11)$$

Finally, the present value of energy generation cost is given by dividing the total life-cycle cost by the total energy generation during the same period, taking into consideration of the expected price of electricity produced by the MG, given the time-value increase in cost:

$$c_0 = \frac{C}{E_{tot} \sum_{j=1}^{n} \left(\frac{1+i_0}{1+i} \right)^j} \qquad (10.12)$$

where c_0 is the system energy generation cost, E_{tot} is the total annual total energy generation of system, and i_0 is the electricity price escalation rate.

Renewable Energy Generation Penetration Generation penetration of the RES is defined as the ratio of the electricity generated by the RES units to the total energy generation of the MG system over a given period. Clearly, the higher the proportion of generation provided by RES units, the more significant the fuel savings, and correspondingly the greater the benefit to reducing emissions:

$$\lambda_{re} = \frac{\sum_{j=1}^{n} E_{re,j}}{nE_{tot}} = \frac{\sum_{j=1}^{n} (E_{wt,j} + E_{pv,j})}{nE_{tot}} \qquad (10.13)$$

where λ_{re} is the utilization ratio of the RES, $E_{re,j}$ is the total annual energy generation of the RES, $E_{wt,j}$ is the total annual WT energy generation, and $E_{pv,j}$ is the total annual PV energy generation.

Pollutant Emission The DE is the main airborne pollution source in the MG system. On the one hand, it is preferable to operate the DE at its point of highest efficiency. More DE generation can reduce the size of BS. On the other hand, more generation derived from the DE normally incurs higher quantities of pollutant emissions. The emissions factors of the DE include NO_x, SO_2, CO_2, CO, and dust, which are weighted according to the corresponding per unit energy generated:

$$Q_{de} = \sum_{j=1}^{n} \left[\sum_{h=1}^{8760} (\omega_1 \zeta_{SO_2} + \omega_2 \zeta_{NO_x} + \omega_3 \zeta_{CO_2} + \omega_4 \zeta_{CO} + \omega_5 \zeta_{Dust}) p_{de,j,h} \right] \qquad (10.14)$$

where Q_{de} is the airborne pollutant emission of DE, $\omega_1, \omega_2, \omega_3, \omega_4$, and ω_5 are the weights of pollution emissions, and $\zeta_{SO_2}, \zeta_{CO_2}, \zeta_{CO}, \zeta_{NO_x}$, and ζ_{Dust} are the airborne pollution emissions per unit energy.

Multiobjective Optimization It is noted that the three expressions given in Equations 10.12, 10.13, and 10.14 have different quantitative units. In order to better carry out the multiobjective optimization, the cost of electricity given in Equation 10.12 and the total emissions quantity given in Equation 10.14 are normalized with respect to a base case. The base case, for the purposes of this analysis, was chosen to represent zero RES generation available and all electricity supply to the MG loads was provided by the mainland power grid. In the base case, c_{st} is the energy generation cost and Q_{st} is the pollutant emission. By taking c_{st} and Q_{st} as the reference base for normalization, the optimization problem can be reformulated as a single-objective optimization, according to

$$\min f = \mu_1 \frac{c_0}{c_{st}} + \mu_2 \frac{1}{\lambda_{re}} + \mu_3 \frac{Q_{de}}{Q_{st}} \tag{10.15}$$

where μ_1, μ_2, and μ_3 are the weights of objectives.

10.4.1.3.3 Optimization Process

GAs are amongst the most commonly used population-based heuristic optimization algorithms and have been applied to a variety of problems [60]. A GA is also used to assist solving the unit-sizing optimization problem for the hybrid MG system (i.e., to select the optimal power ratings for the BS, WT, PV, and DE units), as well as choose the BS energy capacity. The optimization process is shown in Figure 10.13 with the fitness function, expressing the goal of the optimization, given in Equation 10.15. A set of weighting factors reflects the emphasis of each objective within the fitness function; namely, cost reduction, energy/fuel saving, and environmental emissions improvement. As given in Equation 10.16, each solution is constrained by the system safety and feasibility factors, based on (i) adequate power in each operational mode, (ii) high utilization of RESs, (iii) limits of RES penetration, and (iv) sufficient energy regulation capability:

$$\begin{cases} SF_1 = (\xi_{ss} + PoR - 1) \geq SF_{1,min} \\ SF_2 = (\xi_{de} + PoR - 1) \geq SF_{2,min} \\ PoR = (\xi_{wt} + \xi_{pv}) \geq PoR_{min} \\ 0 \leq \xi_{wt}, \xi_{pv}, \xi_{ss}, \xi_{de} \leq 1 + \delta \\ d_{0,min} \leq d_0 \leq d_{0,max} \end{cases} \tag{10.16}$$

where SF_1 and $SF_{1,min}$ are the safety factor and the corresponding lower limit of Mode 1, SF_2 and $SF_{2,min}$ are the safety factor and the corresponding lower limit of Mode 2, PoR is the capacity penetration of RESs and PoR_{min} is its lower limit, δ is the margin of component sizing, and $d_{0,max}$ and $d_{0,min}$ are the limits of the BS duration time.

The operating strategy has a critical influence upon the actual performance of systems given different sizes and configurations. In other words, the optimal sizes of system components can differ based on disparate operating strategies. For a holistic analysis, a lifetime simulation is carried out for each system solution based on each possible operating strategy discussed previously and historical data. During the lifetime simulation, units are operated according to the operational strategy and the corresponding technical limits, expressed as constraints. These constraints include the power limits of

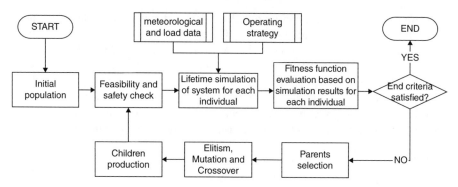

Figure 10.13 Flowchart of the optimization process.

DE ($P_{de,max}$, $P_{de,min}$), the maximum possible power of the WT and the PV generation ($P_{re,max}$) based on historical data, and the charging/discharging power limits of the BS ($P_{c,max}/P_{c,min}$ and $P_{d,max}/P_{d,min}$). The charging/discharging of the BS changes the SOC over its lifetime, as given by

$$\begin{cases} SOC(t+1) = SOC(t) + \frac{Tp_{bat}(t)\eta_{ss}}{E_{ss}}, & \text{Charging} \\ SOC(t+1) = SOC(t) - \frac{Tp_{bat}(t)/\eta_{ss}}{E_{ss}}, & \text{Discharging} \end{cases} \qquad (10.17)$$

The BS efficiency η_{ss} is assumed constant for both the charge and discharge processes. As discussed previously, the system may change its operation mode based on the SOC of the BS.

The performance of each algorithmic population generated as part of the GA operation is evaluated by the fitness function in Equation 10.15. Each new population is produced by the elitism, mutation, and crossover procedures, well defined in the literature. The optimization is considered complete when the end-point criterion is satisfied.

10.4.1.4 System Sizing Optimization

The sizing optimization method and model framework discussed in Sections 10.4.1.1 and 10.4.1.2 have been used to design and build the Dongfushan MG. As introduced in Section 10.4.1.1, the system consists of DE, BS, WT, and PV units, operated according to the strategy and optimization process discussed. The parameters defining the MG optimization problem were based on the actual conditions specific to the system installed on Dongfushan Island. The operation parameters are list in Table 10.1 [3].

Table 10.1 Operation parameters.

Parameters	Percentage of capacity
SOC_{max1}, SOC_{min1}	85, 60
SOC_{max2}, SOC_{min2}	90, 50
SOC_{max3}, SOC_{min3}	100, 50
$P_{c,max}$, $P_{c,min}$	10, 1
$P_{d,max}$, $P_{d,min}$	3, 1
$P_{de,max}$, $P_{de,min}$	100, 30

In addition to residential and commercial load demand, pre-existing sensitive loads on Dongfushan Island were included in the MG system design. Overall, to meet a high level of reliability, the rated capacities of the RES/BS subsystem and the DE subsystem were both required to be larger than the peak demand. Consistent with Equation 10.16 and further enumerated in Equation 10.18, by ensuring that the rated capacities of the RES/BS and the DE subsystems can sustain the peak demand, this ensures that the MG may provide a consistently secure electrical supply to all system loads.

To meet the peak demand as given in Equation 10.16, the capacity penetrations satisfy

$$\begin{cases} \xi_{wt}, \xi_{pv} \geq 0 \\ \xi_{ss} \geq 1 \\ \xi_{de} \geq 1 + \frac{P_{c,con}}{P_{pl}} \end{cases} \tag{10.18}$$

where $P_{c,con}$ is the constant charging power of the BS.

It follows from Equation 10.18 that the BS may meet load demand alone in the case of an emergency (i.e., no output from the RES), presuming that there is enough energy stored (i.e., the SOC of the BS is high enough). As Mode 2 is a backup of Mode 1, the DE is not only required to meet the peak load but also to support additional demand in the form of charging power for the BS. This operational detail ensures the system can get back to Mode 1 as soon as possible.

Figure 10.14 shows a profile of RES availability and the load demand of a typical year based on historical data collected at Dongfushan Island. These data have been used in the lifetime simulation of the system. During the development of MG models, some conservative modifications in design were implemented due to climate and geographical factors of the actual site on Dongfushan Island. Namely, the MG was designed to experience actual RES output that was lower than was expected from the historical data. This result is explained as follows:

1) As a result of territorial limitations of the island and other practical difficulties of installation, the WTs were installed on both sides of the island. However, this deployment scheme significantly reduced the actual capacity of the overall WT units. Experience has shown that when the wind is strong on one side of the island, the opposite side may experience diminished winds due to blockage by the island itself.

2) Moreover, there has been an increase in foggy days experienced on the island compared with the historical record. Interestingly, on one occasion, the MG experienced an entire week without PV output due to fog. Therefore, for the Dongfushan Island MG, low capacity factors of the WTs and the PV generation were used as part of the lifetime simulation, to make conservative estimates of actual climate and geographical factors.

Characteristics of the Dongfushan Island wind and solar resource are given in Table 10.2. Clearly, the wind resource is capable of yielding higher availability than the solar resource. For the demonstrated wind speed profile, 80% of expected winds are below the cut-out speed of the turbines chosen and 78% are below the rated wind speed. On the other hand, the maximum solar irradiance is about 878.8 W/m².

10.4.1.4.1 Sizing Optimization

The optimization goals implemented in this method are flexible to incorporate different needs by adjusting the weighting factors in the objective function (Equation 10.15).

Figure 10.14 Historical data of a typical year: (a) power percentage of the WTs; (b) power percentage of the PVs; (c) load demand.

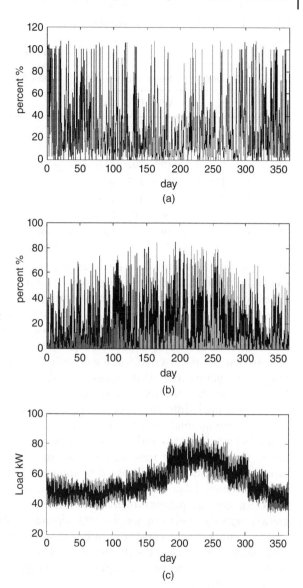

Table 10.2 Characteristics of wind and solar resources.

	Maximum	Average	Available time (h)	Generating capacity (kWh/m^2)
Wind speed	22.70 m/s	6.64 m/s	7010	1931.03
Irradiance	878.80 W/m^2	422.52 W/m^2	2200	929.54

Table 10.3 System sizing with different objectives.

Case	Weight of objectives			Nominal power (kW)				Storage capacity of BS (kWh)	Generation cost (¥/kWh)	RES penetration (%)	Pollutant emission (ton/year)
	μ_1	μ_2	μ_3	WT	PV	DE	BS				
1	1	0	0	194	115	192	86	772	2.55	55.26	55.09
2	0	1	0	198	190	272	156	1398	2.93	61.15	47.90
3	0	0	1	198	202	267	163	1453	2.90	61.07	48.01
4	1/2	1/2	0	193	187	218	103	923	2.71	58.53	51.09
5	1/2	0	1/2	197	199	223	120	1081	2.78	59.56	49.77
6	0	1/2	1/2	194	202	270	165	1482	3.01	61.20	47.82
7	1/3	1/3	1/3	193	115	202	90	801	2.57	55.59	54.69

Some special case studies have been carried out on the Dongfushan MG with different weighting combinations of the objectives. As shown in Table 10.3, the first three cases are equivalent to the optimization considering only one of the three objectives; the last case considers all the three objectives with equal weighting factors (i.e., 1/3); and the remaining three cases are for the optimization for two of the objectives.

The tradeoff between penetration and utilization is one of the key issues for RES exploitation. An oversized RES unit will cause the system to abandon/dump power more frequently than desired, while undersized RES units cannot make the best use of the favorable potential of renewable resources. The optimization results of nominal WT power have small differences for different cases, as shown in Table 10.3. This implies that the same optimal size of WT can be determined even under different considerations. As Dongfushan Island has a very good wind profile, WTs can produce electricity during most of the days, meeting the base load. Compared with the WT, the PV arrays produce electricity only during daytime and make no contribution to the system at night. Hence, the nominal power of the PV arrays in case 2 nearly doubles that in case 1, while the RES generation penetration increases only a little bit less than 6%. When the RE generation capacity level is higher, the MG depends more upon energy storage to transfer energy. Limited by the constraint in Equation 10.18, the DE nominal power increases accordingly as the size of the BS increases. However, the DE run-time correspondingly decreases since, in general, it requires less energy from the DE when higher capacity RESs are available.

As shown in Table 10.3, the minimum generation cost is 2.55 ¥/kW h, which is proximate to the equivalent cost if the electricity was only generated by the DE units on the island. This demonstrates the economic viability of incorporating RESs into the MG, rather than relying upon a system only with DEs, when considering the cost and difficulty of diesel fuel transportation. Among the seven cases listed, the highest RES penetration is 61.15%, which implies a possible savings of about 60% of diesel fuel, when comparing the MG with the base case. As the RES penetration increases from about 55% to 60%, as shown in Figure 10.15, the run-time of the DE is reduced by about 1000 h, while the pollutant emission is reduced by about 8 tons in a year. Compared with the DE-only base case, the RES MG proposed is more economic, reducing the expenditure of fuel and emission of airborne pollutants.

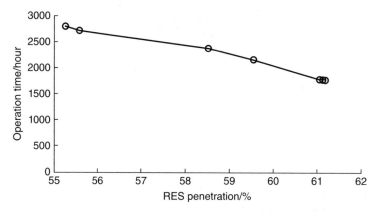

Figure 10.15 The operation time of DE versus the RES generation penetration.

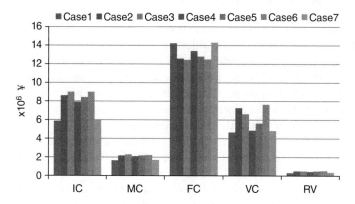

Figure 10.16 Breakdown of the life-cycle cost.

Usually, the generation costs decrease as RES penetration level increases, but at some point this trend reverses and the generation cost will start to increase. While a higher penetration of RESs can reduce fuel cost, RESs come with higher initial capital investment and replacement costs. Although fuel consumption makes up the main proportion of cost during the 20-year life-cycle (see Figure 10.16), the initial and replacement costs of storage system units also have a significant impact on the overall economic analysis. However, a larger size RES unit may decrease its utilization factor and also requires a larger energy storage system. Moreover, beyond a certain point, as the storage system capacity gets larger, there is less extension of system lifetime, relative to smaller capacity BS systems, as shown in Figure 10.17.

The case studies on Dongfushan MG project show that: (i) the life-cycle cost of an MG with RESs is superior to a system only with DE units, especially considering the cost and difficulty of diesel fuel transportation; (ii) there is an optimal level of RES penetration that balances the generation cost and the size of energy storage unit.

10.4.1.4.2 Practical Considerations
On the one hand, the GA is an intelligent computational method and its solutions are influenced by aspects of its formulation; namely, the population size and the generation

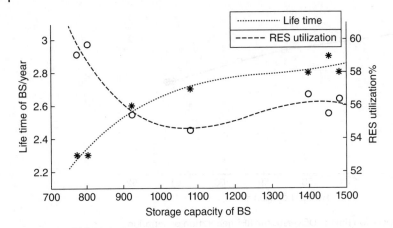

Figure 10.17 RES utilization (and lifetime of BS) versus BS storage capacity.

size. On the other hand, practical capacities (or power ratings) of system components are discrete, which makes the actual solution slightly different from the solutions of a GA. Table 10.3 give the feasible capacity/power rating ranges of the devices in the Dong-fushan MG. By extension, the optimal sizing of components must account for actual limitations and other realistic factors, such as DC bus voltage levels, available inverter ratings, and so on. This reality has a direct effect on the solution-finding for the ratings of the WT, PV, BS, and the DE, which will settle at discrete values.

According to the practical choices given in Table 10.4, there were 72 alternative schemes analyzed, but 18 of them did not meet the constraints of Equation 10.16. The generation cost and the RES penetration of each scheme are shown in Figure 10.18. It is obvious that a larger size of RES, especially the WT, leads to a higher RES penetration. On the other hand, a smaller size of BS leads to a lower generation cost. The best theoretical solution was 210 kW WT + 150 kW PV + 200 kW DE + 960 kW h BS (small orange diamond in Figure 10.18). However, the impact of RES units incorporated into the MG design are based on the historical data in this study. In reality, the actual impact of RES generation will change based upon actual weather conditions and real load demands, inclusive of real-world uncertainties. In addition, the initial capital cost is of paramount concern for customers. Based on this appreciation and other practical

Table 10.4 Available choices of units in the Dongfushan MG.

	WT	PV	DE	BS
Module capacity	30 kW	50 kW (180 W × 278)	–	240 kWh (1000 A h × 240)
Alternative capacity	180, 210 kW	100, 150, 200 kW	200, 220, 280 kW	720, 960, 1200, 1440 kWh
Alternative plan	180 kW WT + 100 kW PV – Purple		200 kW – Small	720 kW h – Circle
	180 kW WT + 150 kW PV – Blue		220 kW – Middle	960 kW h – Diamond
	180 kW WT + 200 kW PV – Green		280 kW – Big	1200 kW h – Square
	210 kW WT + 100 kW PV – Cyan			1440 kW h – Triangle
	210 kW WT + 150 kW PV – Orange			
	210 kW WT + 200 kW PV – Red			

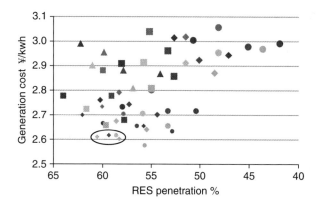

Figure 10.18 Optimization results of different system sizing schemes.

considerations, the actual optimal configuration was chosen as 210 kW WT + 100 kW PV + 200 kW DE + 960 kW h BS (small cyan diamond in Figure 10.18). This selection led to a lower initial cost and generation cost, while invoking a slightly lower RES penetration than the theoretical value.

10.4.1.5 Optimal Configuration and Operation Practice

10.4.1.5.1 Optimal Configuration
As discussed in Section 10.4.1.4, the optimal configuration of Dongfushan MG was chosen as 70 kW×3 (with a total of 210 kW) WT units, 100 kW PV units, a 960 kW h BS unit, and a 200 kW DE unit. Figure 10.19 shows the simulation result of the MG system over a period of 1 week. It can be seen from Figure 10.19a that the system is mostly operated in Mode 1. In the figure, for the RES units, "1" implies that all the RES units are ON and "0" means no RES unit is ON (i.e., all OFF). There are very few power regulation control actions upon the RES units during any given day. For the DE, "1" implies that the unit is ON and "0" means the unit is OFF (shutdown). The simulation results show over 100 h with a high-level output of the RES units. The DE only operates during a short period of time when the load level is continuously high. The RES can meet the load demand with the help of the BS for the majority of the time. Figure 10.19b shows that the BS unit mostly operates with an SOC between 0.6 and 0.85; namely, the excepted charging/discharging state. The DE unit operates in a low power range as a result of a high utilization of the RES. This implies DE operation under 0.55 of its nominal power for more than 90% of operational time, as shown in Figure 10.19c.

10.4.1.5.2 Operation Practice
Dongfushan MG has been in continuous operation for more than 6 years, since July 2011. The reliability of the power supply to island inhabitants has significantly improved, and periods of power outage time have been reduced. Experience shows that the wind power availability is good from October until the following March. During those months, the WT can reduce about 60% of the diesel fuel consumption and 70% of the DE run-time on the island. The wind power is ordinary between April and June, resulting in a 40% reduction in diesel fuel consumption and a 50% reduction in DE run-time. Even when the wind power is typically poor (between July and September),

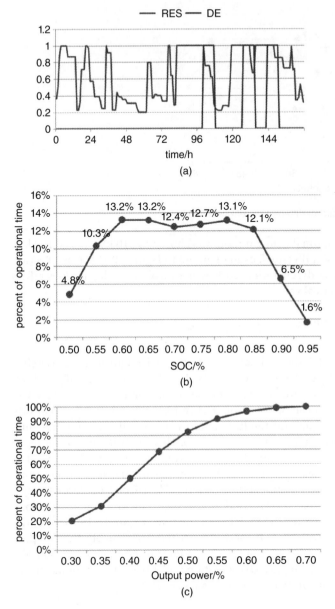

Figure 10.19 Simulation study results of the system with the optimal configuration. (a) Operation state of the RES and the DE. (b) The SOC distribution of the BS. (c) Power output distribution of the DE.

the WT can still reduce 35% of the diesel fuel consumption and 40% of the DE run-time. A selection of actual operational data from 2011 is given in Table 10.5.

Since the completion of the Dongfushan MG project there has been a rapid increase in the air conditioning load on the island. This load growth causes higher short-term load peak spikes that were not originally considered in the MG design. Nevertheless, the system built with the conservative configuration proposed is able to meet the load

Table 10.5 Operation practice of year 2011 (August to December).

	Demand (kW h)	WT (kW h)	PV (kW h)	DE (kWh)	RES penetration (%)	Diesel consumption (ton)
Aug	35 996	11944	3 334	20 718	42.44	4.97
Sep	37 384	9 674	2 937	24 773	33.73	5.95
Oct	33 330	10 349	2 904	20 077	39.76	4.82
Nov	33 257	11 220	2 603	19 434	41.56	4.66
Dec	34 773	14 291	2 889	17 593	49.41	4.22

demand on Dongfushan Island consistently, despite the added air conditioning load. The WT units incorporated into the Dongfushan MG were newly developed products at that time. Their reliability and stability have not been as high as expected. Hence, the wind energy generation has not yet reached the expected design value. Similarly, the PV arrays have been heavily affected by foggy days. On Dongfushan Island, fog can form and disappear relatively faster than on the mainland. This makes the PV output even more intermittent than either the historical data or experience-based expectations predicted. Future projects on geographic islands should give full consideration to the impacts of fog on PV installations.

Owing to high precipitation on the island in the past year, the SWDS, which was also designed for balancing excess RES generation, was not operated often. This change in load operation relative to the design forecast further reduced the actual RES generation penetration within the MG. Given each of these practical reasons, the energy generation of RES units within the MG has not reached its design value yet. Nevertheless, the generation from RES units has resulted in well over 30% of the total energy delivered to the MG, proving the MG system's effectiveness at reducing diesel fuel expenditure and airborne pollution.

10.4.2 Plateau Microgrid: A Multiagent-System-Based Energy Management System

10.4.2.1 System Configuration

This standalone MG is developed in a remote area at an altitude above 4000 m [58]. It utilizes the local RESs (solar and hydro) to form a PV–small hydro hybrid (PVSHH) MG to meet the local electricity needs. In the plateau environment, the low temperature and the shortage of oxygen cause incomplete combustion of fuels, which results in low system efficiency and high pollutant emissions when diesel generators are used. However, the solar energy resource in the area is abundant and fairly stable, which makes it suitable for the development of PV power generation.

The original MG was composed of a small-hydro generation plant (SHGP) with four 1.6 MW small-hydro generators and a diesel generation plant (DGP) with four 2.5 MW diesel generators. The SHGP has a 6.4×10^6 m^3 reservoir and is responsible for primary frequency regulation. In the wet season (i.e., spring and summer) the reservoir is close to full. However, the water in the reservoir is normally not enough to meet the electric power and water demand in the dry season. The DGP works in load peak hours to assist the SHGP for secure system operation. Owing to the high altitude, each diesel generator only generates a maximum of 1.0MW under such environment. The emissions can cause

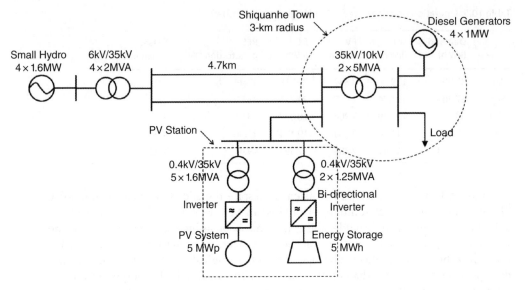

Figure 10.20 Configuration diagram of the plateau MG.

significant adverse impacts on the vulnerable local ecological system. In addition, the transportation cost of fuels to the remote plateau area is very expensive. The high cost and the high emission rate limit the operation time of the diesel generators.

As the load demand increases in the area, the SHGP can no longer meet the needs. Without increasing the generation of the diesel generators, RESs are considered to serve the increasing demand. To utilize the abundant solar resources (with a sunshine duration over 2000 h in a year) in the area, a PV plant (PVGP) is planned to be developed and integrated with the existing system to form a new PVSHH MG, as shown in Figure 10.20. A 5 MWp PV system and a 5 MW h battery energy storage system (BESS) have been installed. The PVSHH MG will be able to enhance system stability, improve efficiency, reduce emissions, and to meet the electric power and water demands of the area.

To achieve a secure and efficient operation, an MAS EMS of the PVSHH MG is designed based on the following considerations:

The PVGP has the highest priority in generation. With the help of the BESS, the extra energy from PV generation, if any, will be stored in the daytime and used to support the peak load in the evening. This increases the utilization of PV generation and reduces the operation time of the DGP. Only when the BESS is fully charged the PV generation will be curtailed to balance the load demand and to avoid overvoltage.

The operation of the SHGP has a high priority to meet the residential water demand, especially in the dry season. For most of the time, the SHGP undertakes the base load, and its total generation has to be above the lower output threshold. The SHGP is responsible for the system frequency regulation.

The usage of the diesel generators should be minimized. The DGP is only used as a peaker to meet peak demands. Centralized control schemes have been widely used in EMSs for MGs. Under a centralized framework, an MGCC can be designed to carry out economic power dispatch for DGSs and energy storage systems at multiple time scales [61]. However, as the time scale increases, the design of a centralized controller struggles with challenges from larger prediction errors in renewable generation and more complexity in solving optimization problems. When the system size and the number of components are getting larger, it also becomes more challenging to realize real-time management. Moreover, reliability is another concern for centralized schemes, since a single fault at the MGCC can cause the whole system to stop working [61].

To avoid the drawbacks of a centralized control scheme, MASs have been proposed for MG energy management. An MAS is a collection of autonomous computational entities (agents), which are [46]: (i) autonomous (i.e., they perform "tasks based on goals") and (ii) intelligent (i.e. they pursue their goals such that they "optimize" given performance measures in an environment, which is hard to define analytically). An agent is a computational entity such as a software program that can act upon its environment as well as interact with other agents with possibly conflicting goals toward an ultimate common goal. An MAS is a distributed control approach based on local information and actions, which is suitable for real-time control and has the following features [44, 58]:

1) **Flexibility.** The MAS structure can integrate various plug-and-play DG sources and adaptively adjust the control of MG according to actual conditions and targets.
2) **Scalability.** It is convenient to extend the functions of an MAS based on the needs of customers.
3) **Fault tolerance.** The localized control of individual DGSs is robust to disturbances and faults in the MG.

The studies of MAS-based MG EMSs have been carried out on the system framework, dispatch algorithms, and coordination strategies of DGSs. Detailed reviews on MASs for power engineering applications are available in the literature [62, 63]. An MAS-based EMS is developed for a PVSHH MG. In the EMS, the optimal dispatch of DGSs is achieved by means of virtual bidding (VBD). Based on the VBD, the system operation schedule and capacity reserve are determined. The DGSs are controlled to follow the schedules according to their individual control strategies and local information. In addition, the real-time power control error (PCE) is compensated by the DGSs with capacity reserve using a model predictive control (MPC) method. The MAS is implemented based on an NI-PXI physical platform to realize local control strategies and agent functions such as dynamic price adjustment. The EMS is tested and validated on an RTDS–PXI real-time simulation platform before it is actually used in the PVSHH MG.

10.4.2.2 Multiagent-System-Based Energy Management Method

The proposed MAS-based EMS carries out multi-time-scale energy management based on a VBD strategy and a dynamic power dispatch algorithm. The VBD determines the system operation schedule and capacity reserve based on a public bidding platform, while the dynamic dispatch algorithm works in real time to compensate the PCE of the DGSs. The proposed method is an EMS model with general applicability. Moreover, the high-altitude environment leads to specific control target and parameters in the model, such as the virtual price adjustment for different DGSs, which reflects the operational considerations mentioned earlier.

10.4.2.2.1 Virtual Bidding

For grid-connected MGs, the electricity price from the utility is normally set as the reference for DGSs to bid. However, the method is not suitable for standalone MGs, not only because a standalone MG is isolated from the main power grid, but also because it fails to recognize the diversity in different DG techniques. Therefore, a VBD strategy is proposed to schedule the operation of the DGSs in the hybrid MG with consideration of operational conditions and DG features, especially the resources utilization and allocation, instead of economic benefit only. It aims to improve system operation stability by meeting the base load via an SHGP and the peak load using other resources, maximize the utilization of PV generation with the help of a BESS, and to reduce the operation time of the DGP in accordance with the system operational priorities/requirements discussed in Section 10.4.2.1.

As shown in Equation 10.19, the virtual price of a DG is composed of a marginal cost and a dynamic compensation cost. The marginal cost correlates with the actual generation cost, while the dynamic compensation cost is updated in real time according to the generation limits, resources allocation, and environmental impacts:

$$C_{bid,t}(P) = C_{mar}(P) + C_{dyn,t}(\cdot) \tag{10.19}$$

where $C_{bid,t}$ ($\$$) is the virtual price, C_{mar} is the marginal cost, and $C_{dyn,t}$ is the dynamic compensation price.

In the PVSHH MG, the VBD first calculates the clearing price $C'_{bid,t}$ and the corresponding power dispatch scheme $P'_{bid,t}$ for the DGSs based on their marginal costs and the load level of the system. The clearing price is the reference of dynamic compensation for the DGSs and is similar to electricity price from the utility in grid-connected mode.

To better utilize the PV generation, a parameter named PV expectation P_{ev} is defined. The actual PV power is expected to be between P_{ev} and the maximum PV power. If the PV clearing power generation $P'_{bid,t}$ is smaller than P_{ev}, then $C_{dyn,v,t}$ (dynamic compensation price of PV plant) is set as a negative number to stimulate more PV generation; in contrast, if $P'_{bid,t}$ is greater than or equal to P_{ev}, then $C_{dyn,v,t}$ is set as zero, as shown in Equation 10.20. P_{ev} is a constant obtained in the system sizing optimization with operational considerations. The extra PV power above P_{ev} can be stored in the BESS and then used to support the peak load in the evening. The stored energy in the BESS has a fixed virtual price that is a little cheaper than that of the DGP, with the aim to reduce the operation time of the DGP.

$$C_{dyn,v,t}(\cdot) = \begin{cases} C'_{bid,t} - C_{mar,v}(P_{ev}) & P'_{bid,t} < P_{ev} \\ 0 & P'_{bid,t} \geq P_{ev} \end{cases} \tag{10.20}$$

where $C_{mar,v}$ is the marginal cost of the PV power.

The participation of the DGP is limited by another parameter named peak load index P_{el}. If load demand $P_{L,t}$ is greater than P_{el} (e.g., during the peak load period), the corresponding $C_{dyn,d,t}$ (dynamic compensation price of diesel power plant) is negative to increase the generation from the diesel generators; otherwise, $C_{dyn,d,t}$ is greater than or equal to zero to limit the generation, as shown in Equation 10.21. The P_{el} is different when the water level and the load level vary in wet and dry seasons.

$$C_{dyn,d,t}(\cdot) = \begin{cases} C'_{bid,t} - C_{mar,d}(P_{L,t} - P_{el}) & P_{L,t} > P_{el} \\ 0 & P_{L,t} \leq P_{el} \end{cases} \tag{10.21}$$

where $C_{mar,d}$ is the marginal cost of diesel power plant.

The power expectation of the SHGP P_{eh} is determined by the residential water demand. If the power generation of the SHGP is smaller than or equal to P_{eh}, the corresponding $C_{dyn,h,t}$ (dynamic compensation price of small hydro plant) is negative to increase the generation from the SHGP; otherwise, the $C_{dyn,h,t}$ is positive to limit the generation and save the water resource:

$$C_{dyn,h,t}(\cdot) = \begin{cases} -k_{h1} & P_{H,t} \leq P_{eh} \\ -k_{h2}(H_t - H_s) & P_{H,t} > P_{eh} \end{cases} \tag{10.22}$$

where k_{h1} is the fixed compensation of small hydro, k_{h2} is the variable compensation of small hydro, H_s is the standard water level, and H_t is the water level.

The flowchart of VBD is shown in Figure 10.21. At the beginning, the schedule agent (SA) sends the information of load, the system clearing price $C'_{bid,t}$ and the corresponding power dispatch scheme $P'_{bid,t}$ to the local agents, which are the "participation information" in Figure 10.21 to help the local agents in decision-making. Based on the information received from the SA and the local information, each local agent will send back its VBD price and power range to the SA. The local agents can easily change their generation intentions by changing the corresponding dynamic compensation prices. If the local agents agree with the schedule, this round of VBD ends. Should any local agents disagree, they would quit this round of VBD. The remaining agents then continue to bid for the portions of those agents that have quit in the operation/reserves schedule until all the remaining agents agree. Based on the real-time information exchange and feedback, the SA and the local agents determine the schedule of operation and capacity reserve all together.

The generation priority of the DG sources is ranked based on their virtual prices; that is, a DG with a lower virtual price has a higher generation priority. The dispatch scheme will be obtained as $[P_{H,t}, P_{D,t}, P_{V,t}]$; that is, the dispatch power for the SHGP, DGP, and the PVGP. $P_{V,t}$ is the overall dispatch power of the PVGP that includes both the output of the PV system and the BESS in the PVGP. Owing to the forecast error of the PVs, the potential minimum power $(P_{V,max} - \delta_V P_{V,max})$ may be smaller than $P_{V,t}$. This means the actual generation of the PV system may be lower than the dispatch requirement, which would cause a power shortage. Although the BESS may give support, an appropriate reserve should be dispatched to guarantee the system's secure operation, as given in Equation 10.23. Meanwhile, the error from the load prediction can also cause a shortage in the power supply. Thus, the reserve should be large enough to compensate the power shortage caused by the prediction error of load and the PVs, as given in Equation 10.24.

$$R_{V,t} = P_{V,t} - (P_{V,max} - \delta_V P_{V,max}) \tag{10.23}$$

$$R = \begin{cases} R_{V,t} + \delta_L P_{L,t} & R_{V,t} > 0 \\ \delta_L P_{L,t} & R_{V,t} \leq 0 \end{cases} \tag{10.24}$$

where $R_{V,t}$ is the reserve power for the PV system, $P_{V,t}$ is the dispatch power for PVGP, $P_{V,max}$ is the power limit of the PV system, δ_L is the power prediction error of the load, and δ_V is the power prediction error of the PVs.

Only SHGP and DGP will then continue to bid for the reserve demand based on the dispatch scheme. The power reserve is vital to guarantee the stable operation of the MG and also is the base of the dynamic power dispatch. The unit commitment plan

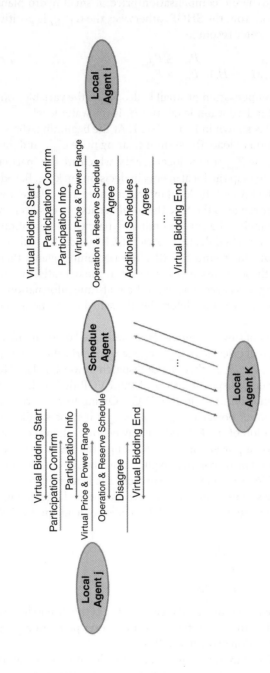

Figure 10.21 Flowchart of virtual bidding.

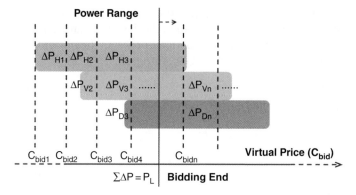

Figure 10.22 Visualization of operation scheduling.

for the small-hydro generators and the diesel generators is obtained according to the corresponding dispatch power and the reserve demand.

The schedule of operation and reserve is shown in Figure 10.22. The DGSs take part in the system operation based on their virtual prices and power ranges. Because the virtual price is nonlinear and segmented, it is difficult to calculate the balance point of the DG output power and load demand. The power ranges of the DGSs will be divided into segments which are the increased generation (ΔP_{Hn}, ΔP_{Vn} and ΔP_{Dn}) by raising a unit price. By pricing low to high, the load demand is met with power segments (ΔP). If the load demand is totally met at the nth segment (C_{bidn}), the dispatch power for SHGP is the sum of increased generations from the first segment to nth segments ($\Delta P_{H1} + \Delta P_{H2} + \cdots + \Delta P_{Hn}$), and PVGP and DGP are the same.

10.4.2.2.2 Dynamic Power Dispatch

The VBD formulates the system operation schedule and capacity reserve for DGSs. However, owing to the complexity of the actual situation, the DGSs may not be able to exactly follow the schedules. For example, a large drop of the solar irradiance can cause the generation of the PV system to be far lower than the schedule. In addition, in the MAS-based EMS, the DGSs are controlled based on local information. Therefore, an operation agent (OA) is needed to compensate actual deviations in real time. Based on the error signals fed back from the local agents, the OA will carry out dynamic power dispatch to keep the system secure and stable.

SHGPs (i.e., the small-hydro generators) are responsible for frequency regulation and their output changes are proportional to the system frequency deviation. The other DGSs are operated as constant-power sources. Hence, the PCE can be calculated as

$$\text{PCE} = \sum_{i \in \theta_f} k_i \Delta f + \sum_{j \in \theta_c} (P_{j,\text{sch}} - P_{j,t}) \tag{10.25}$$

where Δf is the system frequency variation, k_i is the frequency control coefficient, $P_{j,\text{sch}}$ is the dispatch power by the virtual bidding, and $P_{j,t}$ is the real-time output power.

Based on the MPC method and the dynamic characteristics of the DGSs, a dynamic dispatch algorithm is developed to distribute the PCE signal using the capacity reserve for DGSs to balance the load demand and power generation, and to adjust the system dispatch plan in real time. Based on the capacity reserve and the dynamic characteristics of the DGSs, the dynamic power dispatch is also an optimal process. The objective of the algorithm is to minimize the adjustment of the DGSs' outputs:

$$\min \sum |\Delta u_i(t)| \tag{10.26}$$

$$\text{s.t.} \qquad \Delta X(t) = A\Delta X(t-1) + B\Delta U(t) \tag{10.27}$$

$$\sum \Delta x_i(t) = \text{PCE} \tag{10.28}$$

$$\Delta X_{\min} \leq \Delta X(t) \leq \Delta X_{\max} \tag{10.29}$$

$$X_{\min} \leq X(t) \leq X_{\max} \tag{10.30}$$

$$\Delta U_{\min} \leq \Delta U(t) \leq \Delta U_{\max} \tag{10.31}$$

$$U_{\min} \leq U(t) \leq U_{\max} \tag{10.32}$$

where Δu_i is the control variable of the MPC, $\Delta X(t)$ is the vector form of the state variables, A is the state matrix of the prediction model, B is the output matrix of the prediction model, and PCE is the power control error.

The control variables are the adjustment of the DG dispatch power and the quantity of load shedding. For a standalone MG, load shedding is a necessary measure for secure operations. The state variables are the changes in the DG outputs. The dynamic equation given in Equation 10.27, which includes the frequency regulation characteristics and dynamic responses of the constant-power sources, is used to predict the DG outputs. A first-order model derived based on the step response of the reference power is used as the prediction model for DGSs [64]. Therefore, dynamic power dispatch takes advantage of a fast DG for transient load/RES changes and then gradually replaces it with a slow DG. The constraint in Equation 10.28 guarantees the power balance in the next step. The inequality constraints in Equations 10.29–10.32 define the limits of the state and the control variables.

When a local emergency happens, the DGSs will first adjust their outputs based on their local control strategies via the local agents. Meanwhile, the DGSs for frequency regulation will adjust their outputs to respond to the frequency deviation signals. The OA updates the operation schedule to compensate the PCE either in a regular manner or in emergent conditions. The VBD is for determining long-term operation schedules. Figure 10.23 shows the whole time scales of the proposed EMS of the MG.

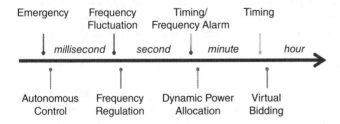

Figure 10.23 Dynamic response process of the proposed EMS in different time scales.

10.4.2.3 Validation of the Microgrid Energy Management System

To test the effectiveness of the proposed EMS for future implementation, the EMS is simulated on an RTDS–PXI real-time simulation platform. In return, the simulation results can be used to improve the management algorithm.

The RTDS–PXI simulates the real-time operation of the MG being studied. The MAS-based framework is realized on an NI–PXI platform. The communication system is designed based on IEC 61850, which is also used in the actual MG. By implementing the practical standards and requirements in the simulation system, the RTDS–PXI platform recreates the practical operational conditions as realistic as possible and verifies the effectiveness of the proposed EMS.

10.4.2.3.1 RTDS-PXI Real-Time Simulation Platform

The real-time RTDS–PXI simulation system is composed of a PXIe-1062Q, RTDS, and its communication and data acquisition accessories, as shown in Figure 10.24. The communication protocol of the whole system is IEC 61850.

The PXI system is the platform for implementing the MAS to carry out functions such as VBD, dynamic power dispatch, local control, and so on. The following features of the adopted PXIe-1062Q system make it suitable for the task (i.e., the implementation of MAS): fast calculation speed for real-time control by utilizing a low-jitter internal 100 MHz reference clock and a large bandwidth up to 3 GB/s; scalable internal CPU, communication and acquisition cards; capable of parallel computing; and visualized code programming.

RTDS is used to model the MG by simulating the dynamic (including frequency and voltage) responses and the real-time control of the system. RTDS is a flexible real-time platform for time-domain simulations. It has an extendable computing rack and supports hardware-in-the-loop simulation/emulation. The PXI, RTDS, DGSs, and the other devices are interfaced through an IEC 61850-based communication system. IEC 61850 is a well-accepted standard for communication in power systems, which enhances the interoperability among the different DGSs in the MG.

Figure 10.24 The RTDS–PXI real-time simulation platform (A: real-time monitoring of PXI; B: real-time simulation results of MG).

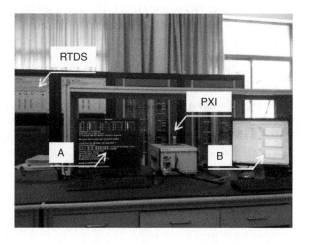

10.4.2.3.2 Structure and Implementation of Multiagent System

The structure of the MAS-based EMS for the PVSHH MG is shown in Figure 10.25, which includes seven types of agents:

1) **SA.** The SA periodically calls for bidding based on the VBD mechanism. The SA makes and releases the operation schedule and reserve according to the bids and the power range of the DGSs. As shown in Figure 10.25, the IEC 61850 based *Communication Module* is an essential module for all the agents. In addition, the SA has a *Virtual Bidding Center* to call for a VBD and gives the operation scheme and a *Reserved Power Check* module to make the reserve plan. In practice, for the MG under study, the local agents agree with the operation/reserves schedule of the SA, because the local agents send the power range and corresponding price to the SA after participating in the VBD. The schedule is carried out within the power ranges; that is, each DG should have the capacity to complete the scheduling.

2) **OA.** Based on the MPC method, the OA regularly adjusts the power dispatches according to the DGSs' dynamic characteristics. Under extreme conditions, the OA launches urgently to secure the system frequency. Therefore, the OA has to monitor the change of frequency and the power flow through a *System State Monitor* and make the power dispatch in the *Dynamic Power Dispatch* module.

3) **Dispatchable DG agent (DDA).** Based on the local information and individual control strategy of the DGP, the agent participates in the VBD and executes the generation schedules. In Figure 10.25, the *VBD Participant* module is used to adjust the virtual price to cooperate with the *Virtual Bidding Center* to participate in the VBD. According to the operation scheme and reserve demand, the unit commitment plan and power dispatch are executed under the local control strategy in the *Local Control* module.

4) **Frequency regulation agent (FRA).** Based on the local information and individual control strategy of the SHGP, the FRA is responsible for system frequency regulation, participates in the VBD, and executes the generation schedules. Except the functions of *VBD Participant* and *Local Control*, there is a *Frequency Regulation* module for the frequency measurement and regulation.

5) **Intermittent DG agent (IDA).** The agent predicts the generation of the PV system. Based on the prediction, local information, and local control strategy, the IDA controls the PVGP to take part in the VBD and follows the generation schedules. Compared with the DDA, the IDA needs an additional *Generation Forecast* module to support decision-making of the *Local Control* and *VBD Participant* as shown in Figure 10.25.

6) **Energy storage agent (ESA).** This agent controls the BESS of the PVGP to smooth the output of the PV system and reinforces the system during the peak time in the evening if the stored energy of the BESS is sufficient. Corresponding to these functions, there are *Smooth Fluctuation* of PV power, *Coordinated Control* with the PV system, and *Peak Shaving* modules in the ESA.

7) **Demand management agent (DMA).** The agent carries out load forecast and sheds loads for secure operation as needed under the dispatch of the OA. In Figure 10.25, the DMA has a *Load Forecast* module to support decision-making of the agents and a *Load Curtailing* module to execute the order of the OA in emergency.

In the PVSHH MG, the SHGP is controlled by the FRA, the diesel generators are controlled by the DDA, the PV system is controlled by the IDA, the BESS is controlled

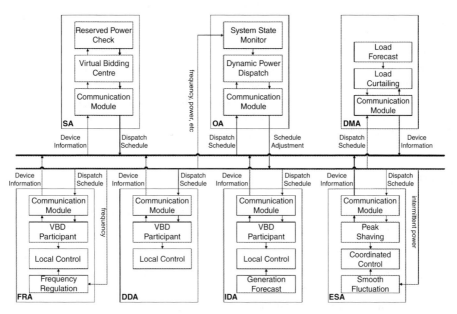

Figure 10.25 Classification and functions of the agents.

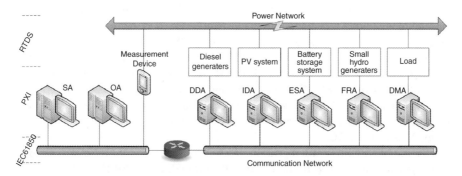

Figure 10.26 Schematic diagram of the real-time PXI–RTDS simulation/validation platform.

by the ESA, and the controllable loads are controlled by the DMA. The control and implementation of the proposed MAS-based EMS on the real-time simulation platform is shown in Figure 10.26. The power network and the devices of MG, including the small-hydro generators, the diesel generators, the PV system, and the BESS, are modeled and simulated in the RTDS. All the agents are implemented on the PXI. The commutation between the RTDS and the PXI is based on IEC 61850. The simulation platform supports real-time simulation, real-time decision-making, and real-time commutation.

10.5 Summary

In this chapter we have discussed the design, control, and implementation of standalone multisource MGs with high penetration of PV generation. Different configurations of MGs (AC, DC, and AC/DC hybrid structures), control schemes (centralized, distributed, and hybrid hierarchical control), and unit sizing methods have been reviewed. Some standalone MG systems around the world have been briefly introduced as MG implementation examples. The development and implementation of two MG projects have been given in detail. Dongfushan Island MG has been introduced to demonstrate how the joint optimization of operation and component sizing is carried out in the MG implementation. A plateau MG at high altitude has been discussed to show the implementation of an MAS for MG energy management.

References

1 Baziar A. and Kavousi-Fard A. (2013) Considering uncertainty in the optimal energy management of renewable micro-grids including storage devices. *Renewable Energy*, **59**, 158–66.

2 López M., Martín S., Aguado J., and de la Torre S. (2013) V2G strategies for congestion management in microgrids with high penetration of electric vehicles. *Electric Power Systems Research*, **104**, 28–34.

3 Zhao B., Zhang X., Li P., *et al.* (2014) Optimal sizing, operating strategy and operational experience of a stand-alone microgrid on Dongfushan Island. *Applied Energy*, **113**, 1656–1666.

4 Zhao J., Graves K., Wang C., *et al.* (2012) A hybrid electric/hydro storage solution for stand-alone photovoltaic applications in remote areas. In *2012 IEEE Power and Energy Society General Meeting*. IEEE. doi: 10.1109/PESGM.2012.6345187.

5 Kroposki B., Basso T., and DeBlasio R. (2008) Microgrid standards and technologies. In *2008 IEEE Power Energy Society General Meeting – Conversion and Delivery of Electrical Energy in the 21st Century*. IEEE. doi: 10.1109/PES.2008.4596703.

6 Lasseter R.H., Akhil A.A., Marnay C., *et al.* (2003) Integration of Distributed Energy Resources: The CERTS MicroGrid Concept. Report LBNL-50829. Department of Energy, Washington, DC.

7 Lasseter R. (2007) CERTS microgrid. In *2007 IEEE International Conference on System of Systems Engineering*. IEEE. doi: 10.1109/SYSOSE.2007.4304248.

8 Lasseter R. (2008) Extended CERTS microgrid. In *2008 IEEE Power Energy Society General Meeting – Conversion and Delivery of Electrical Energy in the 21st Century*. IEEE. doi: 10.1109/PES.2008.4596492.

9 Microgrid Exchange Group. (2011) DOE Microgrid Workshop Report. https://energy.gov/sites/prod/files/Microgrid%20Workshop%20Report%20August%202011.pdf (accessed March 11, 2017).

10 Venkataraman G. and Marnay C. (2008) A larger role for microgrids: are microgrids a viable paradigm for electricity supply expansion? *IEEE Power & Energy Magazine*, **6**(3), 78–82.

11 Kroposki B., Lasseter R., Ise T., *et al.* (2008) Making microgrids work: a look at microgrid technologies and testing projects from around the world. *IEEE Power & Energy Magazine*, **6**(3), 40–53.

12 Marnay C., Asano H., Papathanassiou S., and Strbac G. (2008) Policymaking for microgrids: economic and regulatory issues of microgrid implementation. *IEEE Power & Energy Magazine*, **6**(3), 66–77.

13 Papaefthymiou S.V., Karamanou E.G., Papathanassiou S.A., and Papadopoulos M.P. (2010) A wind–hydro-pumped storage station leading to high RES penetration in the autonomous island system of Ikaria. *IEEE Transactions on Sustainable Energy*, **1**(3), 163–172.

14 Lasseter R., Akhil A., Marnay C., *et al.* (2002) Integration of Distributed Energy Resources: The CERTS Microgrid Concept. Consortium for Electric Reliability Technology Solutions, Berkeley, CA.

15 Hatziargyriou N., Asano H., Iravani R., and Marnay C. (2007) Microgrids: an overview of ongoing research, development, and demonstration projects. *IEEE Power & Energy Magazine*, **5**(4), 78–94.

16 Kyriakarakos G., Dounis A.I., Rozakis S., *et al.* (2011) Polygeneration microgrids: a viable solution in remote areas for supplying power, potable water and hydrogen as transportation fuel. *Applied Energy*, **12**, 4517–4526.

17 Obara S., Kawai M., Kawae O., and Morizane Y. (2013) Operational planning of an independent microgrid containing tidal power generators, SOFCs, and photo-voltaics. *Applied Energy*, **102**, 1443–1457.

18 Khodaei A., Bahramirad S., Shahidehpour M.. (2015) Microgrid planning under uncertainty. *IEEE Transactions on Power Systems*, **30**(5), 2417–2425.

19 Huang C.C., Chen M.J., Liao Y.T., and Lu C.N. (2012) DC microgrid operation planning. In *2012 International Conference on Renewable Energy Research and Applications (ICRERA)*. IEEE; pp. 1–7. doi: 10.1109/ICRERA.2012.6477447.

20 Lu X., Liu N., Chen Q., and Zhang J. (2014) Multi-objective optimal scheduling of a DC micro-grid consisted of PV system and EV charging station. In *2014 IEEE Innovative Smart Grid Technologies – Asia (ISGT Asia)*. IEEE; pp. 487–491.

21 Hossein L. and Amin K. (2017) AC versus DC microgrid planning. *IEEE Transactions on Smart Grid*, **8**, 296–304.

22 Kellogg W.D., Nehrir M.H., Venkataramanan G., and Gerez V. (1998) Generation unit sizing and cost analysis for stand-alone wind, photovoltaic, and hybrid wind/PV systems. *IEEE Transactions on Energy Conversion*, **13**(1), 70–75.

23 Chedid R., Akiki H., and Rahman S. (1998) A decision support technique for the design of hybrid solar-wind power systems. *IEEE Transactions on Energy Conversion*, **13**(1), 76–83.

24 Borowy B. and Salameh Z. (1996) Methodology for the optimally sizing the combination of a battery bank and PV array in a wind/PV hybrid system. *IEEE Transactions on Energy Conversion*, **11**(2), 367–375.

25 Giraud F. and Salameh Z. (2001) Steady-state performance of a grid-connected rooftop hybrid wind–photovoltaic power system with battery storage. *IEEE Transactions on Energy Conversion*, **16**(1), 1–7.

26 Marnay C., Venkataramanan G., Stadler M., and Chandran B. (2008) Optimal technology selection and operation of commercial-building microgrids. *IEEE Transactions on Power Systems*, **23**, 975–982.

27 Lasseter R.H. and Piagi P. (2006) Control and Design of Microgrid Components: Final Project Report. PSERC Publication 06-03.

28 Henry L. and Kai S. (2007) Superconducting magnetic energy storage (SMES) for energy cache control in modular distributed hydrogen–electric energy systems. *IEEE Transactions on Applied Superconductivity*, **17**(2), 2361–2364.

29 Hashem M. and Wang C. (2009) *Modeling and Control of Fuel Cells: Distributed Generation Applications*. IEEE Press–Wiley; Chapter 9.

30 Nehrir M.H., Wang C., Strunz K., *et al.* (2011) A review of hybrid renewable/alternative energy systems for electric power generation: configurations, control and applications. *IEEE Transactions on Sustainable Energy*, **2**, 392–403.

31 Laxman M., Shigenori I., and Hirofumi A. (2008) A transformerless energy storage system based on a cascade multilevel PWM converter with star configuration. *IEEE Transactions on Industry Applications*, **44**(5), 1621–1630.

32 Sood P., Lipo T., and Hansen I. (1988) A versatile power converter for high-frequency link systems. *IEEE Transactions on Power Electronics*, **3**(4), 383–390.

33 Cha H. and Enjeti P. (2003) A three-phase AC/AC high-frequency link matrix converter for VSCF applications. In *IEEE 34th Annual Power Electronics Specialist Conference, 2003 (PESC '03)*, vol. **4**. IEEE; pp. 1971–1976.

34 Nelson D., Nehrir M., and Wang C. (2006) Unit sizing and cost analysis of stand-alone hybrid wind/PV/fuel cell power generation systems. *Renewable Energy*, **31**, 1641–1656.

35 Delson J. and Shahidehpour S. (1994) Linear programming applications to power system economics, planning and operations. *IEEE Transactions on Power Systems*, **7**(3), 1155–1163.

36 Granville S. (1994) Optimal reactive dispatch through interior point methods. *IEEE Transactions on Power Systems*, **9**(1), 136–146.

37 Pruitt K., Braun R., and Newman A. (2013) Evaluating shortfalls in mixed-integer programming approaches for the optimal design and dispatch of distributed generation systems. *Applied Energy*, **102**, 386–398.

38 Koutroulis E., Kolokotsa D., Potirakis A., and Kalaitzakis K. (2006) Methodology for optimal sizing of stand-alone photovoltaic/wind-generator systems using genetic algorithms. *Solar Energy*, **80**, 1072–1088.

39 Yang H., Zhou W., Lu L., and Fang Z. (2008) Optimal sizing method for stand-alone hybrid solar-wind system with LPSP technology by using genetic algorithm. *Solar Energy*, **82**, 354–367.

40 Ekren O. and Ekren B. (2008) Size optimization of a PV/wind hybrid energy conversion system with battery storage using response surface methodology. *Applied Energy*, **85**, 1086–1101.

41 Kashefi A., Riahya G., and Kouhsari S. (2009) Optimal design of a reliable hydrogen-based stand-alone wind/PV generating system, considering component outages. *Renewable Energy*, **34**, 2380–2390.

42 Wang L. and Sigh C. (2009) Multicriteria design of hybrid power generation systems based on modified particle swarm optimization algorithm. *IEEE Transactions on Energy Conversion*, **24**(1), 163–172.

43 Pourmousavi S.A., Nehrir M.H., Colson C.M., and Wang C. (2010) Real-time energy management of a stand-alone hybrid wind–microturbine energy system using particle swarm optimization. *IEEE Transactions on Sustainable Energy*, **1**(3), 193–201.

44 Dimeas A. and Hatziargyriou N. (2005) Operation of a multiagent system for microgrid control. *IEEE Transactions on Power Systems*, **20**(3), 1447–1455.

45 Tsikalakis A.G. and Hatziargyriou N.D. (2011) Centralized control for optimizing microgrids operation. In *2011 IEEE Power and Energy Society General Meeting*. IEEE. doi: 10.1109/PES.2011.6039737.

46 Gerhard W. (1999) *Multiagent Systems: A Modern Approach to distributed Artificial Intelligence*. MIT Press, Cambridge, MA.

47 Kelash H.M., Faheem H.M., and Amoon M. (2007) It takes a multiagent system to manage distributed systems. *IEEE Potentials*, **26**(2), 39–45.

48 Yang Z., Ma C., Feng J.Q., *et al.* (2006) A multi-agent framework for power system automation. *International Journal of Innovation in Energy Systems and Power*, **1**, 39–45.

49 Toroczkai Z. and Eubank S. (2005) Agent-based modeling as a decision-making tool. *The Bridge*, **35**(4), 22–27.

50 Huang K., Cartes D.A., and Srivastava S.K. (2007) A multiagent-based algorithm for ring-structured shipboard power system reconfiguration. *IEEE Transactions on Systems, Man, and Cybernetics, Part C: Applications and Reviews*, **37**(5), 1016–1021.

51 Nagata T., and Sasaki H. (2002) A multi-agent approach to power system restoration. *IEEE Transactions on Power Systems*, **17**(2), 457–462.

52 Jiang Z., and Dougal R. (2008) Hierarchical microgrid paradigm for integration of distributed energy resources. In *2008 IEEE Power Engineering Society General Meeting – Conversion and Delivery of Electrical Energy in the 21st Century*. IEEE; pp. 20–24.

53 Guerrero J., Vasquez J., Matas J., *et al.* (2011) Hierarchical control of droop-controlled AC and DC microgrids – a general approach toward standardization. *IEEE Transactions on Industrial Electronics*, **58**(1), 158–172.

54 Nakken T., Strand L.R., Frantzen E., *et al.* (2006) The Utsira wind–hydrogen system – operational experience. In *European Wind Energy Conference 2006*.

55 Landau M. and Deubler H. (2006) 10 years PV Hybrid system as an energy supply for a remote alpine lodge. In *4th European PV–Hybrid and Mini-Grid Conference*.

56 Yasuhiro K. (2009) Operation results of the Hachinohe microgrid project. In *Microgrid Symposium*, San Diego, CA.

57 Robert W. Galvin Center for Electricity Innovation. *Microgrid at IIT*. http://www.iitmicrogrid.net/ (accessed March 9, 2017).

58 Zhao B., Xue M., Zhang X., *et al.* (2015) An MAS based energy management system for a stand-alone microgrid at high altitude. *Applied Energy*, **143**, 251–261.

59 Kaldellis J., Zafirakis D., and Kavadias K. (2009) Techno-economic comparison of energy storage systems for island autonomous electrical networks. *Renewable and Sustainable Energy Reviews*, **13**(2), 378–392.

60 Goldberg D. (1989) *Genetic Algorithms in Search, Optimization and Machine Learning*. Addison-Wesley, New York.

61 Zhao J., Wang C., and Zhao B. (2014) A review of active management for distribution networks: Current status and future development trends. *Electric Power Components and Systems*, **42**(3–4), 280–293.

62 McArthur S.D., Davidson E.M., Catterson V.M., *et al.* (2007) Multi-agent systems for power engineering applications – part I: concepts, approaches, and technical challenges. *IEEE Transactions on Power Systems*, **22**(4), 1743–1752.

63 McArthur S.D., Davidson E.M., Catterson V.M., *et al.* (2007) Multi-agent systems for power engineering applications – part II: technologies, standards, and tools for building multi-agent systems. *IEEE Transactions on Power Systems*, **22**(4), 1753–1759.

64 Falahi M., Lotfifard S., Ehsani M., and Butler-Purry K. (2013) Dynamic model predictive based energy management of DG integrated distribution systems. *IEEE Transactions on Power Delivery*, **28**(4), 2217–2227.

Index

a

active material 230, 232, 233
active and reactive power-voltage control
    scheme 259
AE (alternative energy) 274
ambient temperature 24, 25, 30–32, 36,
    39, 49, 91, 233, 263
analysis method
  curve 87, 210
  frequency-domain 211, 213
  three-level assessment 198, 207
  time-domain 211
anode 230–233
AS (ancillary service) 243–245
automatic voltage control (AVC) 125, 130,
    139, 147
automobile company C1 114–116
available reactive capacity 130, 131
AVC *see* automatic voltage control (AVC)
average sensitivity value 257, 258

b

battery company C2 114–116
battery energy storage system (BESS) 19,
    20, 230–233, 242, 271, 300–309 *see
    also* energy storage
bidirectional power flow 10

c

CAP (capacity absorption planning) 244
capacitor banks 130, 131, 140, 141,
    144–162 *see also* static capacitor
    (SC)
CASP (capacity/ancillary service program)
    243

cathode 230–233
cell
  flow 229, 232, 233
  heterojunction solar 20
  homojunction solar 20
  multi-junction solar 20
  noncrystalline silicon semiconductor 21
  organic solar 21
  PV 2, 5, 6, 19–26, 29–31, 39, 171
  Schottky junction solar 20
  silicon solar 2, 21–23
  thin-film solar 20
  wet PV 20
charging/discharging controller of battery
    19
CHP generation 17
close-loop control 38
combined heat and power generation *see*
    CHP generation
community partition theory 254
connection topology between inverter and
    PV 26–28
  centralized 26
  control logic 131
  module integrated 26–28
  multiple series 26–28
control principle 132, 133
constrained optimization method 82–84
  algorithms 82
  random scenario 88–94
converter
  DC-DC 33
  flyback 29
  rear-end 29, 34
correlation coefficient 79–81

Grid-Integrated and Standalone Photovoltaic Distributed Generation Systems: Analysis, Design, and Control,
First Edition. Bo Zhao, Caisheng Wang and Xuesong Zhang.
© 2018 China Electric Power Press. Published 2018 by John Wiley & Sons Singapore Pte. Ltd.